2008 IEEE/LEOS International Conference on Optical MEMS and Nanophotonics

Freiburg, Germany
11-14 August 2008

IEEE Catalog Number: CFPO8MOE-PRT
ISBN: 978-1-4244-1917-3

Copyright © 2008 by The Institute of Electrical and Electronics Engineers, Inc.
All Rights Reserved

Copyright and Reprint Permissions: Abstracting is permitted with credit to the source. Libraries are permitted to photocopy beyond the limit of U.S. copyright law for private use of patrons those articles in this volume that carry a code at the bottom of the first page, provided the per-copy fee indicated in the code is paid through Copyright Clearance Center, 222 Rosewood Drive, Danvers, MA 01923.

For other copying, reprint or republications permission, write to IEEE Copyrights Manager, IEEE Operations Center, 445 Hoes Lane, Piscataway, New Jersey USA 08854. All rights reserved.

IEEE Catalog Number: CFP08MOE-PRT

ISBN 13:   978-1-4244-1917-3

LOC: 2007909044

Additional Copies of This Publication Are Available from:

IEEE Service Center
445 Hoes Lane
Piscataway, NJ 08854

Phone:	(800) 678-IEEE
	(732) 981-1393
Fax:	(732) 981-9667
E-mail:	customer-service@ieee.org

TABLE OF CONTENTS

Monday, 11 August 2008

M1	**Nano Particles & Nanowires**	
M1.1	Advances in Quantum Dots and 2D/3D Photonic Crystal Nanocavity based on Micro-Machining Technology	1
M1.2	Spectral Reflectance Measurement of Two–Dimensional Photonic Nanocavities with Embedded Quantum Dots	3
M1.3	Synthesis and Characterization of ZnO Nanorod Arrays and Their Integration into Polymer Solar Cells	5
M1.4	Parallel Assembly of Nanowires Using Lateral-Field Optoelectronic Tweezers	7
M1.5	Sub-30nm Alignment Accuracy between Layered Photonic Nanostructures using Optimized Nanomagnet Arrays	9

M2	**Microfluidics and Biosensors**	
M2.1	How Can We use Nanophotonics to Help in Combating Climate Change?	11
M2.2	Low-Power Consumption Integrated Laser Doppler Blood Flowmeter with a Built-in Silicon Microlens	13
M2.3	Reconfigurable Liquid Micro-Lens System for Variable Fiber Coupling	15
M2.4	High Resolution Integrated Microfluidic Fabry-Perot Refractometer in Silicon	17
M2.5	Pulsed Laser Triggered High Speed Microfluidic Switch	19

M3	**Waveguide Devices**	
M3.1	Optical Interconnects & Nanophotonics	21
M3.2	Silicon Microring Resonator Connected with Submicorn Comb Actuator	23
M3.3	Four Port Nanophotonic Couplers for Dense, Planar Integrated Optics	25
M3.4	A Novel Etching-Oxidation Fabrication Method for 3D Nano Structures on Silicon and Its Application to SOI Symmetric Waveguide and 3D Taper Spot Size Converter	27
M3.5	Coupled-Ring Reflector Laser Diodes Composed of Squared Ring Waveguides	29

M4	**Light Sources and Displays**	
M4.1	MEMS-based Pico Projector Display	31
M4.2	High-Efficiency MEMS Tuneable Gratings for External Cavity Lasers and Microspectrometers	33
M4.3	Optical Wavelength Selection and Amplification by Silica Microcavities and Erbium Doped	35

	Fiber	
M4.4	Micropatterned Complex Optical Surface for Wide Angle Illumination	37
M4.5	Large Linear Micromirror Array for UV Femtosecond Laser Pulse Shaping	39

Tuesday, 12 August 2008

Tu1	**Biomedical Microsystems**	
Tu1.1	Cooling and Amplifying Micro-Mechanical Motion with Light	41
Tu1.2	MEMS based Dual-Axes Confocal Clinical Endoscope for Real Time *in vivo* Imaging	42
Tu1.3	Tunable Multi-Micro-Lens System for High Lateral Resolution Endoscopic Optical Coherence Tomography	44
Tu1.4	Resonant Cantilever Bio Sensor with Integrated Grating Readout	46
Tu1.5	Design and Fabrication of Parylene-Hinged Slow-Scan Optical Scanner for OCT Endoscope Application	48
Tu2	**Periodic-Structure Devices**	
Tu2.1	Photonic Crystal Biosensors and Tunable Resonant Optical Devices	50
Tu2.2	Ultrahigh Sensitivity Slot-Waveguide Biosensor on a Highly Integrated Chip for Simultaneous Diagnosis of Multiple Diseases	52
Tu2.3	Nanostructured Effective-Index Micro-Optical Devices based on Blazed 2-D Sub-Wavelength Gratings with Uniform Features on a Variable-Pitch	54
Tu2.4	Characterization of Silicon-on-Insulator Waveguide Chirped Grating for Coupling to a Vertical Optical Fiber	56
Tu3	**Microfabrication**	
Tu3.1	Microspectrometer: From Ideas to Product	58
Tu3.2	Fabrication of Aberration-Corrected Tunable Micro-Lenses	60
Tu3.3	Three-Dimensional Integration of Optical Multi-Chips using Surface-Activated Bonding for High-Density Microsystems Packaging	62
Tu3.4	Pulsed Thermal Excitation of Luminescent Microparticles for Radiation Dosimetry	64
Tu3.5	Widely Tunable Fabry-Perot Optical Filter using Fixed-Fixed Beam Actuators	66
Tu4	**Scanners and Applications**	
Tu4.1	Impact of High Optical Power on Optical MEMS	68
Tu4.2	A High-Power Handling MEMS Optical Scanner for Display Applications	70

	Fast and High-Precision 3D Tracking and Position Measurement with MEMS Micromirrors	72
Tu4.4	In-situ Single Cell Electroporation using Optoelectronic Tweezers	74
Tu4.5	Large Area High-Reflectivity Broadband Monolithic Silicon Photonic Crystal Mirror MEMS Scanner	76
Tu4.6	3D Imaging using Resonant Large-Aperture MEMS Mirror Arrays and Laser Distance Measurement	78

Wednesday, 13 August 2008

W1	**Metamaterials and Photonic Crystals**	
W1.1	Photonic Metamaterials: Optics Starts Walking on Two Feet	80
W1.2	Nanophotonics	82
W1.3	Clarification of Electromagnetic Responses in Split-Ring Resonators from Electric Excitation	84
W1.4	Large-Area Monolithic Photonic Crystal Mirrors with High Reflectivity in the 1250-1650nm Band Patterned by Optical Lithography	86
W2	**Adaptive Optics**	
W2.1	Applications of LCoS-based Adaptive Optical Elements in Microscopy	88
W2.2	MEMS Deformable Mirrors for Adaptive Optics using Single Crystal PMN-PT	90
W2.3	A Varifocal Micromirror with Pure Parabolic Surface using Bending Moment Drive	92
W2.4	Simulation and Characterization of Tunable Achromatic Micro-Lenses	94
W2.5	Optical Scanner with Deformable Mirror Fabricated from SOI Wafer	96
P	**Poster Session**	
P1	A Micromachined Vibratory Sub-Wavelength Diffraction Grating Laser Scanner	98
P2	ZnO Nanorod-based Polymer Solar Cells with Optimized Electrodes	100
P3	Wafer Level Batch Fabrication and Assembly of Small Form Factor Optical Pickup Head	102
P4	Evaluation of X-ray Reflectivity of a MEMS X-ray Optic	104
P5	Design and Fabrication of CMOS-Integrated Thermoelectric IR Microsensors	106
P6	Performance Improvement of a Two-Axis Radial-Vertical-Combdrive Scanner by using a Symmetric Spring Design	108
P7	Assembly of Micro Mirrors on SOI Wafers using SU-8 Mechanisms and One-Push	110

	Operation	
P8	Spatially Resolved Optical Characterization of Photonic Crystal Slabs using Direct Evaluation of Photonic Modes	112
P9	Micro-Mirror Array for Multi-Object Spectroscopy in Cryogenic Environment	114
P10	Low Operation Voltage Non Self-Emissive MEMS Color Filter Pixels	116
P11	White-Light Electroluminescence from ZnO Nanowires/Polyfluorene Heterojunction Diodes	118
P12	Stabilization of Temperature Characteristics of Micromirror for Low-Voltage Driving using Thin Film Torsion Bar of Tensile Poly-Si	120
P13	Simulation-based Design of a MEMS X-ray Optic for X-ray Astronomy	122
P14	Fabrication of Sub-Micrometer Si Spheres with Atomic-Scale Surface Smoothness using Homogenized KrF Excimer Laser Reformation System	124
P15	Space Instruments based on MOEMS Technology	126
P16	Polymer Biochips with Micro-Optics for LIF Detection	128
P17	3D Modeling of Photonic Devices using Dynamic Thermal Electron Quantum Medium Finite-Different Time-Domain (DTEQM-FDTD) Method	130
P18	High-Density Piezoelectric Actuator Array for MEMS Deformable Mirrors Composed of PZT Thin Films	132
P19	Monolithic Integration of a Tunable Photodetector Based on InP/Airgap Fabry-Pérot Filters	134
P20	Polarization Singularities and Local Field Symmetries in Photonic Crystals	136
P21	Polymer Deformable Membrane Mirrors for Focus Control using SU-8 2002	138
P22	Mechanically Coupled Comb Drive MEMS Stages	140
P23	Photonic Crystal Rods utilizing Fano Resonance for Tunable Filter Applications	142
P24	A Glass Cantilever Beam Sensor Combined with a Spherical Reflecting Mirror for Sensitivity Enhancement	144
P25	A Micromirror Scanner with Vertical Combs Tilted by Assembly Process	146
P26	Design and Fabrication of Etched Diffraction Grating Demultiplexers based on a-Si Nanowire Technology	148
P27	Planar Centering Mechanism for Dielectrically Liquid Lens	150
P28	The Two-Axis Magnetostatic-Drive Single-Crystal-Si Micro Scanner Driven by Back-side Electroplating Ni Film	152
P29	All-Optical Ultra-Compact Photonic Crystal Controllable Logic Gate	154
P30	Conductive Pattern Forming Method on Vertical Wall using Spray Coating and Angled Exposure Technologies	156
P31	A Novel Lens Formation Technology to Implement a 3D Spherical Polymer Lens	158

	Fabrication of 2D and 3D Photonic Quasi-Crystals with High Rotation Symmetry by Holographic Lithography Technique	160
P33	Fabrication of Wall-Coated Cs Vapor Cells for a Chip-Scale Atomic Clock	162
P34	The Torque-Enhancement Design for Magnetostatic Scanner	164
P35	Six Port Waveguide Filter based on Circular Photonic Crystal	166
P36	Towards Integration of Glass Microlens with Silicon Comb-Drive X-Y Microstage	168
P37	Optical Add/drop Filter Based on Dual Curved Photonic Crystal Resonator	170

Thursday, 14 August 2008

Th1	**Detectors and Imaging Systems**	
Th1.1	Cavity-Optomechanics in Nanoscale Photonic Crystals	N/A
Th1.2	CMOS-SOI-MEMS Transistor (TMOS) for Infrared Imaging	172
Th1.3	Mid-Infrared Tunable Resonant Cavity Enhanced Detectors Employing Vertically Moving Comb Drive Actuated MEMS Micromirrors	174
Th1.4	A Dynamic Subwavelength Pitch Grating Modulator for Continuous Time-of-Flight Ranging with Optical Mixing	176
Th1.5	Radiation Heat Transfer Dominated Microbolometers	178
Th2	**Nanofabrication and Characterization**	
Th2.1	Near-Field Scanning Nanophotonic Microscopy	180
Th2.2	The Dependence of Poly-Crystalline SiC Mid-Infrared Optical Properties on Deposition Conditions	182
Th2.3	Impact of an Air Barrier on the Electron States of Etch-Released Quantum Heterostructures	184
Th2.4	Bandgap Tuning of Photonic Crystals by Polymer Swelling	186
Th2.5	Formation of a Nitrified Hafnium Oxide Buffer Layer on Silicon Substrate and GaN Quantum Well Crystal Growth for GaN-Si Hybrid Optical MEMS	188

This page intentionally left blank.

Advances in Quantum Dots and 2D/3D Photonic Crystal Nanocavity based on Micro-Machining Technology

Yasuhiko Arakawa

Institute for Nano Quantum Information Electronics (INQIE)
The University of Tokyo
4-6-1 Meguro-ku, 153-8505 Tokyo, Japan
arakawa@iis.u-tokyo.ac.jp

ABSTRACT — **We discuss recent advances in quantum dots for nanophotonics and quantum information devices, including high-performance quantum dot lasers and single photon emitters. Moreover, controlled light emission from quantum dots embedded in 2D/3D photonic crystal nanocavity is demonstrated.**

Keywords — quantum dots, quantum dot lasers, photonic crystal nanocavity, single photon emitters, micro-machining

Following Esaki's pioneering work on superlattices and quantum wells, the concept of quantum dots (QDs) was proposed by Arakawa *et al.* in 1982 for application to semiconductor lasers. with the theoretical prediction of temperature-insensitive threshold current characteristics [1]. Full confinement of electrons in the QDs has brought up unique features of artificial atoms, such as discrete energy states and quantum correlation due to spin/charging effects.

After the proposal of the QDs, most of significant characteristics of the QD lasers were predicted in 1980's, including low threshold current [2,3], enhanced modulation bandwidth[4], small linewidth enhancement factor[4], p-doping[5] and a scheme of tunneling structures for enhancing carrier injection[6]. Discovery of formation of a three dimensional small structure in lattice mismatched InAs/GaAs by Goldstein *et al.* in 1985 triggered development of fabrication of self-assembled QDs [7]. This pioneering work as well as subsequent many investigations in 1990's was led to successful demonstration of QD lasers[8-11].

Recent remarkable progress in the p-doped layers which increase optical gain and differential gain in QD lasers has realized 1.3-μm p-doped QD lasers with temperature-insensitive eye-opening under 10-Gb/s modulation without current adjustments with a high extinction ratio between 20 and 70°C[12]. This successful result has brought launch of a curve-out company (a new type of venture company) "QD Lasers Corp.", capitalized by Fujitsu and Mitsui Venture Capitals in 2006.

The commercialization of the QD lasers for the telecom market requires the fabrication of efficient lasers emitting at 1.3 μm, via a process suitable for mass-production. We have recently reported the fabrication of the first QD lasers above 1.3 μm, grown by metal organic chemical vapor deposition (MOCVD). This was achieved thanks to the beneficial effects of the antimony-mediated growth of InAs/Sb:GaAs QDs. We demonstrated for the first time ground-state (GS) lasing above 1.3 μm from MOCVD-grown GaAs-based QD lasers [13,14], with values of maximum GS modal gain of 12.5 and 19.3 cm^{-1} for five- and ten-stack lasers, respectively. Lasing occurs at the wavelength of 1.35 μm, which corresponds to the GS emission.

The photonic crystal (PhC) nanocavity laser is considered one of the best candidates for ultra-low threshold lasers due to their small mode-volume of the order of cubic wavelength and high quality factor Q. The first PhC laser was reported with multi-quantum well structures in 1999 [15]. In the past decade, some groups have been challenging to achieve laser operation in PhC nanocavities with QD gain material[16]. Recently, we achieved continuous-wave (cw) laser operation at low temperature [17] and room temperature [18] has been reported in PhC nanocavities with QDs. Moreover, a high-Q 3D photonic crystal embedding QDs based on micro-manipulation technique inside a secondary electron microscope (SEM) chamber was successfully demonstrated.

In addition to nanophotonic device applications based on an ensemble of QDs, single or coupled QDs are promising for quantum information devices, such as single photon emitters, and quantum-bit devices. In the quantum information devices, single photon-electron interaction and quantum entangled states are manipulated based on electron spins, charges, and nuclear spins. Even for classical light sources, a single dot with photonic nanocavities has been led to the concept of "single artificial atom lasers". Recently we demonstrated a nearly-single QD laser with a PhC nanocavity.

Regarding single photon emission, there have already been various demonstrations using single atoms, single molecules and single QDs. In particular, semiconductor QDs are the most promising for practical applications. However, the experiments on single photon emission from QDs were mainly at liquid helium temperature and emitting light was at the wavelength less than 1 μm. Recently, emission of single-photon pulses in the C-band (1.55-μm band: the highest transmittance in optical

978-1-4244-1917-3/08/$25.00 ©2008 IEEE

telecommunication bands) from a single InAs/InP QD were demonstrated [19,20]. Moreover, by establishing growth technology of high-quality GaN/AlN QDs, triggered single-photon emission at higher temperatures up to 200 K was demonstrated, which is the highest temperature in all semiconductor single photon emitters[21].

The QDs are promising for realizing various quantum information devices such as entangled photon pair generators and quantum logic gate. Moreover, the QDs can be applied to bio-markers and solar energy technologies. As indicated in Fig.1, through the

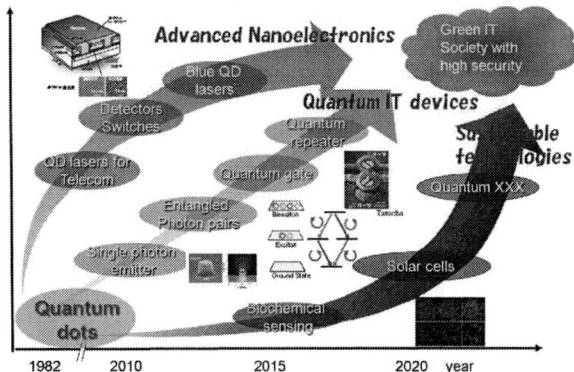

investigation of physics and device technologies of the QDs, we could contribute to "green society" which can be realized by highly efficient IT and global eco-technologies.

Fig1. Quantum dots toward green society

ACKNOWLEDGEMENTS

This work was partly supported by Special Coordination Funds for Promoting Science and Technology. The authors would like to thank S. Iwamoto, D. Guimard, M. Nomura and Kako at the University of Tokyo and N. Yokoyama, M. Sugawara and T. Usuki at Fujitsu laboratories for their intensive collaborations.

REFERENCES

1. Y. Arakawa and H. Sakaki: Appl. Phys. Lett., 40, 939 (1982)
2. M. Asada, Y. Miyamoto, and Y. Suematsu, IEEE J. Quantum Electron., 22, 1915--1921, (1986).
3. A. Yariv, Appl. Phys. Lett., 53, 1033 (1988)
4. Y. Arakawa and A.Yariv, IEEE J. of Quantum electron., QE-22, 1887 (1986)
5. Ref, 1 and see also T. Takahahsi and Arakawa, Optoelectronics-Devices and Technologies, 3, 155 (1988)
6. Y. Arakawa, *Solid-State Electronics* 37, 523 (1994)
7. L. Goldstein, F. Glas, J. Y. Marzin, M. N. Charasse, and G. Le Roux, Appl. Phys. Lett. 47, 1099 (1985)
8. N. Kirstaedter, N. Ledentsov, M.Grundmann, D. Bimberg, V. Ustinov, S. Ruvimov, M. Maximov, P. Kop'ev, Z. Alferov, U. Richter, P. Werner, U. Gosele, J. Heydenreich, Electronics Letters 30, 1416 (1994)
9. L. Huffaker, G. Park, Z. Zou, O. B. Shchekin, and D. G. Deppe, Appl. Phys. Lett. 73, 2564 (1998).
10. K. Mukai, Y. Nakata, K. Otsubo, M. Sugawara, N. Yokoyama, and H. Ishikawa, IEEE Photon. Tech. Lett. 11, 1205 (1999).
11. J. A. Lott, N. N. Ledentsov, V. M. Ustinov, N. A. Maleev, A. E. Zhukov, A. R. Kovsh, M. V. Maximov, B. V.Volovik, Zh. I. Alferov, and D. Bimberg, Electron. Lett. 36, 1384 (2000).
12. K. Otsubo, N. Hatori, M. Ishida, S. Okumura, T. Akiyama, Y. Nakata, H. Ebe, M. Sugawara and Y. Arakawa Jpn. J. of Appl. Phys. Vol. 43, L1124 (2004)
13. D. Guimard, M. Nishioka, S. Tsukamoto and Y. Arakawa, Appl. Phys. Lett., 89, 183124 (2006).
14. D. Guimard, M. Ishida, S. Tsukamoto, M. Nishioka, Y. Nakata, H. Sudo, T. Yamamoto, M. Sugawara, and Y. Arakawa, Appl. Phys. Lett., 90, 241110 (2007).
15. O. Painter, R.K. Lee, A. Scherer, A. Yariv, J.D. O'Brien, and P. D. Dapkus, Science 284, 1819 (1999).
16. S. Strauf, K. Hennessy, M.T. Rakher, Y.-S. Choi, A. Badolato, L.C. Andreani, E.L. Hu, P.M. Petroff, and D. Bouwmeester, Phys. Rev. Lett. 96, 127404 (2006).
17. M. Nomura, S. Iwamoto, K. Watanabe, N. Kumagai, Y. Nakata, S. Ishida, and Y. Arakawa, Opt. Express 14, 6308 (2006).
18. M. Nomura, S. Iwamoto, K. Watanabe, N. Kumagai, Y. Nakata, S. Ishida, and Y. Arakawa, Phys. Rev. B, 75, 195313 (2007)
19. K. Takemoto, Y. Sakuma, S. Hirose, T. Usuki, N. Yokoyama, T. Miyazawa, M. Takatsu, and Y.Arakawa , Jpn. J. Appl. Phys. 43, 7B, L993, (2004).;
20. T. Miyazawa, K. Takemoto, Y. Sakuma, S. Hirose, T. Usuki, N. Yokoyama, M. Takatsu and Y. Arakawa, JJAP vol.44, L620 (2005).
21. S. Kako, C. Santori, K. Hoshino, S. Goetzinger, Y. Yamamoto, Y. Arakawa, Nature Material, 5, 887 (2006)

Spectral Reflectance Measurement of Two-Dimensional Photonic Nanocavities with Embedded Quantum Dots

Wolfgang C. Stumpf, Takashi Asano, Takanori Kojima, Masayuki Fujita, Yoshinori Tanaka, and Susumu Noda

Department of Electronic Science and Engineering, Kyoto University, Kyoto-Daigaku-Katsura, Nishikyo-ku, Kyoto 615-8510, Japan
Phone +81-75-383-2319, Fax +81-75-383-2320
E-mail:wolfgang@qoe.kuee.kyoto-u.ac.jp, snoda@kuee.kyoto-u.ac.jp

Abstract: The resonant wavelength spectra of photonic crystal nanocavities containing quantum dots measured by reflectivity are consistent with photoluminescence. The reflectance method provides Q-value estimation of both active and passive nanocavities in good agreement with theory.

1. Introduction

In recent years, quantum dots (QDs) combined with photonic crystal (PC) nanocavities [1] as shown in Fig. 1(a) have generated increasing interest due to their potential application as ultra-low threshold lasers, solid-state quantum information processing devices and integrated optical circuits, as well as their use in fundamental research of interaction phenomena [2]. PC nanocavities have extremely small mode volumes V and can reach ultrahigh quality factors Q up to 2 500 000 [3]. For any experiment or application it is essential to know the experimental Q value. Commonly used measurement methods such as photoluminescence (PL) from embedded emitters like QDs are influenced by emission related phenomena that can make analysis difficult [4]. However, such complicated effects can be largely avoided by employing this new method based on reflectivity [5]. This technique can measure the Q factor of both passive and active cavities and has been independently demonstrated to control a nanocavity's reflectivity [6].

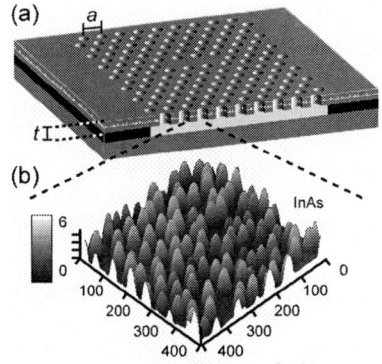

Fig. 1 (a) L3 type PC nanocavity with (b) embedded QDs imaged by AFM.

2. Sample fabrication methods

The substrate was grown by molecular beam epitaxy on a semi-insulating GaAs wafer. First, 200 nm of GaAs were grown as a buffer layer at 600°C. This was followed by 1200 nm sacrificial layer of $Al_{0.7}Ga_{0.3}As$ and 100 nm GaAs, i.e.: the first half of the PC slab at the same temperature. At 520°C and 5.6×10^{-2} mPa arsenic pressure, an amount corresponding to 2.6 monolayers (MLs) of InAs was supplied at a growth rate of 0.033 ML/s. The second half of the PC slab, a GaAs capping layer of 100 nm was then grown on top of the QDs at 480°C. As confirmed by a substrate grown under the same conditions but without capping layer, this yields an InAs QD density of about 400 QDs/μm^2 according to atomic force microscopy (AFM) as shown in Fig. 1(b) [4]. The growth mechanism of these self-assembled QDs (Stranski–Krastanow), results in a size distribution i.e.: the QDs as shown in Fig. 1(b), vary from 20 to 40 nm in lateral size and are on average 6 nm in height. The triangular lattice PC pattern with air-holes of radius $r = 0.3a$ where a is the PC lattice constant was introduced into the slab layer using HI/Xe inductively coupled plasma etching. The sacrificial layer underneath was removed by HCl wet etching. The resulting symmetrically air-clad structure confines light vertically by total internal reflection and in plane by its two-dimensional (2D) photonic bandgap. The nanocavities consist of a line of three missing air-holes [L3 donor type, shown in Fig. 1(a)]. A series of samples was prepared with a ranging from 350 to 490 nm. The physical thickness t of the 2D PC slab is constant 200 nm, so the thickness relative to the lattice constant varies between samples. This causes the theoretical Q factors to range from 3400 for $t = 0.4a$ ($a = 490$ nm) to 5000 for $t = 0.6a$ ($a = 350$ nm) according to three-dimensional (3D) finite-difference time-domain (FDTD) simulations.

3. Experimental work

Figure 2(a) shows a schematic of the experimental setup used for reflectance measurements. Light from an intensity stabilized, tunable, and continuous wavelength laser source is focused onto a L3 nanocavity area by a microscope objective. The wavelength resolution of the setup is 0.3 pm. The focus diameter is about 1 μm. The incident light is linearly polarized at a 45° angle relative to the polarization of the nanocavity's zeroth order mode. The reflected light is collected by the same objective, but only the portion which is polarized 90° to the incident beam is detected by the photo diode. This cross-polarization configuration greatly improves the visibility of the nanocavity mode and successfully suppresses unwanted background, i.e.: light which is not reflected by the nanocavity, as clearly seen in

Fig. 2(b). In Fig. 3 the reflectance spectra of nanocavities ranging linearly from $a = 350$ to 390 nm in 10 nm steps are compared to their conventional PL spectra. Figure 3(a) shows the reflectance measurement results, where a distinct peak was observed for each of the nanocavities' zeroth order modes with an almost constant wavelength interval between them. Comparing Fig. 3(a) and (b), good correspondence of the peak wavelengths between the reflectance and PL data can be seen, which also supports the validity of the reflectance measurements. The QD PL emission from an area without PC at room temperature (RT) is shown in the inset of Fig. 3(b). The ground state has an inhomogeneous broadening of 30 meV resulting from the growth mechanism and ranges up to 1350 nm at RT. This means that QD emission can couple to the nanocavity mode if a is smaller or equal to 380 nm and therefore the behavior of the nanocavities will be active, whereas for a greater than 390 nm, the nanocavities can be assumed to be passive as they are completely detuned from QD related emission. This data demonstrates that the reflectance method can appraise both active and passive nanocavities unlike PL methods that can probe only active nanocavities. To show the validity of this concept as a proof of principle, measurements were extended to lattice constants a ranging up to 490 nm, as shown in Fig. 4. The error bars correspond to standard deviation (SD). The experimental data was fitted by a Lorentzian-like function that also accounts for Fabry–Pérot resonances and the Fano effect. There is a certain fluctuation in Q factor among samples originating from the fabrication process, but also a thin film effect. Around 1365 nm the PC slab layer acts as a thin film of t of about half wavelength where the coupling between the nanocavity and the light beam vertical to the slab becomes minimal. This effect is responsible for a spread in Q values over some wavelength range where the coupling constant remains small (1360–1480 nm), as indicated in Fig. 4. In this range, the ratio of cavity-resonant peak maximum to background noise floor due to surface scattering is much smaller for the reflectance spectra and thus there is greater deviation from the theoretical Q than elsewhere. However, the trend of Q values agrees well with the theoretical prediction, i.e.: Q of about 3400 for $a = 490$ nm and of about 5000 for $a = 350$ nm due to the relative thickness t.

4. Conclusion

We have demonstrated that the reflectance method has the ability to probe both of active and passive nanocavities over a wide range of wavelengths. The overall trend of the nanocavities' Q values measured agrees well with the theoretical prediction according to 3D FDTD calculations. We think that this method can also be used to gain insight into the fundamentals of the QD–nanocavity system's interaction mechanism and believe that this will contribute to the development of applications and encourage experiments in the future.

5. References

[1] Y. Akahane, T. Asano, B.-S. Song, and S. Noda: *Nature* **425**, 944 (2003).
[2] S. Noda, M. Fujita, and T. Asano: *Nat. Photon.* **1**, 449 (2007).
[3] Y. Takahashi, H. Hagino, Y. Tanaka, B.-S. Song, T. Asano, and S. Noda: *Opt. Express* **15**, 17206 (2007).
[4] W. Stumpf, M. Fujita, M. Yamaguchi, T. Asano, and S. Noda: *Appl. Phys. Lett.* **90**, 231101 (2007).
[5] W. Stumpf, M. Fujita, M. Yamaguchi, T. Asano, and S. Noda: *Jap. Soc. of Appl. Phys. Conf.*, 7p-R-14, Sapporo, Japan, Sept. 7 (2007).
[6] D. Englund, A. Faraon, I. Fushman, N. Stoltz, P. Petroff, and J. Vučković: *Nature* **450**, 857 (2007).

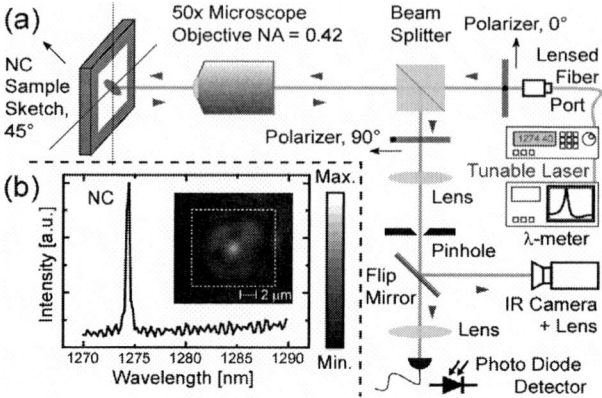

Fig. 2 (a) Simplified schematic of the measurement setup. (b) Experimental reflectance spectrum for a L3 nanocavity (NC) of Q ~4000 ($a = 350$ nm). Inset of (b): Infrared reflection in false colors at the nanocavity's resonant wavelength at RT.

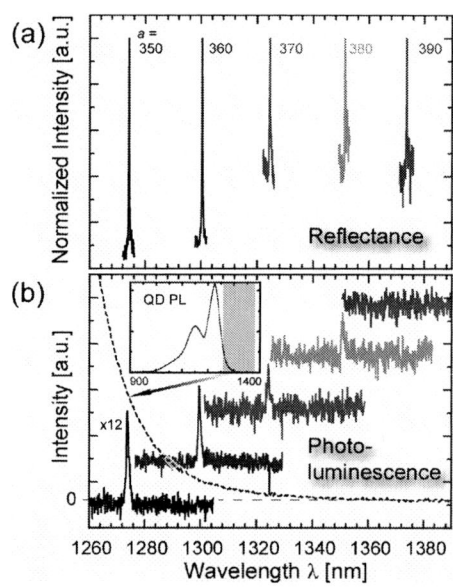

Fig. 3 (a) L3 nanocavity reflectance spectra. (b) The corresponding mode PL spectra at RT with offset added for clarity. Inset: QD PL without PC at RT.

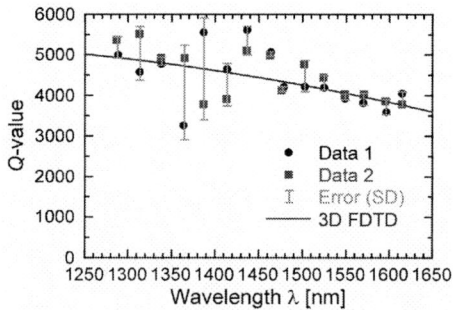

Fig. 4 Survey of Q-values for nanocavities where $a = 350$ - 490 nm.

Synthesis and characterization of ZnO nanorod arrays and their integration into polymer solar cells

Jing-Shun Huang[1], Chen-Yu Chou[1], Chun-Yu Lee[1], and Ching-Fuh Lin[1,2], *Senior Member, IEEE*

[1]Institute of Photonics and Optoelectronics, National Taiwan University; [2]Graduate Institute of Electronics Engineering, and Department of Electrical Engineering, National Taiwan University
Taipei, 10617 Taiwan, Republic of China
cflin@cc.ee.ntu.edu.tw

Abstract—ZnO nanorod arrays (ZNAs), grown from aqueous solutions, were integrated into polymer solar cells. Morphological, crystalline and optical properties of ZNAs were studied. Infiltration-improved polymer/ZNAs solar cells by adding fullerene were assembled and characterized.

Keywords-polymer solar cell, nanostructures, zinc oxide

I. INTRODUCTION

Organic solar cells (OSCs), based on soluble conjugated polymer blends are among the most promising ones of many recent photovoltaic device architectures that have great potential to provide efficient solar-to-electrical power conversion, as well as many advantages such as mechanical flexibility, light weight, and low cost [1], [2]. For practical application, it is still limited by several factors such as low-charge mobility and short exciton diffusion lengths, etc [3], [4]. One promising solution to such impasse is to employ inorganic nanostructures with high electron affinity and high electron mobility.

The environmental friendly and low-cost ZnO nanorod array (ZNAs) is particularly well suited for this application as they can be grown using a hydrothermal method [5], [6]. It had been demonstrated that OSCs could be fabricated using the combination of the P3HT:PCBM blend and ZNAs [7], [8]. These studies indicate that the ZNAs play an important role in collecting and transporting electrons. However, the photogenerated current of the devices was not high in these reports. For example, the highest short-circuit current density (Jsc) in [7] is 10.0 mA/cm^2, and the highest Jsc in [8] is 9.6 mA/cm^2. Both have not exceeded 10 mA/cm^2. It is suggested that the poor infiltration of polymer into ZnO nanorod results in low Jsc.

In the beginning of this work, we synthesize and characterize ZNAs with different heat treatments. Our investigation shows that ZNAs with the lowest-temperature heat treatment have the smallest diameter, and are aligned most vertically to the substrates. Then, the ZNAs were integrated into polymer solar cells. We report a new method to improve both the short-circuit current density and the infiltration in polymer/ZNAs solar cells by introducing fullerene as the electron transport layer. We have demonstrated that the Jsc can be increased (larger than 16 mA/cm^2), and concluded that the fullerene can help the polymer infiltrate into ZnO nanorods well.

This work was supported in part by the National Science Council, Taiwan, Republic of China, with Grant Nos. NSC96-2221-E-002-277-MY3 and NSC96-2218-E-002-025

II. CURRENT RESULTS

A. ZnO nanorod arrays

Hydrothermal growth of the ZNAs on the ZnO sol-gel thin films was investigated, where the ZnO thin films were heated at different temperatures before the growth of ZNAs. Fig. 1(a) shows relation between the diameter of ZnO nanorods and the heating temperature of ZnO thin films. As the heating temperature decreases from 900 to 130 °C, the diameter of ZnO nanorods decreases from 250 to 60 nm. Fig. 1(b) shows the X-ray diffraction (XRD) patterns of the ZNAs corresponding to those shown in Fig. 1(a). The insets show that the surface morphologies of corresponding ZNAs. The ZNAs are grown together at the same temperature of 90 °C, while the thin films were heated from 130 to 900 °C. It is noticeable that for the sample at 130 °C, the XRD pattern shows only the (002) diffraction peak. In addition, the intensity of (002) diffraction peak is strongest, compared to other samples at higher temperatures, implying that the ZNAs are high-quality single crystals growing along the c-axis direction with a preferential orientation perpendicular to the substrates. Fig. 1(c) shows the photoluminescence spectrum of ZNAs at 130 °C. An evident ultra-violet near band edge emission peak at 380 nm is observed, which originates from the excitonic recombination through an exciton-exciton collision process [9]. These vertical ZNAs are highly suitable for use in ordered nanorod-polymer solar cells as described in the following section.

B. Polymer/ZNAs solar cells

We have fabricated ZnO/ poly(3-hexylthiophene) (P3HT) : (6,6)-phenyl C61 butyric acid methyl ester (PCBM) solar cells successfully with an inverted sandwich structure of indium tin oxide (ITO) / ZNAs / PCBM / P3HT:PCBM blend / Poly(3,4-ethylenedioxylenethiophene):polystyrene sulfonic acid (PEDOT:PSS) / Pt. It is notable that a PCBM layer between ZNAs and P3HT:PCBM blend is introduced in the device. For comparison, another sample without this layer was prepared. The results for the both devices parameters measured at 100 mW/cm^2 from a 150 W Oriel solar simulator using an air mass 1.5 global (AM 1.5G) filter are summarized in Table 1. The device without the PCBM layer exhibited an open circuit voltage (Voc) of 0.24 V, a short circuit current density (Jsc) of 10.34 mA/cm^2, a calculated fill-factor (FF) of 0.38, and a power conversion efficiency (PCE) of 0.94 %. Interestingly, the device with the PCBM layer exhibits a significant improvement in the photovoltaic performance. The Jsc and

PCE are found to increase significantly to 16.26 mA/cm² and 1.33 %, respectively. The introduction of the extra PCBM layer results in efficient collection of charges and a factor of 1.57 increase in Jsc over the deivce without the extra PCBM layer. With the extra PCBM layer, the polymer in the device was found to infiltrate well. In contrast, the infiltration of the polymer into the ZnO nanorods became poor for the devices without the extra PCBM layer. Moreover, without the extra PCBM layer, there were some void spaces between the ZnO nanorods. This resulted in the reduction of the interfacial area between the ZnO nanorods and the polymer.

TABLE I.   DEVICES PARAMETERS WITH AND WITHOUT THE PCBM LAYER

Devices	PCE (%)	Voc (V)	Jsc (mA/cm²)	FF (%)
Without PCBM	0.94	0.24	10.34	38
With PCBM	1.33	0.245	16.26	33

III.   CONCLUSION

This work provides a route to fabrication of low-cost highly oriented ZNAs at low temperature. At the temperature of 130 °C, the ZnO nanorod arrays align very vertically with growth along the c-axis direction. Our investigations show that the extra PCBM layer improves the infiltration of the photoactive layer into ZNAs. After introducing the extra PCBM layer, the short circuit current density of the solar cell increases to 16.26 mA/cm². This value is higher than previous reports in this system. The extra PCBM layer can reduce void spaces between ZnO nanorods and increase the interfacial area between the ZnO nanorods and photoactive layer. This results in the increase in the photocurrent and PCE.

REFERENCES

[1]   G. Yu, J. Gao, J. C. Hummelen, F. Wudl, and A. J. Heeger, "Polymer Photovoltaic Cells: Enhanced Efficiencies via a Network of Internal Donor-Acceptor Heterojunctions," *Science,* vol. 270, pp. 1789-1791, Dec 1995.

[2]   C. J. Brabec, N. S. Sariciftci, and J. C. Hummelen, "Plastic Solar Cells," *Advanced Functional Materials,* vol. 11, pp. 15-26, Feb 2001.

[3]   N. C. Greenham, X. Peng, and A. P. Alivisatos, "Charge separation and transport in conjugated-polymer/semiconductor-nanocrystal composites studied by photoluminescence quenching and photoconductivity," *Physical Review B,* vol. 54, p. 17628, Dec 1996.

[4]   J. E. Kroeze, T. J. Savenije, M. J. W. Vermeulen, and J. M. Warman, "Contactless determination of the photoconductivity action spectrum, exciton diffusion length, and charge separation efficiency in polythiophene-sensitized TiO2 bilayers," *Journal of Physical Chemistry B,* vol. 107, pp. 7696-7705, Aug 2003.

[5]   L. Vayssieres, "Growth of Arrayed Nanorods and Nanowires of ZnO from Aqueous Solutions," *Advanced Materials,* vol. 15, pp. 464-466, Mar 2003.

[6]   G. Hua, Y. Zhang, C. Ye, M. Wang, and L. Zhang, "Controllable growth of ZnO nanoarrays in aqueous solution and their optical properties," *Nanotechnology,* vol. 18, p. 145605, Mar 2007.

[7]   D. C. Olson, J. Piris, R. T. Collins, S. E. Shaheen, and D. S. Ginley, "Hybrid photovoltaic devices of polymer and ZnO nanofiber composites," *Thin Solid Films,* vol. 496, pp. 26-29, Feb 2006.

[8]   K. Takanezawa, K. Hirota, Q. S. Wei, K. Tajima, and K. Hashimoto, "Efficient charge collection with ZnO nanorod array in hybrid photovoltaic devices," *Journal of Physical Chemistry C,* vol. 111, pp. 7218-7223, May 2007.

[9]   C. Bekeny, T. Voss, H. Gafsi, J. Gutowski, B. Postels, M. Kreye, and A. Waag, "Origin of the near-band-edge photoluminescence emission in aqueous chemically grown ZnO nanorods," *Journal of Applied Physics,* vol. 100, p. 104317, Nov 2006.

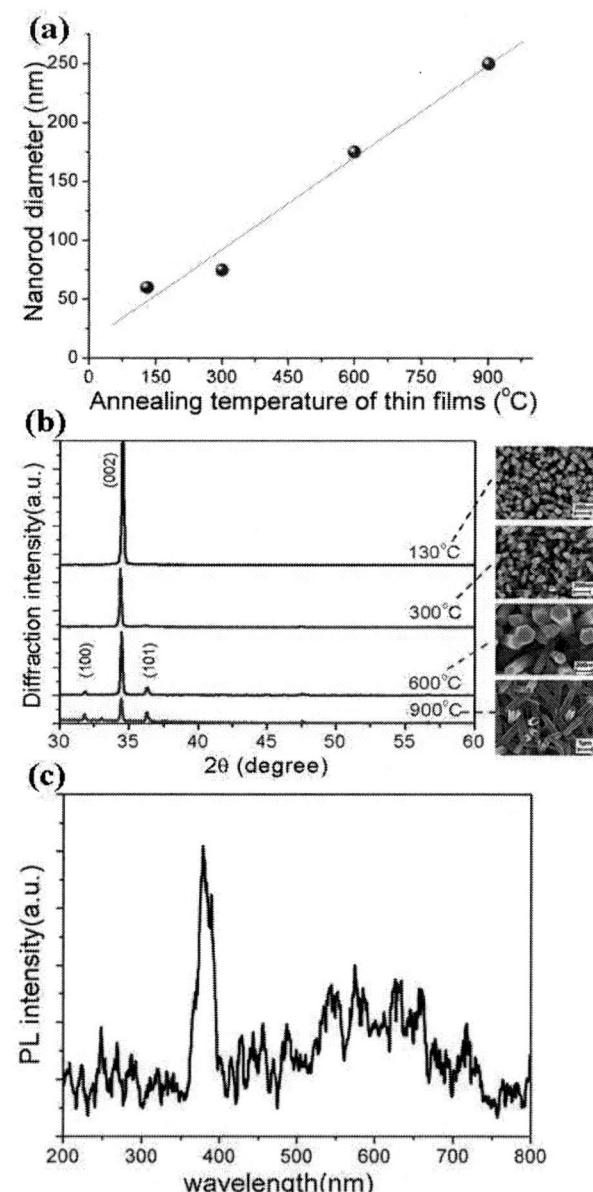

Figure 1.   (a)Dependence of diameter of ZNAs on annealing temperature; (b) XRD spectra of ZNAs with ZnO thin films heated from 130 to 900 ℃. Insets show the corresponding ZNAs; (c) PL spectra of ZNAs with heting temperature of ZnO thin film at 130 ℃ (excitation wavelength: 266nm)

PARALLEL ASSEMBLY OF NANOWIRES USING LATERAL-FIELD OPTOELECTRONIC TWEEZERS

Aaron T. Ohta, Steven L. Neale, Hsan-Yin Hsu, Justin K. Valley, and Ming C. Wu

Berkeley Sensor & Actuator Center (BSAC) and Department of Electrical Engineering and Computer Sciences,
University of California, Berkeley, CA 94720, USA, E-mail: aohta@eecs.berkeley.edu

ABSTRACT

We report on the parallel manipulation and assembly of nanowires using paired virtual optical tips projected on lateral-field optoelectronic tweezers. Precise position and angular control has been demonstrated on four 80-nm-diameter silver nanowires.

INTRODUCTION

An ongoing challenge for the mass production of nanowire-based electronics is the controlled assembly of single nanowires. Nanowire fabrication using "bottom-up" approaches presents difficulties to integration with heterogeneous material systems. Post-synthesis integration circumvents these issues, but has its own limitations. Post-synthesis assembly of individual nanowires and carbon nanotubes has been demonstrated using mechanical manipulators [1, 2] and optical tweezers [3, 4]; however, the parallel processing capabilities of these tools are limited. Single nanowires have also been assembled using the dielectrophoretic forces produced by microfabricated metal electrodes, but the trap locations and trapping patterns are static [5]. Optoelectronic tweezers uses optically-induced dielectrophoretic forces to assemble single nanowires in parallel, but is limited to aligning nanowires in a direction normal to the photoconductive surface of the device [6]. Using a lateral-field optoelectronic tweezers device (LOET), we demonstrate the parallel assembly of single nanowires, with control over the nanowire position and rotational orientation. This technology has the potential to fabricate nanowire electronics in a parallel assembly process.

LATERAL-FIELD OPTOELECTRONIC TWEEZERS (LOET)

Nanowires in an electric field experience a torque on their induced dipole, aligning them in parallel to the electric field lines. Thus, in order to use optoelectronic tweezers to assemble nanowires parallel to the plane of a substrate, electric fields must be generated laterally across the device surface. This is achieved using a lateral-field optoelectronic tweezers device. We have previously demonstrated nanowire trapping using an LOET device [7, 8]; however, this device did not provide control over the rotational orientation of the nanowires.

A new version of the LOET device has been fabricated that provides control of the in-plane orientation of nanowires. This device consists of an interdigitated array of 100-nm-thick aluminum electrodes on an oxidized silicon wafer (Fig. 1a). A 0.75-μm-thick amorphous silicon (a-Si) layer is blanket deposited over the aluminum electrodes using plasma-enhanced chemical vapor deposition. The electrode fingers are separated by gaps of 10 or 25 μm. The gap between the electrodes forms the active area for nanowire assembly.

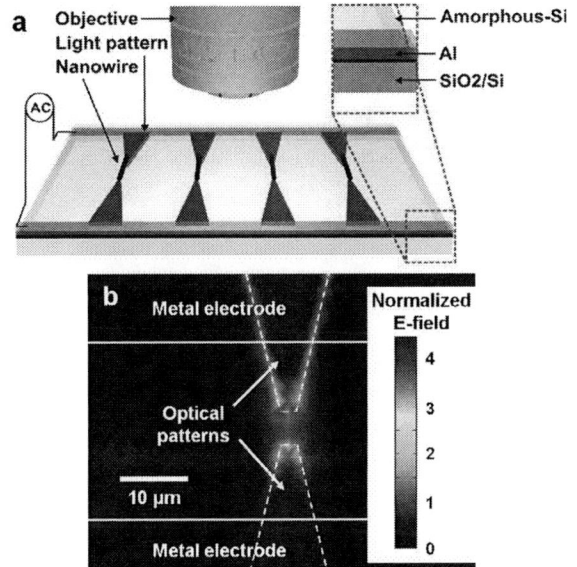

Fig. 1. (a) Schematic of lateral-field optoelectronic tweezers (LOET) for nanowire assembly. The device consists of an unpatterned amorphous silicon layer over an aluminum electrode array, fabricated on an oxidized silicon wafer. Paired triangular optical patterns create nanowire traps in between the metal electrodes. (b) Finite-element simulation of the electric field profile across the LOET electrodes.

The principle of operation for this version of the LOET is different from previous devices. An AC bias is placed across the electrode arrays, resulting in a uniform electric field between electrode fingers. To create a nanowire trap, optical patterns created by a digital micromirror device (DMD) are projected onto the LOET device. The optical patterns act as virtual electrodes by lowering the impedance of the a-Si in the illuminated areas. Thus, the optical patterns function as extensions of the metal electrodes. Paired triangular patterns are used to create a strong electric field gradient at the tips of the illuminated areas (Fig. 1b). The nanowires are attracted to these areas of strong electric field, and align between the triangular patterns. A typical triangular optical pattern has a tip diameter of 2 μm and a taper angle of 14 degrees. The gap between

978-1-4244-1917-3/08/$25.00 ©2008 IEEE

paired triangular patterns is adjusted to approximate the length of the trapped nanowire, which is typically 5 μm.

NANOWIRE ASSEMBLY

Silver nanowires with diameters of 80 to 100 nm and lengths of 1 to 10 μm were suspended in ethanol, and introduced into the LOET device. The optical patterns can be used to transport nanowires parallel to the plane of the LOET device, and can control the nanowire orientation in both the *x*- and *y*-directions (Fig. 2). Rotational control is achieved by adjusting the relative alignment of the triangular trapping patterns. Continuous rotation control has been performed over a range of ±28 degrees. Further optimization should result in a larger range of orientation control.

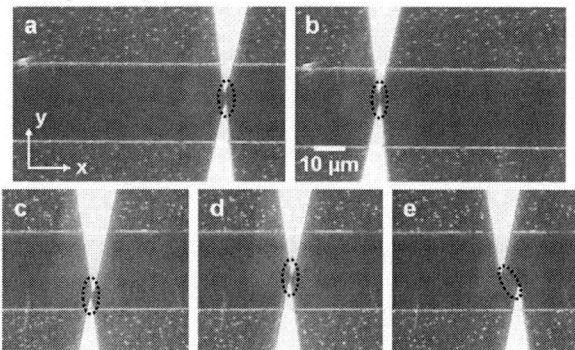

Fig. 2. Trapping of silver nanowires using lateral-field optoelectronic tweezers. Aluminum electrodes are visible underneath the amorphous silicon layer, at the top and bottom of each image. Optical patterns, visible as the bright areas, are used to create nanowire traps. The nanowire positions are indicated by dashed circles. (a, b) Transport of a nanowire in the negative x-direction at a rate of approximately 3 μm/s. (c, d) Transport of a nanowire in the positive y-direction. Other nanowires are weakly trapped at the edges of the electrodes, but can be moved by the LOET trap patterns. (e) Rotational control of a nanowire is achieved by moving the tips of the optical patterns relative to each other.

One of the advantages of LOET is its parallel manipulation capabilities. The proof-of-concept is demonstrated here with the assembly of four individual silver nanowires into a regularly-spaced array (Fig. 3). Each nanowire is trapped in parallel by a pair of optical patterns. The nanowires are positioned at with a 200-kHz AC signal at a voltage of 300 mVpp. Once the nanowires are trapped in the desired locations, the voltage is increased to 2.8 Vpp to anchor the nanowires to the surface.

CONCLUSION

We have demonstrated the parallel assembly of individual silver nanowires on a lateral-field optoelectronic tweezers device. Nanowires can be positioned with control over their *x*- and *y*-positions and rotational orientations. This technique enables parallel nanowire assembly for nanowire electronics.

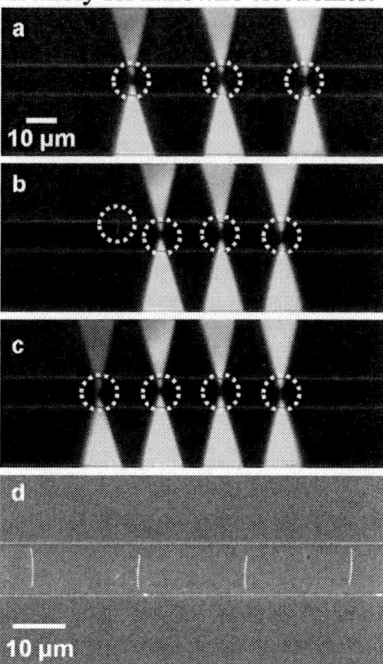

Fig. 3. Fabrication of a silver nanowire array. (a) Three nanowires have been trapped in parallel. (b) The nanowires are arranged in a closer, more regular spacing. Another nanowire can be seen to the left of the array. (c) The fourth nanowire is trapped by another set of optical patterns. The applied voltage is increased to assemble the nanowires on the surface of the device. (d) An SEM image of the nanowire array.

ACKNOWLEDGMENT

The authors thank Jiaxing Huang for providing the silver nanowires.

REFERENCES

[1] C. Thelander and L. Samuelson, "AFM manipulation of carbon nanotubes: realization of ultra-fine nanoelectrodes," *Nanotechnology*, vol. 13, pp. 108-113, 2002.

[2] D. J. Sirbuly, et al., "Optical routing and sensing with nanowire assemblies," *Proc. Natl Acad. Sci. U. S. A.*, vol. 102, pp. 7800-7805, 2005.

[3] R. Agarwal, et al., "Manipulation and assembly of nanowires with holographic optical traps," *Optics Express*, vol. 13, pp. 8906-8912, 2005.

[4] P. J. Pauzauskie, et al., "Optical trapping and integration of semiconductor nanowire assemblies in water," *Nature Materials*, vol. 5, pp. 97-101, 2006.

[5] H. W. Seo, C. S. Han, S. O. Hwang, and J. Park, "Dielectrophoretic assembly and characterization of individually suspended Ag, GaN, SnO2 and Ga2O3 nanowires," *Nanotechnology*, vol. 17, pp. 3388-3393, 2006.

[6] A. Jamshidi, et al., "Dynamic manipulation and separation of individual semiconducting and metallic nanowires," *Nature Photonics*, vol. 2, pp. 85-89, 2008.

[7] A. T. Ohta, et al., "Optically Controlled Cell Discrimination and Trapping Using Optoelectronic Tweezers," *IEEE J. Selected Topics in Quantum Electronics*, vol. 13, pp. 235-243, 2007.

[8] A. T. Ohta, et al., "Trapping and transport of silicon nanowires using lateral-field optoelectronic tweezers," in *Proc. Conference on Lasers and Electro-Optics*, pp. 828-829, 2007.

Sub-30nm alignment accuracy between layered photonic nanostructures using optimized nanomagnet arrays

Anthony J. Nichol, George Barbastathis

Department of Mechanical Engineering – Massachusetts Institute of Technology

We present experimental results indicating fine alignment of stacked nanopatterned silicon nitride membranes to better than 30nm. The initial coarse alignment via membrane folding was $2\mu m$, which gives a coarse-to-fine alignment improvement of more than 60×. Alignment was achieved by using the attractive force between arrays of nanomagnets patterned on the membranes before folding. Optical gratings and circular voids forming a 3D photonic crystal-like structure were patterned before membrane release and self-aligned after folding, as shown in Figures 1 and 2. Alignment performance was maintained during multiple folding and unfolding operations.

Layer-to-layer alignment of nanopatterned membranes can be achieved by active or passive methods. Passive methods do not require feedback, and include mechanical [1], surface-tension [2], and magnetic forces [3]. Surface-tension and mechanical features work well for chip placement and wafer-bonding alignment. However, magnetic forces show the most promise for ultra-precise multi-layered nanophotonic and optical MEMS applications because of favorable scaling. We use the long-distance interaction between the nanomagnets and an external field for actuation and coarse alignment during assembly, and the interaction between nanomagnets on the almost-folded layers for fine alignment just before folding is complete. Nanomagnets can easily and accurately be integrated into most membrane assemblies via additive or subtractive thin-film processes. Since the magnet bond can be released and re-aligned, our approach is favorable for optical MEMS devices that require repeated alignment of moving components. The passive alignment technique can be further improved to sub-10nm by extending the strategies described here, making the method readily applicable to layered nanophotonic devices such as 3D photonic crystals.

The process shown in Figure 3 was used to fabricate the membranes. A magnetic tri-layer of 3Cr\50Co\7Au (nm) was evaporated onto e-beam patterned PMMA on nitride membranes, followed by lift-off. Three nanomagnet sizes, 200nm×500nm, 0.3um×1um, and 1um×3um with in-plane magnetization preference were patterned in various distributions. Our theoretical analysis indicated two competing trends. On the one hand, a few larger magnets provide initial coarse alignment, while hundreds of smaller magnets provide the fine alignment. On the other hand, the small magnets are subject to lithographic errors and reduced in-plane magnetization preference from shape anisotropy; moreover, small magnets are subject to displacement errors equal to shift by one period (nanomagnet spacing). Therefore, the purpose of our experiment was to check whether the variable size distribution does indeed result in improved alignment or the advantage is cancelled by the smaller nanomagnets magnetizing out-of-plane or aligning by one-period shift errors.

The experimental results of Figure 4 prove that the second is the case. Alignment was documented using an SEM to measure the position error between fiducial marks on the two layers. We also tracked the dynamics of magnetic alignment (i.e. the trajectory of the folding surfaces) under an optical microscope. We found that the alignment accuracy was mostly limited by unwanted out-of-plane magnetization resulting in deviations of the folding trajectory from its optimal shape. The case of all-large magnets (labeled as "D" in Fig. 4) yielded average alignment of 30nm between layers, even though the statistical variation of alignment using other nanomagnet size distributions was not significant.

In future work, we will investigate higher external magnetic fields, reduction of the nanomagnets' size, and optimization of the nanomagnets' shape to enforce a single magnetic domain within each nanomagnet and thus improve layer-to-layer alignment even further.

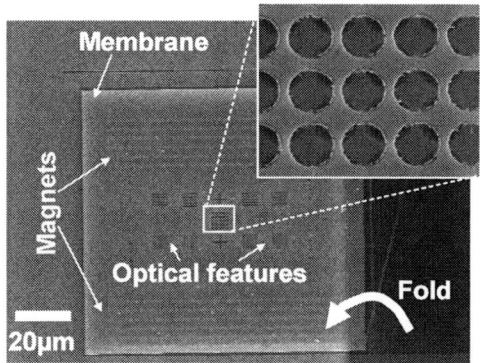

Fig. 1: Folded membrane with optical features aligned from nanomagnet attraction. Inlet shows aligned voids as precursor to 3D photonic crystal.

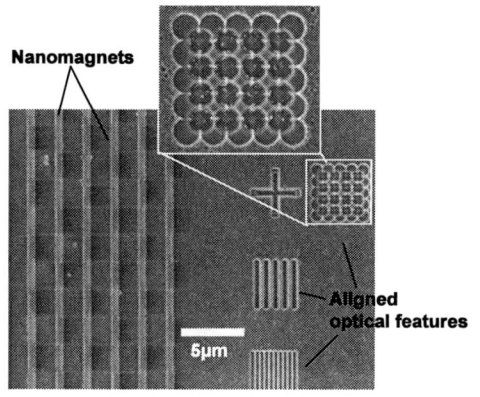

Fig. 2: Alignment of the nanomagnets and optical features shown after etching through the top membrane.

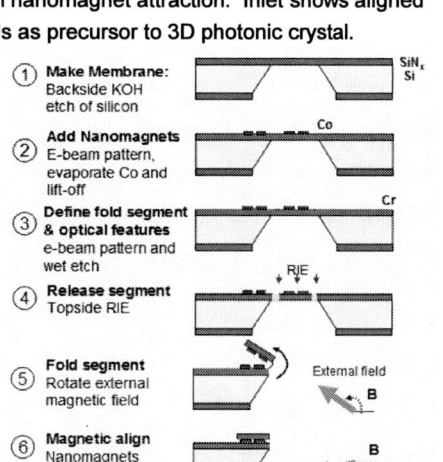

Fig. 3: Process flow for membranes patterned with nanomagnets, released, folded and aligned via nanomagnet attraction.

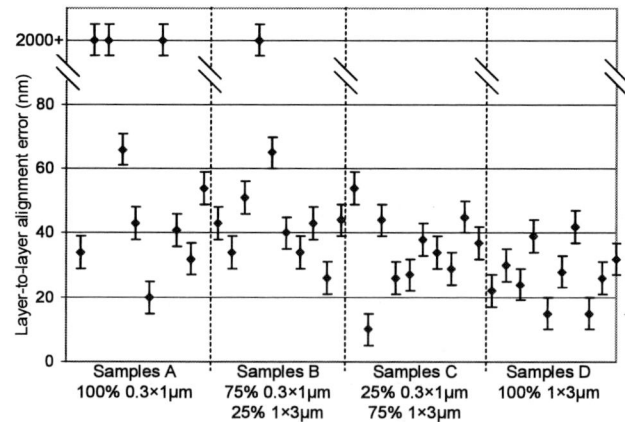

Fig. 4: Layer-to-layer alignment error for 4 different nanomagnet size distributions.

[1] A.H. Slocum. "Precision passive mechanical alignment of wafers." *J. MEMS.* Vol 12, Issue: 6. 2003. pp 826-834

[2] J.M. Kim. "3-D Highly Precise Self-Alignment Process Using Surface Tension of Liquid Resin Material." *IEICE Trans. on electronics.* Vol.E85-C, No.7(20020701) pp. 1491-1498

[3]] A.J. Nichol, P.S. Stellman, W.J. Arora, and G. Barbastathis, "Two-step magnetic self- membranes for 3D nanomanufacturing," *Microelectronic Engineering*, vol. 84, no. 5-8, pp. 1168-1171, Feb. 2007

How can we use nanophotonics to help in combating climate change?

Hilmi Volkan Demir

Department of Physics and Department of Electrical and Electronics Engineering
Nanotechnology Research Center and Institute of Materials Science and Nanotechnology
Bilkent University, Ankara, Turkey
www.bilkent.edu.tr/~volkan
volkan@bilkent.edu.tr

Climate change is one of the major issues that we face in our century [1]. Today as a scientific fact, it is observed that the emission rate of greenhouse gasses including carbon dioxide and nitrous oxide is on an alarmingly rapid rise [2]. Looking forward to the era of nanotechnology, one natural question to pose is "can nanotechnology help in combating this major problem?" According to recent studies [3-5], the answer is yes, but it is a cautious one. Nanotechnology and other technologies need to work hand in hand. A variety of possible solutions that nanotechnology offers include examples in photovoltaic technology for solar cells, hydrogen economy and fuel cells, fuel additives to increase the efficiency of diesel engines, batteries and supercapacitors for energy storage, and improved insulation for buildings. As a matter of fact, nanoparticles are already in industrial use for this purpose. For example, molybdenum disulphide nanocrystals are used as catalysts to remove potentially harmful sulphur compounds from crude oil. [3]

In nanophotonics, as a subfield of nanotechnology, one of our tasks is to incorporate custom-design nanomaterial into photonic devices and systems for the realization of different optical functionalities favorably controlled with external effects. In my research group, to this end, we work on the development and demonstration of new hybrid nanophotonic devices and systems that consist of different combinations of nanomaterials and nanostructures put together in hybrid architectures (e.g., [6-18]). These devices and systems operate in a wide range of optical spectrum from the ultraviolet (UV) to the visible to the infrared (IR) and serve a variety of optical functions for a long range of applications including light generation, modulation, sensing, imaging, displays, communications, alternative energy generation, and environmental decontamination. Examples of these nanomaterials employed in our devices include epitaxially grown quantum structures; in-solution synthesized semiconductor nanocrystals, metal nanoparticles, and polymer nanoparticles; electroplated, evaporated, or patterned metal nanostructures, and grinded semiconductor nanoparticles [6-18].

In particular to help combat climate change, in our research group, we ask ourselves "How can we use nanophotonics?" This question can of course be tackled in many different approaches. Among the others, we investigate three specific questions to seek answers for "how": *1.)* Can we generate high quality white light in solid state lighting based on LED platforms to reduce the global energy consumption? *2.)* Can we produce more efficient solar cells based on Si for alternative energy generation? *3.)* Can we offer means to reduce the amount of emitted CO_x, NO_x and other greenhouse gases?

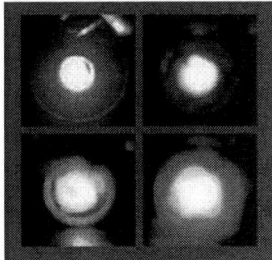

Fig. 1. Nanocrystal emitters hybridized on LEDs to generate and tune white light [6].

For the first question, we are working on nanocrystal hybridized white LEDs (Fig. 1 [6]) for the generation of high-quality, tunable white light [6-15]. Currently, about 20% of global energy production is consumed for lighting; LED based solid state lighting potentially offers 50% reduction in the energy consumption, corresponding to a significant carbon emission reduction (as large as 300 million tons annually). Currently available commercial LEDs, however, typically provide cool white light with a low color rendering index (~70). Using multiple combinations of nanocrystals, we achieved warm white light with high color rendering index (~80) meeting the requirements of future lighting [7]. Also, we successfully demonstrated white light generation and tuning using polymer-nanocrystal assemblies [8, 9], white emitting nanocrystal luminophores [10], and multiple color emitting heteronanocrystals [11, 12], both using blue [6, 7] and n-UV LED platforms [13, 15]. We also utilized metal nanostructures to control and enhance spontaneous emission from nanocrystal emitters [14].

For the second question, to produce alternative efficient energy sources, we developed photovoltaic nanocrystal scintillators hybridized on Si solar cells to extend solar conversion toward UV [16, 17]. Our nanocrystal hybridized scintillators led to significant enhancement in optical responsivity in UV (experimentally by two orders of magnitude enhancement) [16], and two fold enhancement in the solar conversion efficiency under white light illumination [17].

Finally, for the third generation, to help reducing greenhouse gas levels, we are working on photocatalytic nanocomposite systems for massive environmental decontamination (in collaboration with our industrial partner DYO) [18]. We demonstrated time evolution of optical spectral efficiency curves and strong size effect in our UV and visible photocatalytic organic-inorganic nanocomposites.

In conclusion, as emerging technologies enabled by nanophotonics to combat climate change, we developed and demonstrated *1.)* nanocrystal hybridized white light sources and those also embedded with plasmonic metal nanoparticles for the reduction of global energy consumption, *2.)* photovoltaic nanocrystal scintillators hybridized on solar cells for the generation of efficient green energy sources, and *3.)* photocatalytic nanoparticle and nanocomposite systems for massive environmental decontamination, all to contribute to the reduction of greenhouse gas amount.

References:
1. IPCC Climate Change 2007, "Is the Current Climate Change Unusual Compared to Earlier Changes in Earth's History?," FAQ chapter, p. 114 (2007).
2. IPCC Climate Change 2007, Human and Natural Drivers of Climate Change, Summary for Policymakers, p. 2 (2007).
3. Nature Nanotechnology, "Combating Climate Change", editorial, 2(6), p. 325 (2007).
4. Oakdene Hollins, "Enviromentally Beneficial Nanotechnologies, Barriers and Opportunities; a report for the Department of Environment, Food, and Rural Affairs," UK, May 2007.
5. Karen F. Schmidt, "Green Nanotechnologies," Woodrow Wilson International Center for Scholars, USA, April 2007.
6. S. Nizamoglu, T. Ozel, E. Sari, and H. V. Demir, Nanotechnology, 18(6), 065709 (2007).
7. S. Nizamoglu, G. Zengin, and H. V. Demir, Applied Physics Letters, 92, 031102 (2008).
8. H. V. Demir, S. Nizamoglu, T. Ozel, E. Mutlugun, I. O. Huyal, E. Sari, E. Holder, and N. Tian, New Journal of Physics, 9, 362 (2007).
9. I. O. Huyal, T. Ozel, U. Koldemir, S. Nizamoglu, D. Tuncel, and H. V. Demir, Optics Express, 16(2), 1115-1124 (2008).
10. S. Nizamoglu, E. Mutlugun, T. Özel, H. V. Demir, S. Sapra, N. Gaponik, A. Eychmüller, Applied Physics Letters 92, 113110 (2008).
11. S. Nizamoglu and H. V. Demir, Optics Express 16(6),3515-3526 (2008).
12. S. Nizamoglu, E. Mutlugun, O. Akyuz, N. Kosku Perkgoz, H. V. Demir, L. Liebscher, S. Sapra, N. Gaponik, A. Eychmüller, New Journal of Physics 10, 023026 (2008).
13. S. Nizamoglu and H. V. Demir, Nanotechnology, 18(40), 405702 (2007).
14. I. M. Soganci, S. Nizamoglu, E. Mutlugun, O. Akin, and H. V. Demir, Optics Express, 15(22), 14289-14298 (2007).
15. S. Nizamoglu and H. V. Demir, J. Opt. A: Pure Appl. Opt., Special Issue on Nanophotonics, 9(9), S419-S424 (2007).
16. E. Mutlugun, I. M Soganci, and H. V. Demir, Optics Express 16(6), 3537-3545 (2008).
17. E. Mutlugun, I. M. Soganci, and H. V. Demir, Optics Express, 15(3), 1128-1134 (2007).
18. S. Tek, E. Mutlugun, I. M. Soganci, N. Kosku Perkgoz, D. Yucel, G. Celiker, and H. V. Demir, Journal of Nano Photonics, 1, 011685 (2007).

Low-power Consumption Integrated Laser Doppler Blood Flowmeter with a Built-in Silicon Microlens

Yoshinori Kimura[1,3], Atsushi Onoe[1], Eiji Higurashi[2] and Renshi Sawada[3,4]

[1]Corporate Research and Development Laboratories, Pioneer Corporation,
6-1-2 Fujimi, Tsurugashima-shi, Saitama 350-2288, Japan
Tel: +81-49-279-2300, Fax: +81-49-279-1511, E-mail: yoshinori_kimura@post.pioneer.co.jp
[2]Research Center for Advanced Science and Technology, University of Tokyo, Komaba, Meguro-ku, Tokyo 153-8904, Japan
[3]Graduate School of Systems Life Science, Kyushu University, Motooka, Nishi-ku, Fukuoka 819-0395, Japan
[4]Department of Intelligent Machinery and Systems, Kyushu University, Motooka, Nishi-ku, Fukuoka 819-0395, Japan

Abstract

We propose a new structure for the integrated laser Doppler blood flowmeter which consists of an Si cavity and a photodiode and a laser diode within the cavity. A silicon cover formed with a converging microlens also functions as the package. This structure has a reduced rate of optical power loss in the sensor, resulting in significantly low power consumption.

Keywords: laser Doppler, blood flowmeter, microlens, AR coating,

1 INTRODUCTION

Continuous health monitoring using biomedical sensors has been widely studied to prevent some diseases. The detection of a car driver's drowsiness using biomedical sensors is also studied to prevent traffic accidents.[1] We think a wearable integrated laser Doppler blood flowmeter is the most effective candidate for such continuously monitoring biomedical sensors, since the blood flowmetry monitoring peripheral blood flow just beneath the skin is intimately correlated with health conditions and nervous systems such as sympathesis and parasympathesis related to feeling, emotion and drowsiness. The integrated laser doppler blood flowmeter has been fabricated based on micromachining techniques of polyimide waveguide formation and surface mounting by some of the authors.[2] The blood sensor is only approximately 3mm square, one-hundredth the size of conventional probe type instruments, however it can provide blood flow measurements as precise as those from conventional optical fiber type sensors.

In this paper, we propose a new structure of blood flow sensors with a silicon cover and a converging microlens which also functions as a package. The structure does not suffer from power loss based on coupling with light waveguide, leading to a drastically low operating consumption of power.

2 DEVICE DESIGN

As shown in fig.1, our integrated laser doppler blood flowmeter emits a laser beam almost vertically upward. A laser diode (LD) and a photodiode (PD) are mounted on the bottom surface of cavities fabricated by using anisotropic silicon etching technique. These are covered with a silicon-cover-plate of which the undersurface is partially processed into microlens by using photoresisit reflow technique. The mounted LD and PD are electrically connected with external electrode pads through silicon vias (TSV). Distributed feedback bragg-reflector LD (DFB-LD) with a wavelength in the 1300nm band is employed as the light source, and InGaAs based PD is used to detect the

optical signal backscattered from body tissue. The laser beam emitted from DFB-LD is reflected upward by an etched mirror formed by depositing gold onto the silicon (111) facet. The reflected laser beam incidents the internal microlens so that its divergence angle become narrower, and subsequently the laser beam emits to the body tissue through the laser emission window of the sensor chip. The integrated laser doppler blood flowmeter shown in fig.1 is not a hermetically-sealed structure, but this can be achived by some structural modification.

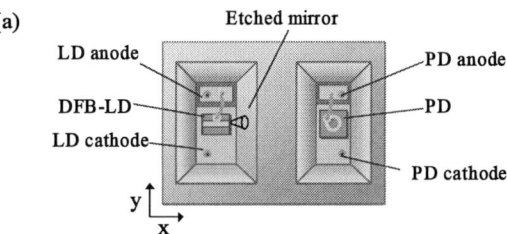

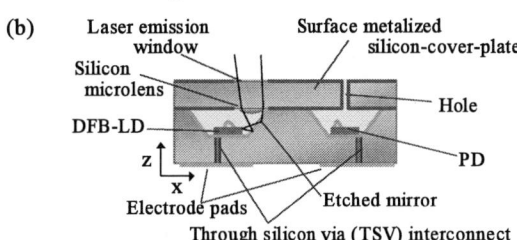

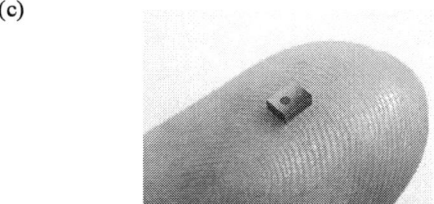

Figure 1. Schematic top view (a), cross sectional-view (b) and photograph (c) for the fabricated integrated laser Doppler blood flowmeter. The silicon cover plate is not shown in figure (a).

Figure 2 shows photograph of silicon microlens built into our sensor chip. Its diameter and height are both 400μm and 42μm, respectively. The diameter was designed larger than the laser beam diameter ($1/e^2$) emitted from DFB-LD, so that almost all laser power would be recieved by the microlens.

Figure 2. Silicon microlens with anti-reflection (AR) coating.

Figure 3 shows the beam profile at 7 mm apart from the surface of the sensor chip. The full width at half maximum (FWHM) of the beam profile was estimated to be 43% compared to the specification of DFB-LD. Accordingly, it was proved that the divergence angle of the laser beam emitted from DFB-LD was narrowed by the internal microlens.

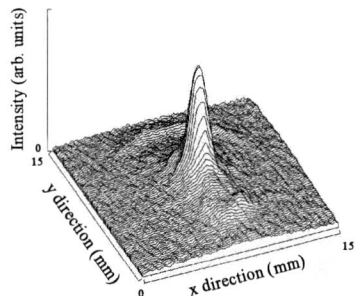

Figure 3. Laser Beam profile emitted from sensor chip.

Figure 4(a) shows the calculated reflectivity spectrum of the silicon surface, the reflectivity at a wavelength of 1300nm was estimated to be up to around 30%. Accordingly we applied an anti-reflection (AR) coating composed of a single layer of SiO_2 to both the silicon-microlens and the laser emission window to reduce the surface reflection loss.

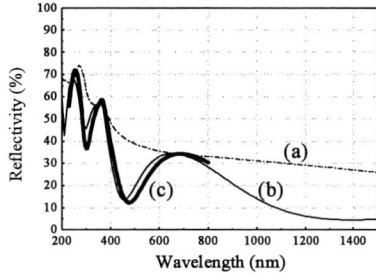

Figure 4. Calculated reflectivity without AR coating (a), calculated reflectivity with AR coating (b) and measured reflectivity with AR coating (c) from a silicon surface.

Calculated reflectivity spectrum of silicon covered with AR coating is shown in fig.4(b). Since the measured reflectivity spectrum, as shown in fig.4(c), was well fitting to the calculated reflectivity spectrum in the wavelength range of 250 to 800nm, it was confirmed that the AR coating was successfully fabricated. therefore, the reflectivity of one

silicon surface was expected to be reduced from 30% to 5% at the wavelength of 1300nm by AR coating. As a result, we could successfully realize such a low power consumption that the required laser power for blood flow measurements was only around 1mW, and the drive current for DFB-LD was 21mA.

3 RESULTS

Figure 5 shows the peripheral blood flow signal at a finger measured by our integrated laser Doppler blood flowmeter, it clearly shows the pulse wave pattern. The fluctuation of the blood flow could be also observed, which occurs due to the motion of hand-up and hand-down. In this way, it was confirmed that the peripheral blood flow could be measured.

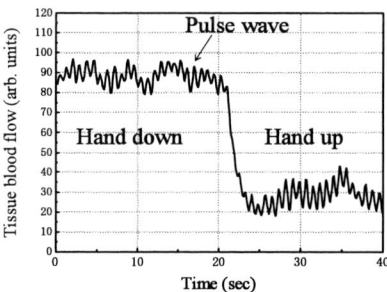

Figure 5. Blood flow signal measured by our integrated laser doppler blood flowmeter.

4 CONCLUSIONS

We proposed and fabricated a new structure of the integrated laser Doppler blood sensor with a silicon cover incorporating a converging microlens which also functions as the package and without power loss based on coupling with light waveguide. The silicon converging microlens functioning as a cover improved the efficiency of laser power extraction, and reduced the divergence angle of the laser beam emitted from the sensor chip. The AR coating applied to the surface of the silicon microlens reduced the reflection optical loss, and contributed to a reduction of the drive current for DFB-LD. The necessary laser power for blood flow measurements was only around 1mW, and the drive current for DFB-LD was 21mA. These are excellent characteristics from the viewpoint of the low power consumption and the safety of the laser.

ACKNOWLEDGEMENTS

This study was supported in part as research in the Core Research for Evolutional Science and Technology (CREST) program of the Japan Science and Technology Agency.

REFERENCES

[1] M. Yasushi, and M. Yanagidaira, "Estimating sleepiness during expressway driving", Journal of International Society of Life Information Science, Vol.21, No.2, pp.442-444, 2003.

[2] E. Higurashi, R. Sawada, and T. Ito, "An integrated laser Doppler blood flowmeter for wearable health monitoring system", Proc. IEEE/LEOS International Conference on Optical MEMS 2001, pp.49-50, 2001.

Reconfigurable liquid micro-lens system for variable fiber coupling

Riyaz P. Shaik, Wolfgang Mönch, Holger Krause, and Hans Zappe

Laboratory for Micro-optics, Department of Microsystems Engineering – IMTEK, University of Freiburg
Georges-Köhler-Allee 102, 79110 Freiburg, Germany
riyaz.shaik@imtek.uni-freiburg.de

Abstract—**A system capable of both two-dimensional lateral positioning and focusing of a liquid micro-lens for variable fiber coupling is presented. Lateral positioning and focus adjustment is accomplished by electrowetting-on-dielectrics. The chip is fabricated on a transparent Pyrex substrate and is equipped with a two dimensional array of electrodes featuring a well-defined microstructure which allows accurate positioning of the liquid micro-lens. The system is used as a variable optical fiber coupling element, and measurements of optical power coupled to a fiber as a function of the applied voltage are presented. Applications such as adaptive wave front sensing, optical tweezers, or three-dimensional focus control for microscopy are envisaged.**

Index Terms—**Liquid micro-lens, electrowetting on dielectrics, fiber coupling, variable optical attenuator**

I. INTRODUCTION

ELECTROWETTING-on-dielectrics (EWOD) has proven to be an attractive means for the fabrication and control of tunable liquid micro-lenses. Both focal length tuning [1] and lateral positioning [2] of liquid lenses has been demonstrated using this technique and, using a complex substrate design, both actuation possibilities have been integrated in a single system [3].

We present here a system for two-dimensional (2D) liquid lens repositioning. Furthermore, we demonstrate the applicability of the system to variable fiber coupling based on its dynamic focusing capability. The system may ultimately be used for further applications requiring three-dimensional focus control.

II. THEORY

The theory behind EWOD is well known [4]: the contact angle formed by a liquid droplet on a dielectric surface may be varied by applying a bias between the droplet and the substrate, thereby changing the focus of the spherical liquid lens. If an array of electrodes is used instead of a single electrode, the contact angle may be reduced only in parts of the droplet, thereby inducing lateral movement of the liquid lens. We employ both mechanisms here.

For using a liquid lens to couple light into a fiber, the lens should provide the maximum optical output power for a given focal length. Coupling efficiency is maximized if the numerical aperture (NA) of the lens is smaller than the NA of the optical fiber, the the lens area fully overlaps the lateral extent of the illumination and the absorption losses are minimized. We address these considerations below.

III. FABRICATION AND PACKAGING

A schematic view of the 2D liquid micro-lens system is shown in Fig. 1. The transparent chip features an array of 64 (8×8) electrodes and is fabricated on a Pyrex wafer. The 64 electrodes are separated from each other by a finger structure to facilitate movement of a liquid lens across the electrode boundaries. A grid structure with a mesh size of $50 \, \mu m \times 50 \, \mu m \times 300 \, nm$ is etched into the top dielectric layer for precise positioning of the liquid lens [5]. On top of this grid, an indium tin oxide (ITO) electrode is sputtered for electrical contact to the liquid lens. For focal length tuning or lateral repositioning of the liquid lens, a voltage is applied between this ITO droplet contact electrode and one or more of the buried electrodes.

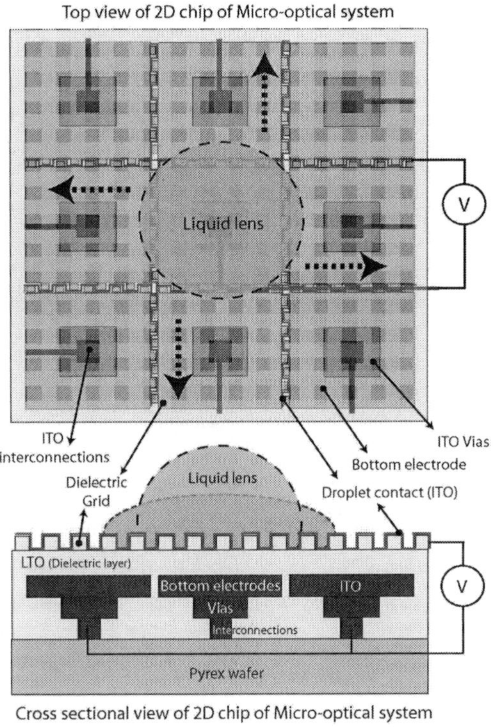

Fig. 1: Schematic with top view and cross sectional view of 2D liquid micro-lens system.

An overview of fabrication process steps is shown in Fig. 2. An ITO multi-layer is used for defining the electrical interconnections for the inner array of electrodes. After deposition and patterning of low temperature oxide (LTO) for each layer, a layer of ITO is sputtered into the patterned LTO structures. Three layers of ITO are required for definition of electric interconnections, vias, and the buried electrodes ($500\,\mu m \times 500\,\mu m$), respectively, as seen in steps 1-3.

On top of the buried ITO electrodes, an 800 nm LTO layer representing the dielectric layer for EWOD is deposited. Finally, a hydrophobic layer is deposited on top of the dielectric grid and the ITO droplet contact.

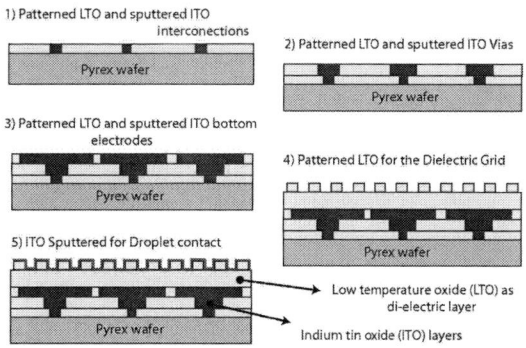

Fig. 2: Fabrication process of the transparent substrate chips for liquid micro-lens repositioning and focal length tuning.

The chip is packaged in a polyurethane-based housing and used a glass lid to result in a hermetically sealed liquid filled system. The liquids used are an aqueous solution for the lens and silicone oil for the ambient liquid. Polyurethane has superior performance when compared to silicone elastomers for packaging which comes into contact with of silicone oil.

IV. OPTICAL POWER MEASUREMENT

This 2D liquid micro-lens system was used for variable fiber coupling. As a source, a laser beam ($\lambda = 633\,nm$) with 1 mm beam diameter was used, such that the intensity of the beam was homogeneous over the micro-lens area; diameter of the micro-lens was $780\,\mu m$. At the output, a multimode optical fiber (core/cladding: $62.5\,\mu m/125\,\mu m$) was mounted above the system and positioned at the focal length of the liquid micro-lens. Starting from its initial state ($V = 0\,V$, with back focal length (BFL) = 4.57 mm), the optical power of the light collected by the fiber was measured using an optical power meter as shown in Fig. 3.

Without the lens in position (i.e., the illumination remains collimated), the optical power measured in the optical fiber was $18\,\mu W$. Placing the lens at biased $V = 0\,V$ into the system such that the the focus was on the fiber facet, a power of $166\,\mu W$ was coupled into the fiber. By increasing the applied voltage on the lens, the focus moved beyond the fiber facet, coupling efficiency was reduced and decreased with applied bias as shown in Fig. 4.

We see that a power variation of -10.63 dB was achievable using this approach. The configuration may be expanded to variably couple input optical power into an array of output fibers or waveguides, such that a novel approach for optical switching is suggested.

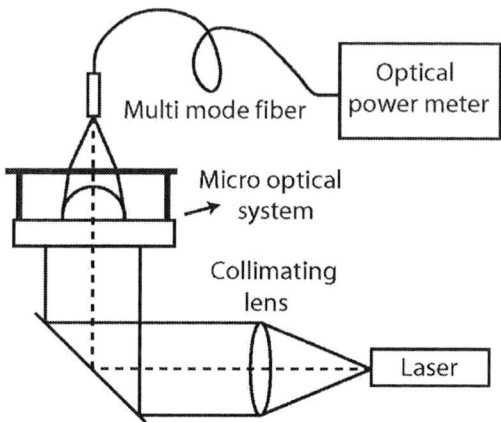

Fig. 3: Setup for demonstration of variable fiber coupling using the 2D liquid micro-lens system.

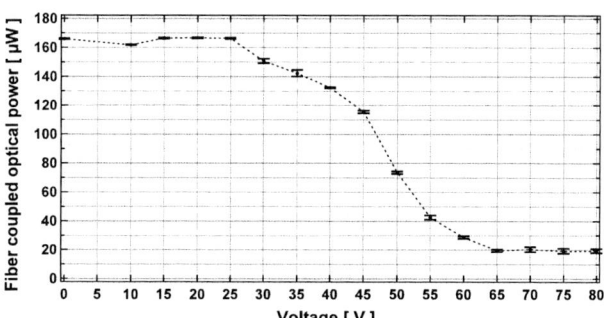

Fig. 4: Measured optical power coupled into the fiber by the liquid lens as a function of applied actuation voltage.

V. SUMMARY

Using repositionable variable-focus liquid micro-lenses, on a 2D electrode array, variable coupling of an optical signal into an output fiber was demonstrated. The technology, when further developed, may be of interest for adaptive wave front sensing, optical tweezers, or three-dimensional focus control.

ACKNOWLEDGEMENTS

The authors gratefully acknowledge financial support from the Landesstiftung Baden-Württemberg and Bernd Aatz for help in building the setup for optical power measurements.

REFERENCES

[1] B. Berge and J. Peseux, "Variable focal lens controlled by an external voltage: An application of electrowetting," *European Physical Journal E*, vol. 3, pp. 159–163, 2000.

[2] P. Paik, V. Pamula, and R. Fair, "Rapid droplet mixers for digital microfluidic systems," *LabChip*, vol. 3, pp. 253–259, 2003.

[3] R. Shaik, L. Lasinger, F. Krogmann, W. Mönch, and H. Zappe, "Fabrication and characterization of a repositionable liquid micro lens system," *Proceedings of IEEE Optical MEMS 2007*, pp. 139–140, 2007.

[4] B. Berge, "Electrocapillarité et mouillage de films isolants parl'eau," *C.R.Acad.SciParis*, vol. 317, pp. 157–163, 1993.

[5] F. Krogmann, R. Shaik, L. Lasinger, W. Mönch, and H. Zappe, "Reconfigurable liquid micro-lenses with high positioning accuracy," *Journal of Sensors and Actuators A*, vol. 143, pp. 129–135, 2008.

High Resolution Integrated Microfluidic Fabry-Perot Refractometer in Silicon

R. St-Gelais, J. Masson and Y.-A. Peter

Ecole Polytechnique de Montréal, Engineering Physics Department

P.O. Box 6079, Station Centre-Ville, Montréal (QC), H3C 3A7 CANADA

Email: {raphael.st-gelais, yves-alain.peter}@polymtl.ca

Tel + 1 514 340 4711 x 3100, Fax + 1 514 340 3218

Abstract—We present a high resolution, robust and low cost integrated refractometer for microfluidic systems. The device is made of two Bragg reflectors vertically etched in silicon to form an in-plane Fabry-Perot filter. Liquids are injected in the cavity through a microfluidic channel and the variation of the refractive index induces a shift of the resonance wavelength. A sensitivity of 920nm/RIU (Refractive Index Units) and a resolution of less than 10^{-3} are obtained and are in good agreement with optical simulations. This resolution is, to our knowledge, the lowest ever reported for an integrated microfluidic Fabry-Perot refractometer.

I. INTRODUCTION

The measurement of the refractive index of gases and liquids has been an important field of research during the last century and still attracts a lot of attention since it is related to many physical parameters of materials. Recently, a lot of work has been done toward the integration of refractive index sensors to microfluidic systems [1]. Many sensors measure the interaction of an evanescent wave at the interface between the sample and a metal (surface plasmon resonance) or a dielectric (integrated wave guide, microcavities). These methods allow very high resolution measurements, sometimes better than 10^{-8}, but the interaction depth at the interface with the sample is typically smaller than $1\mu m$, limiting the possibilities to measure the refractive index of bigger biological specimens such as cells. Single cell measurement is of great interest since it is low-cost, label free, and can be linked to the state or nature of a cell. For example, the effective refractive index can be related to the size and protein level of normal versus cancerous cells [2], or to the structure of a specific cell through the excitation of transverse modes in a Fabry-Perot cavity [3].

These applications require a Fabry-Perot cavity with mirror spacing of the order of a cell diameter, as well as integration to microfluidic systems. The proposed sensor fills these requirements and compares advantageously to similar devices [2], [4], [5], due to its simple fabrication, easy alignment, high resolution and robustness. The Bragg mirrors, microfluidic patterns and alignment grooves are fabricated simultaneously in one conventional microfabrication process. The grooves allow easy alignment of input and output fibers, which do not need any preparation steps such as reflective coating or splicing to collimating optics components. The length of the cavity can be designed from one to more than a hundred microns, depending on the targeted application. The measures yield high resolution

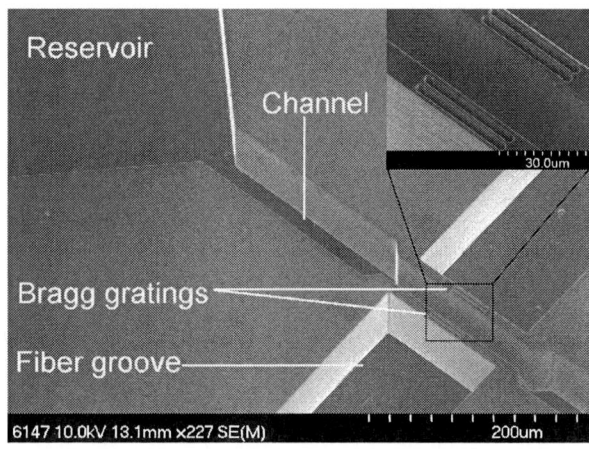

Fig. 1. SEM photograph of the Fabry-Perot sensor integrated with fiber alignment grooves, microfuidic channel and reservoir.

and high sensitivity: $2 \cdot 10^{-3}$ and 920nm/RIU respectively, as well as great repeatability.

II. SENSOR FABRICATION AND PRINCIPLE OF OPERATION

The sensor is presented in Fig. 1. The whole structure is defined by a single photolithography step. The silicon is then etched by deep reactive ion etching (DRIE). The etch depth is $70\mu m$ to allow the use of $125\mu m$ diameter conventional SMF28 optical fibers. The low roughness and high verticality of the optical surfaces are ensured by optimization of the BOSCH process, as previously reported [6].

Liquids placed in the reservoir flow in the microchannel by capillary effect and reach the Fabry-Perot cavity formed by the two Bragg reflectors. Light from a broadband light source (1520nm - 1620nm) is incident on the Fabry-Perot through one of the optical fibers placed in the alignment grooves. The transmitted light is collected by the second optical fiber, which is connected to an optical spectrum analyzer.

Each of the Bragg reflectors is made of three layers of silicon and two layers of air. The photomask is designed with $2.6\mu m$ and $1.7\mu m$ thick walls for silicon and air respectively. These values will be modified by the fabrication process because of the diffraction effects through the small openings of the photomask and because of a typical 300nm undercut

978-1-4244-1917-3/08/$25.00 ©2008 IEEE 17

of silicon during the DRIE process. The length of the cavity (25μm) is chosen such that the phase shift experienced by the reflexion on a Bragg reflector is negligible compared to the phase shift experienced during a round trip in the cavity, thus improving the sensitivity of the device [7].

We can easily design a Fabry-Perot having a cavity of any length between one and more than a hundred microns. A smaller length would decrease the sensitivity and increase the free spectral range. This could be useful for applications where measurements are to be performed on a large range of refractive indexes. The flexibility of photolithography could also allow the variation of the microchannel width to trap cells in the Fabry-Perot cavity, for example.

III. RESULTS AND DISCUSSION

The experimental and simulated results are presented in Fig. 2. The measurements are performed with certified refractive index liquids (Cargille Labs, series AA) with low temperature dependence (0.0004 RIU/°C). The simulated results are obtained by the transfer matrix method, considering a plane wave incident on perfectly parallel and flat surfaces [6]. Simulations considering the effects of a Gaussian beam and of verticality deviation of the walls could explain the transmission losses, as well as the broadening of the resonance peaks [8]. For now we approximate these effects by subtracting 23dB to the simulated results. The dimensions of the walls are measured in top view with a scanning electron microscope, showing an increase of about 1.1μm of the thickness of air layers caused by diffraction during photolithography. For the simulations, we also reduced the thickness of the silicon walls by 300nm at each interface to take into account the undercut of the DRIE process. These values are then slightly adjusted, respectively to 1.086μm and 314nm, to fit with the experimental results. Therefore, we propose that similar devices could be used as test structures to monitor precisely the thickness of vertically etched optical multilayers.

The sensor has a sensitivity of 920nm/RIU and was able to detect experimentally variations of $\Delta n = 2 \cdot 10^{-3}$ of the refractive index, as shown in Fig. 2 (b). For this value of Δn, the peaks were very easily distinguished. Therefore, a resolution of less than 10^{-3} can be reached. Such resolution is, to our knowledge, the highest ever reported for an integrated microfluidic Fabry-Perot refractometer. The sensor is very robust and produces highly reproducible measurements. The optical fibers were removed in order to clean the sensor with acetone and IPA between each measurement. This was done more than 30 times, out of cleanroom environment, and the sensor still produced the same spectrum for a given liquid.

IV. CONCLUSION

We reported a high resolution, robust and potentially low cost integrated microfluidic refractive index sensor in silicon. The sensor allows bulk refractive index measurement of liquids with a resolution of less than $\Delta n = 10^{-3}$ which is, to our knowledge, the lowest resolution ever reported for an integrated microfluidic Fabry-Perot refractometer. The integration of the sensor with microfluidic channels and reservoirs

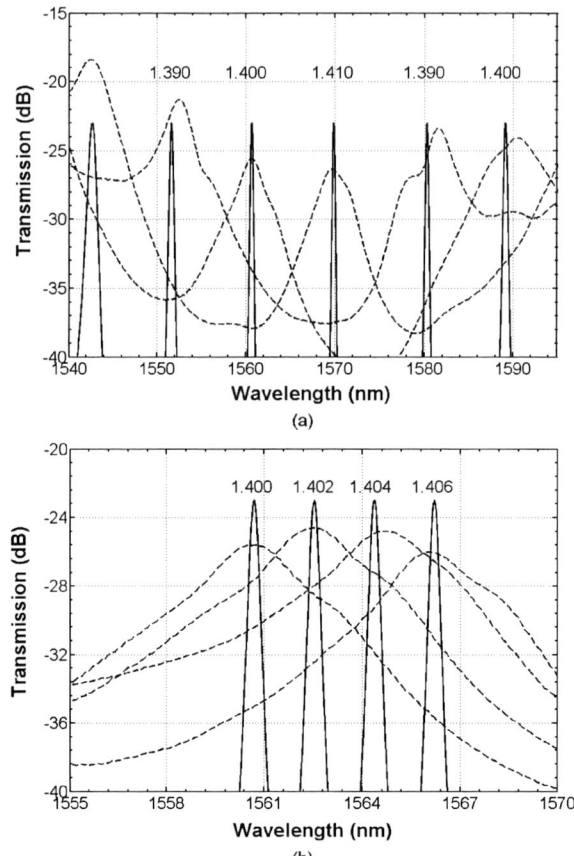

Fig. 2. Measured (dashed lines) and simulated (plain lines) transmission of the Fabry-Perot filter filled with calibration liquids. The curves are labeled by the refractive index of the measured liquid. (a) Variation of the refractive index by $\Delta n = 0.010$. (b) Variation of the refractive index by $\Delta n = 0.002$.

was demonstrated and might be of great interest in practical applications such as intracavity single-cell characterization [2,3].

REFERENCES

[1] H. Hunt and J. Wilkinson, "Optofluidic integration for microanalysis," *Microfluidics and Nanofluidics*, vol. 4, no. 1, pp. 53–79, 2008.
[2] W. Song, X. Zhang, A. Liu, C. Lim, P. Yap, and H. Hosseini, "Refractive index measurement of single living cells using on-chip Fabry-Pérot cavity," *Applied Physics Letters*, vol. 89, pp. 203–901, 2006.
[3] H. Shao, W. Wang, S. Lana, and K. Lear, "Optofluidic intracavity spectroscopy of canine lymphoma and lymphocytes," *Photonics Technology Letters, IEEE*, vol. 20, no. 7, pp. 493–495, 2008.
[4] P. Domachuk, I. Littler, M. Cronin-Golomb, and B. Eggleton, "Compact Resonant Integrated Microfluidic Refractometer," *Applied Physics Letters*, vol. 88, p. 093513, 2006.
[5] H. Shao, D. Kumar, S. Feld, and K. Lear, "Fabrication of a Fabry–Pérot cavity in a microfluidic channel using thermocompressive gold bonding of glass substrates," *Journal of Microelectromechanical Systems*, vol. 14, no. 4, pp. 756–762, 2005.
[6] J. Masson, F. Koné, and Y.-A. Peter, "MEMS tunable silicon Fabry-Perot cavity," *Proceedings of SPIE*, vol. 6717, pp. 671–705, 2007.
[7] S. Weidong, L. Xiangdong, H. Biqin, Z. Yong, L. Xu, and G. Peifu, "Analysis on the tunable optical properties of MOEMS filter based on Fabry–Perot cavity," *Optics Communications*, vol. 239, no. 1-3, pp. 153–160, 2004.
[8] A. Lipson and E. Yeatman, "Low loss 1D photonic band gap filter in (110) silicon," *Opt. Lett*, vol. 31, pp. 395–397, 2006.

Pulsed Laser Triggered High Speed Microfluidic Switch

Ting-Hsiang Wu[1], Lanyu Gao[3], Kenneth Wei[2] and Eric Pei-Yu Chiou[2]

[1]Department of Electrical Engineering, University of California, Los Angeles (UCLA)
[2]Department of Mechanical and Aerospace Engineering, UCLA
48-121 ENG. IV, 420 Westwood Plaza, Los Angeles, CA 90095-1597, USA
[3] Department of Mechanical and Energy Engineering, Zhejiang University
Hangzhou, 310027, P.R. China
Tel +1-310-825-9091, E-mail tsw2008@ucla.edu

ABSTRACT

We report a high speed microfluidic switch capable of achieving a switching time of 70 μsec. The switching mechanism is realized by exciting dynamic vapor bubbles with focused laser pulses in a microfluidic PDMS channel. A time-resolved imaging system has been constructed to capture the nanosecond transient channel wall deformation and the particle flow pattern. A switching of polystyrene microspheres in a Y channel has also been demonstrated. This ultrafast laser triggered switching mechanism has the potential to advance the sorting speed of the state-of-the-art microscale fluorescent activated cell sorting devices for two orders of magnitude.

INTRODUCTION

Microfluidic cell sorters have several advantages over the conventional fluorescence-activated cell sorters (FACS) such as device miniaturization, integration of multiple functions on one chip and achieving high yields for small sample sizes [1]. However, the limitation for microfluidic cell sorters remains the lack of a switching mechanism that can achieve fast switching time and at the same time preserve the cell viability after sorting. Demonstrated sorting speeds range from 25 cells/sec with a thermal reversible polymer to 300 samples/sec using dielectrophoretic focusing [2,3]. The shortest switching time was reported by Want *et al.* [4]. A switching time of 2-4 msec was demonstrated by applying an optical force on the cell, and the recovered cells remained viable and unstressed. We propose here a microfluidic switch triggered by a pulsed-laser microbeam that can achieve a switching time of 70 μsec, nearly two orders of magnitude shorter than that of the optical force switching mechanism.

Pulsed laser mirobeams have been used to create disruption and actuation on the microscale and even nanoscale. Examples include laser drilling [5], laser cell surgery [6], and mixing of fluids in a microchannel [7]. When the laser pulse is focused in a liquid medium (e.g. water), the intense optical field generates heated plasma within the focal volume. The heat dissipates into the surrounding liquid and induces cavitation bubbles. At the initial stage of the expansion, the bubble grows at very high speeds, hundreds of m/s, within hundreds of nanoseconds [6]. In our proposed device, the bubble expansion is used to deform the elastic microchannel wall, serving as the mechanical switch to alter the flow

pattern in the adjacent particle channel. Because of the rapid bubble dynamics, a fast switching time on the order of microseconds can be achieved by controlling the bubble size and the pulsed channel dimensions.

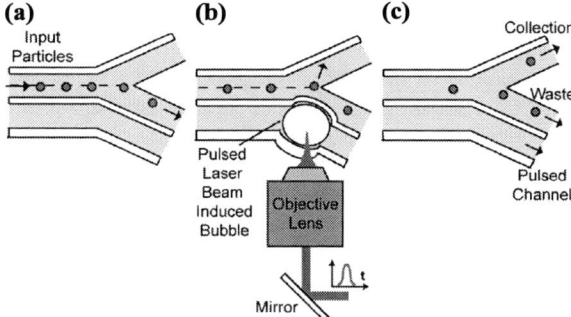

Fig.1 Schematic of the pulsed laser triggered microfluidic switch. (a) Before (b) during and (c) after switching.

PRINCIPLE AND DEVICE STRUCTURE

Figure 1 shows the schematic of the pulsed laser triggered microfluidic switch. The device consists of a Y-shaped microchannel with two outlets, collection and waste. Using hydrodynamic focusing, the input particles are focused slightly off the centerline of the main channel and go into the waste channel after the Y junction. Separated by a thin wall, another channel runs in parallel with the waste channel. This channel is positioned above the objective lens, which focuses the laser pulse to induce optical breakdown of the liquid medium and the subsequent cavitation bubble. As the bubble expands, the channel wall, being elastic, is deformed and squeezes the waste channel on the opposite side. This transient deformation of the channel wall alters the flow pattern in the main channel. A particle approaching the fork of the Y junction is thus switched to flow in the collection channel.

Rapid switching based on deforming and restoring of the channel wall takes place on the time scale of few microseconds to hundreds of microseconds, depending on the bubble size, pulsed channel dimensions and the channel wall width. On the other hand, the switched particles are shielded from the direct impact of the explosive expansion and collapse of the bubble. This way minimum stress is exerted on the particle during switching, and a high level of sample integrity and viability can be maintained. This device has the

978-1-4244-1917-3/08/$25.00 ©2008 IEEE

potential to achieve high-speed and high-efficiency switching on fragile biological samples.

EXPERIMENT AND RESULT

Fabrication of the microchannels employed the conventional casting method. A master mold was made by patterning SU-8 photoresist on a silicon wafer. Polydimethylsiloxane (PDMS) was poured onto the master mold to replicate the channel structures. After curing, the PDMS layer was peeled off from the master mold and bonded to a glass cover slide after oxygen plasma treatment. The main channel had a width of 58 μm and a height of 53 μm. The pulsed channel was 158 μm in width and 53 μm in height. The thickness of the PDMS wall between the pulsed channel and the waste channel was 14 μm. All channel flows were controlled by syringe pumps (KD Scientific). Polystyrene microspheres of diameter 7 μm were pumped through the main channel. Two other converging channels at the beginning of the main channel provided the sheath flows to focus the particle flow, which had a speed of around 0.3 m/s in the main channel.

The pulsed laser source was a Q-switched, frequency-doubled Nd:YAG laser (Continuum, Minilite I). The pulse was 6 nanoseconds in pulsewidth and 532 nm in wavelength. The beam was directed into the fluorescence port of the microscope (Zeiss, AxioObserver) and was focused with an objective lens (Zeiss, Epiplan-Neofluar 50×, 0.8 NA). The pulse energy used was 390 μJ, and the focus was positioned in the pulsed channel. The induced cavitation bubble and PDMS wall deformation were captured using a high-speed Intensified CCD camera (Princeton Instrument, PIMAXII), providing exposure times as short as 500 ps.

Fig. 2 shows the transient flow pattern change in the microchannel due to the bubble induced by a single nanosecond laser pulse. The particle flow was focused slightly off the centerline of the main channel and went into the waste channel on the right after the Y junction (Fig. 2(a)). In Fig. 2(b), a cavitation bubble induced by the focused laser pulse (taken 400 nanoseconds after the laser pulse arrival) deformed the PDMS channel wall and pushed the particle flow towards the left. In Fig. 2(c), 1 sec after the laser pulse, the cavitation bubble collapsed and was flushed away by the fluid flow in the pulsed channel. PDMS wall as well as the particle flow pattern resumed to the original state as before the laser pulse.

Fig. 3 demonstrates particle switching in the Y microchannel triggered by a single laser pulse. Before switching, the particles flew to the waste channel on the right (Fig. 3 (a)). Fig. 3 (b) shows some particles were switched to the collection channel on the left 250 μsec after the laser pulse. The PDMS wall had restored its original shape in this image, which showed a switching time less than 250 μsec. A separate experiment was conducted to measure the dynamics of deforming and restoring of the PDMS wall (images not shown). The

deformation of the channel wall had returned to about 10% of the maximum deformation (less than 1 μm away from the original position) 70 μsec after the laser pulse. By 200 μsec, the wall deformation was indiscernible.

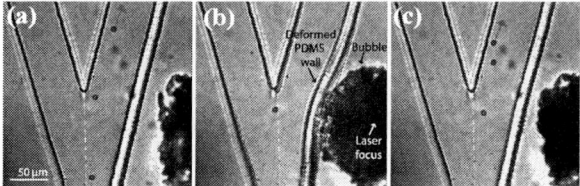

Fig.2 Transient change of particle flow pattern in the microchannel due to a pulsed-laser induced bubble. (a) Particle flow without laser pulses. (b) 400 ns after the laser pulse. (c) 1 s after the laser pulse.

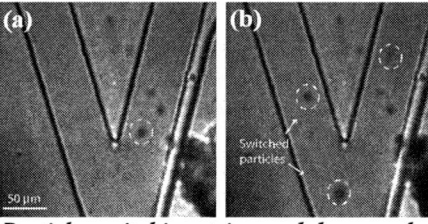

Fig.3 Particle switching triggered by a pulsed laser microbeam. (a) Before switching (b) 250 μs after the laser pulse.

CONCLUSION

A microfluidic switch with a switching time as short as 70 μsec is demonstrated for the first time. Microspheres of 7 μm, size of most mammalian cells, in diameter have being successfully switched in a Y-shaped microchannel triggered by the pulsed laser microbeam. This device has the potential to achieve high-speed and high-efficiency cell sorting on a microfluidic chip.

ACKNOLEDGEMENT

This project is supported by NSF Career Award (ECCS-0747950) and Center for Cell Mimetic Space Exploration (CMISE), a NASA University Research, Engineering and Technology Institute (URETI).

REFERENCE

[1] Huh, A. et al., Physiol. Meas. **26**, R73-R98, 2005.
[2] Holmes, D., Sandison, M.E., Green, N.G., Morgan, H., IEE Proc.-Nanobiotechnol., **152**(4), p.129-135,2005.
[3] Shirasaki, Y. et al., Anal. Chem., **78**, p.695-701, 2006.
[4] Wang, M.M. et al., Nature Biotech. **23**(1), p.83-87, 2005.
[5] Chichkov, B.N. et al., Appl. Phys. A **63**, p.109-115, 1996.
[6] Vogel, A., Noack, J., Huttman, G. and Paltauf, G., Appl. Phys. B **81**, p. 1015-1047, 2005.
[7] Hellman, A.N. et al., Anal. Chem. **79**, p. 4484-4492, 2007.

Optical Interconnects & Nanophotonics

B.J. Offrein

Abstract—. Optical interconnect technology will play an increasingly important role in servers and supercomputers. High density and low-cost optical packaging concepts are required. We consider optical interconnects for board-level and chip-level communication.

Index Terms— Electro-optical packaging, optical interconnections, optical printed circuit boards, silicon photonics.

I. INTRODUCTION

The processor performance continues to grow at a rate of over 50% per year. The IO bandwidth to and from the processor has to scale accordingly in order to take advantage of this development at the system performance level [1]. So far, developments in electrical signaling could keep up with this performance increase, therefore, today servers and supercomputers are purely electrical systems. However, with increasing data-rates, electrical interconnects start to suffer from increased attenuation, distortion, cross-talk and resonant effects. Optical interconnects offer some basic advantages that are essential to continue system performance scaling over the next decade. The transmission behavior in optical waveguides is data-rate independent, originating from the ultra-high optical carrier frequency. There is no cross-talk between adjacent waveguides for the waveguide spacing we consider and the propagation losses are low. The result is a much higher bandwidth density capability of optical interconnects as compared to electrical buses [2].

II. OPTICAL INTERCONNECTS IN FUTURE SERVER AND SUPERCOMPUTING SYSTEMS

In fact, these arguments have propelled optical communication as the interconnect medium of choice for long distance links since the early 80's. With increasing bandwidth demand, optics becomes viable for ever shorter links and today optics is already employed in rack-to-rack interconnects in server and supercomputing applications in the form of parallel fiber-optic links. There are a number of challenges optical technology has to overcome to make the step towards intra-system interconnects. Technological developments are required to fulfill the intra-system interconnect requirements of future server systems. The bandwidth to and from the

B.J. Offrein is with IBM Research GmbH, Säumerstrasse 4, 8803 Rüschlikon, Switzerland.
Phone: +41447248572, e-Mail: ofb@zurich.ibm.com

processor will reach several to ten's of terabit per second, board and backplane level interconnects will have to scale accordingly. The line data-rate will be 10 or 20 gigabit per second, which implies that hundreds to thousand of optical channels will be required to enable the dataflow within the system. Such requirements can not be fulfilled with the components available for today's parallel fiber-optic links. This holds for example for the transceivers, where a much higher channel count and density will be necessary with much lower power consumption, i.e. less than 5 mW per Gb/s.

The first applications of optical technology in server systems will certainly be based on parallel fiber-optic technology. In the long run however, the application of fibers as the interconnect medium for board-level interconnects would lead to extremely tedious routing and assembly procedures. A more integrated approach is required that should incorporate the integration of the optical channels, transceivers, connectors as well as the interfacing to the processor. Only such a higher state of integration will bring the cost of board-level optical technology in the target range of 1 $ per Gb/s or below. In fact, it is this kind of integration is what we know from today's electrical printed circuit board interconnect and assembly technology.

III. BOARD-LEVEL OPTICAL INTERCONNECT TECHNOLOGY

At IBM Research, we have been developing optical printed circuit board technology based on polymer waveguides that are located at the board surface or embedded within the board stack. These multimode waveguides have a height and width in the range between 35 µm and 50 µm. The propagation loss at a wavelength of 850 nm is below 0.05 dB/cm [3]. Passive alignment concepts are available that enable a simple and low-cost interface between the waveguides and transceivers, fibers or other optical elements such as lenses or a mirror [4]. High density and low-power transceivers were developed [5-7].

It is a challenge to integrate a disruptive technology like polymer waveguide-based optical interconnects in computing systems. Only mature technologies have a chance to be considered for such type of systems. Alternative applications can help to mature the technology and to establish the vendor eco-system. For these reasons, our technology developments target a generally applicable optical interconnect system. Furthermore, we envision a step-by step integration of polymer waveguide technology. Polymer waveguide flex sheets appear to be a potential first entry point as an alternative to fiber flex sheets.

Optical printed circuit board technologies enable a tremendous increase of the IO bandwidth at the board-to-board link level. Still, this technology platform is inherently

978-1-4244-1917-3/08/$25.00 ©2008 IEEE

hybrid with discrete processor and transceiver elements that are assembled onto printed circuit boards that carry both electrical and optical signals.

IV. CHIP-TO-CHIP AND INTRA-CHIP LEVEL OPTICS

A much more integrated packaging approach is required to bring the advantages of optical communication to the chip-to-chip or intra-chip level. Here we consider especially the power advantages that optics may bring. Silicon photonics has become a tremendous field of research as it provides the ability to integrate optical communication with silicon logic [8]. The ultra-high index contrast between the silicon waveguide core and the silicon-dioxide cladding facilitates bending radii in the μm range. This leads to extremely small optical devices and the ability to realize ultra-high-density interconnects. Three-dimensional (3D) packaging approaches are under development as a means to continue increasing the processor package performance. In the future, one of the layers in the 3D stack could be dedicated to the optical communication between the many processor cores in the logic-level. Many optical and electro-optical components have to be developed to make this happen. Various examples of silicon waveguide-based optical components and subsystems were demonstrated [8-11].

V. CONCLUSIONS

Optical interconnects offer the potential for much higher bandwidth intra-system communication. Today, servers and supercomputers are still purely electrical systems. Compared with today's fiber-optic technology, a much higher state of integration is required to bring optical link technology into such systems. The different intra-system link levels have specific requirements on the bandwidth, density and technological constraints. We discussed the potential of optical interconnects for board-level and chip-level links.

REFERENCES

[1] A. Benner, M. Ignatowski, J. A. Kash, D. M. Kuchta, M. B. Ritter, "Exploitation of optical interconnects in future server architectures," *IBM J. Res. & Dev.* **49** (4/5), pp. 755-775 (2005).

[2] D. A. B. Miller, "Physical Reasons for Optical Interconnection", *Int. J. Optoelectron.*, **11** pp. 155-168 (1997).

[3] R. Dangel, R. Beyeler, F. Horst, T. Lamprecht, N. Meier, B.J. Offrein, B. Sicard, M. Moynihan, P. Knudsen, E. Anzures, „Waveguide Technology Development based on Temperature- and Humidity-Resistant Low-Loss Silsesquioxane Polymer for Optical Interconnects", *Paper OThH2, OFC/NFOEC 2007, Anaheim, CA, 2007.*

[4] T. Lamprecht, F. Horst, R. Dangel, R. Beyeler, N. Meier, L. Dellmann, M. Gmur, C. Berger, B.J. Offrein, „Passive alignment of optical elements in a printed circuit board", *Electronic Components and Technology Conference*, 2006. Proceedings. 56ᵗʰ ECTC, p.7, 30 May-2 June 2006

[5] R. Dangel, C. Berger, R. Beyeler, L. Dellmann, M. Gmür, R. Hamelin, F. Horst, T. Lamprecht, N. Meier, T. Morf, S. Oggioni, M. Spreafico, R. Stevens, B. J. Offrein, "Polymer-Waveguide-Based Board-Level Optical Interconnect Technology for Datacom Applications", *IEEE Trans. Advanced Packaging*, in Press.

[6] L. Schares *et al.*, "Terabus: Terabit/second-class card-level optical interconnect technologies," *IEEE J. Select. Topics Quantum Electron., Special Issue on Optoelectronic Packaging*, Vol.12, No. 5, pp. 1032-1044, Sept./Oct. 2006.

[7] L. Dellmann, C. Berger, R. Beyeler, R. Dangel, M. Gmur, R. Hamelin, F. Horst, T. Lamprecht, N. Meier, T. Morf, S. Oggioni, M. Spreafico, R. Stevens, B.J. Offrein, "120 Gb/s Optical Card-to-Card Interconnect Link Demonstrator with Embedded Waveguides", *Electronic Components and Technology Conference*, 2007. Proceedings. 57ᵗʰ ECTC, pp. 1288-1293, May 29 2007-June 1 2007.

[8] R. Soref, "The Past, Present, and Future of Silicon Photonics," *IEEE Sel. Topics in Quantum El*, 12, pp. 1678-1687, 2006.

[9] Fengnian Xia, Lidija Sekaric and Yurii Vlasov, "Ultracompact optical buffers on a silicon chip," Nature Photonics 1, pp. 65-71, 2006.

[10] W. M. Green, M.J. Rooks, L. Sekaric, Y.A. Vlasov, "Ultra-compact, low RF power, 10 Gb/s silicon Mach-Zehnder modulator," Optics Express, 15, pp. 17106-17113, 2007.

[11] Y.A. Vlasov, W. M. J. Green, F. Xia, "High-throughput silicon nanophotonic wavelength-insensitive switch for on-chip optical networks," *Nature Photonics 2*, pp. 242 - 246 (16 Mar 2008).

Silicon microring resonator
connected with submicorn comb actuator

K.Takahashi, Y.Kanamori, K.Hane

Department of Nanomechanics, Tohoku University, Sendai, 980-8579, Japan

Tel: +81-22-795-6962, Fax: +81-22-795-6963, E-mail: hane2@hane.mech.tohoku.ac.jp

ABSTRACT

A 500nm wide 260nm thick 63.4μm long silicon microring waveguide was suspended in air by connecting to a comb actuator with the low-loss suspension arms. The airgap between the microring and the input/output waveguides was adjusted by a voltage applied to the comb actuator to vary the coupling efficiency. Transmittance from the input to the drop waveguides was varied from 0% at the gap of 750nm to 50% maximum value (about 30dB) at the voltages from 0V to 28V. Under the drop condition, the bandwidth of the drop signal was 0.5nm, corresponding to the Q-value of 3100.

Keywords: microring, silicon photonics, silicon waveguide, comb actuator

Introduction

Submicron wide silicon waveguides are promising for the dense integration of optical circuits with the silicon electronics in the fields of optical telecommunication and interconnections [1]. As an efficient narrow band filter, microring resonators have been studied intensively. In some advanced researches on microring resonators, the tunable functions were introduced. MEMS technology is also promising as a tunable mechanism. A microtroidal resonator with integrated MEMS tunable coupler was reported [2]. A high Q-value (5400-11000) and a wide tuning range (22.4dB extension ratio) of waveguide coupling were obtained.

In this paper, a silicon microring resonator with an electrostatic micro comb actuator is reported as a coupling variable drop filter. A microring resonator is suspended in air by connecting to an actuator with low-loss suspension arms. All the parts of the device are fabricated from a top layer of a silicon on insulator (SOI) wafer with a single electron beam mask process. The optical properties of the proposed device have been experimentally measured.

Principle, design and experimental

Figure 1 shows the schematic diagram of the proposed wavelength filter using a microring resonator combining with a microactuator. The coupling between the buslines and microring waveguides can be changed by moving the microring. The input wave is selectively transmitted to the drop-port when the microactuator makes the microring approach to the buslines under the condition that the wavelength of the input matches to one of the resonant wavelengths of the microring. On the other hand, when the actuator is not operated, the microring is apart form the buslines so as to make the coupling efficiency very small. The input wave passes to the throughput port without interacting the microring resonator. Therefore, the proposed device works as a wavelength selective add/drop filter.

As shown in Fig.1, The proposed device consists of three submicron silicon waveguides i.e., two bus lines and ring resonator with a microactuator. One busline waveguide is used for the input and throughput ports, respectively. The other busline waveguide is for a drop port. The two busline waveguides are bent 90 degrees with the bent waveguides to be located closely to the microring waveguide as shown in Fig.1. The microring waveguide is a close-loop waveguide having a rounded square shape with four 5μm radius corners. The waveguides are 500nm wide and 260nm thick in design. The microring waveguide is totally 63μm long. The region of the ring parallel to the buslines are 8μm long and the initial gap between the ring and the buslines are 775nm. All the waveguides are supported in air by the narrow suspension arms. In order to minimize the optical loss generated by the suspension mechanism, a gradual s-bend profile for the increase in the waveguide geometry is used in the elliptical bridge as shown in Fig.1 [3].

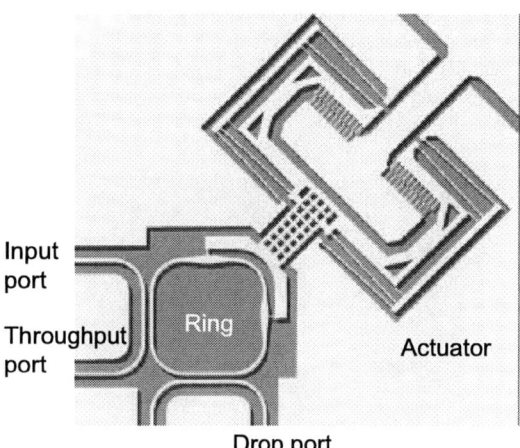

Fig.1 Microring resonator combined with a submicron comb actuator.

The microring can be translated in the plane of the waveguides at the angle of 45degrees against the two buslines. The two buslines are vertical around the microring in order to keep the two gaps between the microring and the buslines equal when the position of the microring is changed. A microactuator is connected to the microring with the suspension arms of the elliptical bridge

978-1-4244-1917-3/08/$25.00 ©2008 IEEE

to move the microring for coupling the buslines.

An electrostatic comb-drive actuator is used to translate the suspended microring. The area of the actuator is 25μm wide and 50μm long. The comb finger is 250nm wide, 260nm thick, and 2μm long and the gap between the fingers is 350nm. The comb area is 2.5μm wide and 8μm long. Double folded springs are utilized in the microactuator.

In the fabrication of the designed device, electron beam lithography was utilized. A 2.0μm SiO_2 layer of SOI wafer was used as the sacrificial layer. After the development of the resist, the top silicon layer was etched by a fast atom beam, which was generated by the neutralization of the ions extracted from a dc SF_6 plasma (Ebara, FAB-60ML). Finally, the hydrogen fluoride vapor generated by heating hydrofluoric acid solution was used to etch the SiO_2 layer underneath the etched top silicon layer [4].

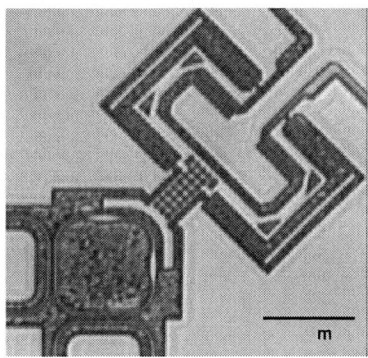

Fig.2 Optical micrograph of the fabricated device.

Results and discussion

Figure 2 shows the optical micrograph of the whole view of the fabricated device. The submicron silicon waveguides and the electrostatic comb actuator are well fabricated and they are suspended in air although the springs and comb are too thin to be imaged by the optical microscope.

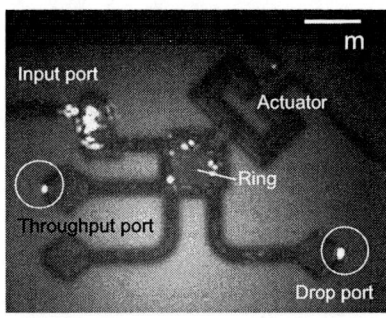

Fig.3 IR micrograph of the signal drop experiment.

The channel drop experiment was carried out using the fabricated device. In optical testing of the fabricated device, a tunable laser (Agilent 81682A) is used as the light source. TE-like polarized light at 1.55 wavelength is coupled into the input waveguide by an objective lens. The fabricated device is observed from the top with an infrared (IR) camera (GoodRich SU320KTS-1.7RT). Figures 3

shows the IR camera images obtained under the experiment. As shown in Fig.3, a bright light spot is observed at the drop port under the condition that 50μW laser light at the ring resonant wavelength of 1.557.94nm is incident on the input port and the actuator is activated at the voltage of 28.2V. A small spot at the throughput port is also seen. Therefore, the drop efficiency is larger than that of throughput in this case.

Quantitative optical properties of the proposed device are also investigated. Figure 4 shows the drop output efficiency measured as a function of the input laser wavelength under the condition that the drop efficiency becomes the maximum. The bandwidth of the drop signal was 0.5nm, corresponding to the Q-value of 3100.

At the resonant wavelength, the actuator was moved and the drop light intensity was measured as a function of voltage. Figure 5 shows the result. At the voltage from 0 to 23V, the coupling efficiency is small and thus the drop intensity is low. Further increasing the voltage, the drop intensity increases as shown in Fig.5. The maximum drop efficiency is approximately 50% at the voltage of 28V.

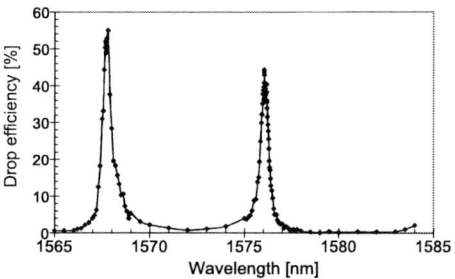

Fig.4 Drop output efficiency measured as function of wavelength.

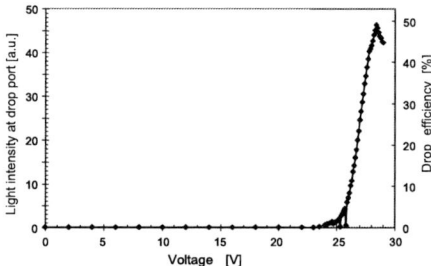

Fig.5 Drop output measured as a function of actuator voltage.

References

[1] B. Jalali, et.al., J. Lightwave. Technol. 24 (2006) 4600.

[2] J. Yao, et.al. IEEE J.Sel. Topics Quantum. Electron. 13 (2007) 202.

[3] L. Martinez, et.al., Opt. Express 14 (2006) 6259.

[4] Y. Fukuta, et.al. Jpn. J. Appl. Phys. 42 (2003) 3690.

Four Port Nanophotonic Couplers for Dense, Planar Integrated Optics

Duncan L. MacFarlane, Manasi Peshave, Wei
Zhou, and Nahid Sultana
Department of Electrical Engineering
The University of Texas at Dallas
Richardson, Texas 75083
972-883-2165 (w) 972-883-2710 (f)
dlm@utdallas.edu

Marc P. Christensen, Nathan R. Huntoon and
Gary A. Evans
Department of Electrical Engineering,
Southern Methodist University
Dallas, Texas 75275

Abstract— **Four port (4x4) couplers will be useful components for dense, planar photonic integration. A compact, high efficiency implementation based on evanescent coupling across a nanoscale gap is discussed.**

I. INTRODUCTION

A useful building block for dense, planar photonic integrated circuits is the four port coupler. In this context, a four port coupler splits and combines four input signals into four output signals, and finds application in photonic filters and packet switches. A signal flow diagram for this component is shown in Fig. 1. In operation, four input signals enter the coupler from directions labeled N, E, W and S, and four output signals exit into the same directions. In general there are 16 field coefficients that describe the coupler, and for each port these are a reflection, ρ, a transmission, τ, a coupling to the right, α, and a coupling to the left, β. If energy is conserved it may be shown that there are 6 independent parameters that describe the coupler, and interesting composition methods have been derived for the four port coupler under this assumption [1,2].

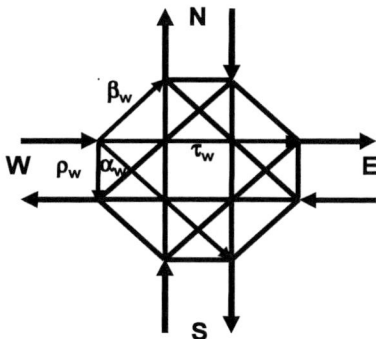

Figure 1. Signal flow diagram of a 4-direction coupler that reflects, transmits, routes left and routes right. In our notation, the 4 ports are labeled W, N, E and S. At each port, there is a reflected component, ρ, a transmitted component, τ, a right directed component, α, and a left directed component, β. Shown explicitly in this diagram are ρ_w, τ_w, α_w, and β_w, the coupling coefficients at the W port.

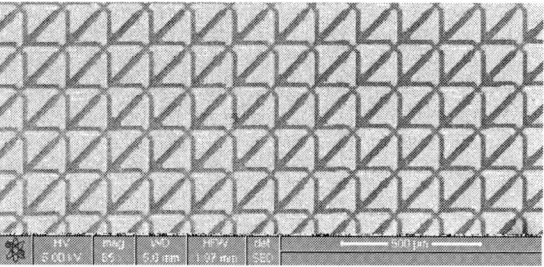

Figure 2. Scanning electron micrograph (SEM) of a two dimensional grid of AlGaAsP semiconductor laser amplifiers. Shown in this figure are triangular bond pads, each connected to a 120 um amplifying waveguide ridge. Where four amplifying ridges meet, a four directional coupler can be etched in order to form a two dimensional active lattice filter.

To illustrate the application for a four port coupler, a photonic integrated circuit is shown in Fig. 2. In this scanning electron micrograph (SEM) is shown a planar, two dimensional field of semiconductor optical amplifiers. This array of semiconductor optical amplifiers forms a two-dimensional active lattice filter that has the potential to be a powerful signal processing engine. A key component is a compact 4-port coupler placed at the intersection formed by the meeting of four ridge waveguide semiconductor optical amplifiers.

In this paper we describe a novel, frustrated total internal reflection based coupler can be greater than 95% efficient. Nanoscale fabrication techniques are also discussed for InP applications.

II. EVANESCENT WAVE BASED FOUR PORT COUPLER

One approach to a compact, high efficiency four port coupler is to use frustrated total internal reflection. Consider a wave incident on a boundary from a high refractive index to a low refractive index. If the angle of incidence is above the critical angle there will be total internal reflection at the interface. A complete analysis includes exponentially decaying evanescent fields that penetrates across the interface into the forbidden low index material. This evanescent field is a non-propagating inductive field with a skin depth that is a fraction of the free space wavelength; in our InP material system this depth is on the order of 100nm. If the low index material

978-1-4244-1917-3/08/$25.00 ©2008 IEEE

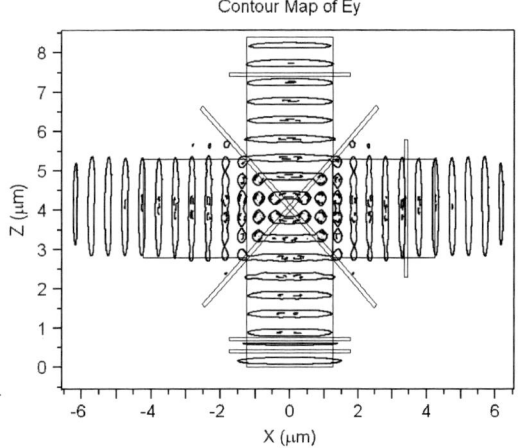

Figure. 3. FDTD analysis of an evanescent field based four port nanophotonic coupler.

is thin compared to this evanescent decay then the field can couple across the gap – it may be said that the photons tunnel across the gap – and there will be power both transmitted and reflected. This is frustrated total internal reflection (F-TIR) and can be used to design a very compact high efficiency coupler, as shown in the FDTD analysis of Fig. 3. In this structure is included two thin, deep, perpendicular air gaps arranged at 45 degrees to the intersection of the four waveguides. The input waves are incident on this lower refractive index gap at 45 degrees, which is well above the critical angle for the 3.2 to 1.0 refractive index transition. For equal air gaps of approximately 110 nm, the coupling will be symmetric and balanced in the sense that the input to any port will be evenly distributed among the outputs.

As can be seen in the contour plot of Fig. 3, there is very little scatter into the non-waveguiding regions, and the modes are not significantly perturbed by the F-TIR process. The principle of operation of this evanescent wave based F-TIR coupler is quite different from the

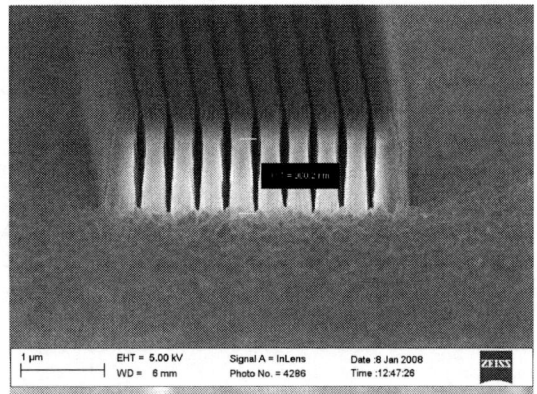

Figure. 4. 0.1x1 micron trenches in InP etched in a Cl2:N2 mix using ICP. These trenches are the approximate dimension required for the evansecent wave coupler.

photonic crystal coupler, and in fact yields higher predicted efficiencies.

As long as the TIR condition is satisfied, the index of refraction of the gap may be increased, and this will increase the required width of the gap for a given coupling percentage. In our calculations, this dependence can be quite strong; for a balanced coupler, an increase of 0.1 in the refractive index will increase the gap by 30 nm. A larger gap filled with a alternate material may be easier to fabricate.

III. FABRICATION

In the coupler discussed, a challenging, high aspect ratio trench must be etched in the InP. For example, in the evanescent wave based device, a trench 140nm wide and 2.5 um deep is required. This trench is filled with ZrO_2 using atomic layer deposition (ALD) [3]. Shown in Fig. 4 are trenches of approximately this size that were fabricated in InP using a Cl2:N2 mix in an inductively coupled plasma. Other etch gases such as HBr and CF4:CHF4 also provide high aspect ratio InP etches.

IV. CONCLUSION

A fundamental building block of photonic integrated circuitry is the coupler used to split and combine optical signals. Traditional linear, one dimensional, lasers and lattice filters are enabled by 2-directional (2-port) couplers characterized by reflection and transmission coefficients. Because of their wide utility, the design and analysis of structures comprising large numbers of these partially reflecting, partially transmitting mirrors are well studied [4-6]. In planar architectures, the four-directional (4-port) coupler enables lasers, filters and hybrid circuits with increasing sophistication and functionality [1,2]. In this paper we have presented two candidate approaches for compact, high efficiency nanophotonic couplers.

REFERENCES

[1] Duncan L. MacFarlane, Jian Tong, Chintan Fafadia, Vishnupriya Govindan, L. Roberts Hunt, and Issa Panahi "Extended Lattice Filters Enabled by Four Directional Couplers" Applied Optics 43, 6124 (2004).

[2] Tiberiu Constantinescu, Viswanath Ramakrishna, Nicholas Spears, L. Roberts Hunt, Jian Tong, Issa Panahi, Govind Kannan, Duncan L. MacFarlane, Gary A. Evans and Marc P. Christensen, "Composition Methods for Four-Port Couplers in Photonic Integrated Circuitry," Journal of the Optical Society A 23, 2919 (2006).

[3] S. X. Lao, R. M. Martin, and J. P. Chang "Plasma enhanced atomic layer deposition of Hfo2 and ZrO2 high-k thin films," J. Vac. Sci. Technol. A 23, 488-496 (2005).

[4] A. H. Gray, Jr., and John D. Markel, "Digital lattice and ladder filter synthesis," IEEE Transactions on Audio and Electroacoustics, AU-21, 491-500 (1973).

[5] H. van de stadt and J. M. Muller, "Multimirror Fabry-Perot interferometers," J. Opt. Soc. Am. A 2, 1367-1370 (1985).

[6] D. L. MacFarlane and E. M. Dowling, "Z-domain techniques in the analysis of Fabry-Perot etalons and multilayer structures," J. Opt. Soc. Am. A, 11, 236-245 (1994).

A NOVEL ETCHING-OXIDATION FABRICATION METHOD FOR 3D NANO STRUCTURES ON SILICON AND ITS APPLICATION TO SOI SYMMETRIC WAVEGUIDE AND 3D TAPER SPOT SIZE CONVERTER

Ling-Han Li[1], Akio Higo[1], Masanori Kubota[2], Masakazu Sugiyama[2] and Yoshiaki Nakano[1]

[1]Research Center for Advanced Science and Technology, The University of Tokyo, Japan

[2]Institute of Engineering Innovation, The University of Tokyo, Japan

Abstract

A novel etching and oxidation method utilizing space effect of dry etching for three dimensional silicon structure is presented. Testing devices of SOI symmetric waveguide with ultra thick SiO_2 cladding and silicon waveguide structure integrated with 3D taper spot size converter are fabricated using this method

Key Word: Sub micron etching and oxidation, Ultra thick SiO_2 layer, Spot size converter, Symmetric SOI waveguide

1. Introduction

Excessive coupling loss between SMF (single mode fiber) and silicon nano wire waveguide caused by the mismatch of their optical mode field diameter causes great research interests in three dimensional SSC (Spot Size Converter) to reduce the vertical and lateral mismatch. However, conventional fabrication methods for vertical SSC include quite complicated method such as grey-scale lithography and precise alignment [1]. Even in some cases ultra thick SiO_2 cladding which is rather difficult for plane oxidation and CVD is also necessary [2]. In this paper, we are reporting a novel fabrication method for making three dimensional silicon structure and ultra thick SiO_2 cladding layer [3]. The process mainly utilizes the dry etching speed difference between varying line width patterns [4]. The method is very simple, flexible and compatible to other silicon device fabrication. The oxidation turns the etched comb shape silicon terrace to homogeneous SiO_2 cladding. Therefore, three dimensional silicon structures, like terrace and taper are realized. By using this process, SOI ridge waveguide with thick SiO_2 cladding as well as the three dimensional taper SSC can be fabricated together in one time.

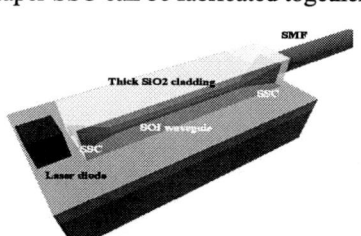

Fig.1 Schematic of Silicon high mesa waveguide with thick SiO_2 cladding layer and 3D SSCs for coupling SMF and compound-semiconductor laser diode

2. Etching-Oxidation fabrication method for three dimensional silicon structures

The fabrication process is shown in Fig.2. Space effect of ICP dry etching in etching speed is the principle of the whole process. By arranging different sub micron scale line patterns for etching, which is called ravine as seen in Fig.2, we can alter the shape of the structure in vertical direction. The patterns are drawn on silicon by Electron Beam lithography. Then it is etched using SF_6 ICP dry etching and the terrace and ravine as in Fig.2 are formed. The etching power is 1500W under 5°C. Finally the sample is wet oxidized under 1050°C for 4 hours to turn the comb shape Si to SiO_2 layer, and the three dimensional structure are `grown` under the cladding SiO_2.

Although several mathematical models, like Deal-Grove model are used for simulating the oxidation thickness, the tridimensional oxidation and process miss like over etching in this case makes it necessary to determine the proper design ratio of terrace and ravine first so that the following oxidation can turn the etched comb shape silicon mesa to SiO_2 layer. Different arrangements of terrace and ravine width as in Table 1 were tested. Their results are shown in Fig.3

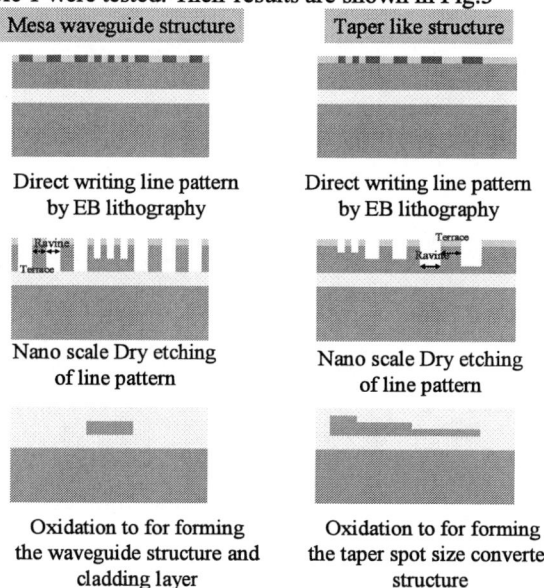

Fig.2 Fabrication process of the Etching Oxidation method for step waveguide structure and taper like structure

Table 1 Sets of Terrace and Ravine with different Ratio

Index	1	2	3	4	5	6	7
Terrace (μm)	0.8	0.7	0.65	0.6	0.55	0.5	0.4
Ravine (μm)	0.2	0.3	0.35	0.4	0.45	0.5	0.6

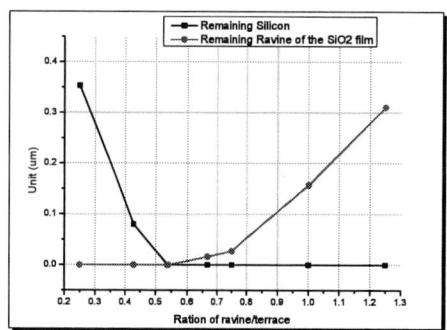

Fig.3 Remaining silicon width, SiO_2 layer gap of different Terrace and Ravine width ratio after oxidation

978-1-4244-1917-3/08/$25.00 ©2008 IEEE

It can be concluded from Fig.3 that with the ratio of 0.65:0.35, comb shape silicon mesa after etching was completely oxidized, making a homogenous SiO_2 layer

According to the above conclusion, the ration of terrace and ravine of 0.65/0.35 was adopted for further fabrication. And by changing the width of the ravine, the height of silicon mesa alters, as in Fig.4

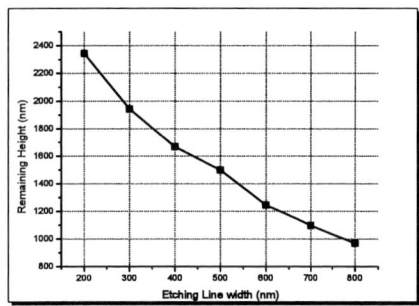

Fig.4 Relationship between the height of fabricated silicon mesa and ravine width (Etching time is 80seconds)

With the fabricating conditions discussed above, silicon terrace, mesa waveguide like structure and three dimensional taper like structure were fabricated and some results are shown in Fig.5.

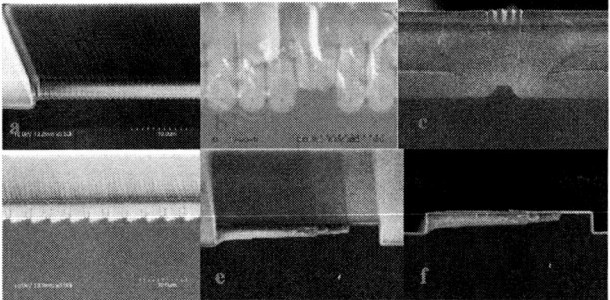

Fig.5 (a) Ultra thick SiO_2 homogeneous cladding layer by etching and oxidation method (b) Silicon terrace (c) Waveguide like structure (d) Silicon 3D triangle (e) Bird view of 3D silicon taper like structure (f) Cross view of silicon taper like structure.

3. Application to silicon photonic devices

Testing silicon photonic devices of ridge waveguide and SSC are fabricated by the etching-oxidation methods on SOI.

For the case of SOI symmetric ridge waveguide, the relationship between the ridge height of waveguide and etching time is investigated and shown in Fig.6. With different ridge width, the `growing` speed of the waveguide is almost the same around 5.3nm/s. Fig.7 illustrates an example of ridge waveguide. The etching time is 120seconds. The ridge height is 770nm, which generally agrees with the predicted height according to Fig.6.

On the other side, a 1000μm long vertical testing SSC, with height changing from 4.5um to 575nm was fabricated, connecting to a symmetric waveguide with silicon core of 8μm wide, 575nm high. As a comparison, the symmetric waveguide connected with an 8μm wide and 4.5μm high 1000μm long silicon block was also done. The insertion loss between two single mode fiber (mode radius of 3.2um) of 3

different set up (waveguide only, silicon block, SSC) are measured and shown in Fig.8 It is known that one SSC prototype can improve the insertion loss around 4.8dB comparing to direct coupling to symmetric waveguide.

However, the large propagation loss in the fabricated waveguide caused by the bad side profile requires process improvement such as introducing cryogenic etching and multiple times oxidation. Moreover, Optimization of waveguide shape and SSC design is currently undergoing

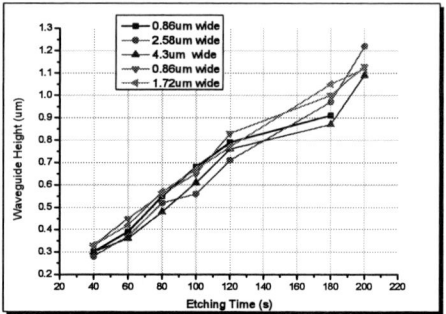

Fig.6 Relationship between waveguide growing speed and etching time, Terrace width 0.65μm and ravine width 0.35μm

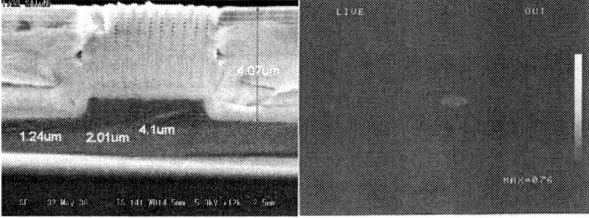

Fig.7 Fabricated symmetric silicon ridge waveguide and it's optical mode near field image (wavelength 1550.5nm)

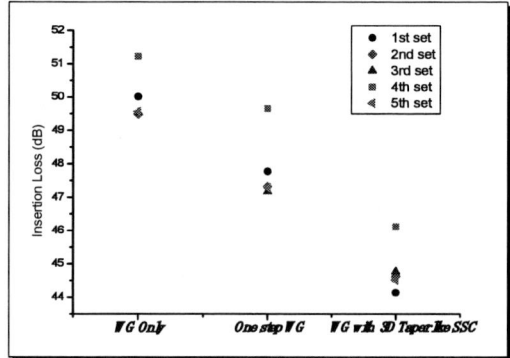

Fig.8 Insertion loss of silicon symmetric mesa waveguide (1.1cm long 8um wide 575nm high), waveguide with silicon block, waveguide with 3D taper SSC between two SMF .

REFERENCES:

[1] K. Shiraishi, H.Yoda, A.Ohshima C.S.Tsai and H.Ikedo, Applied physics leters 91, 141120 2007

[2] L. Vivien, Xavier Le Roux, S. Laval, E. Cassan and D. Marris-Morini IEEE Journal of selected topic in quanum electronic VOL 12. No.6 Nov. 2006

[3] C.Zhang and K.Najafi, Journal of Micromechanics and Microengineering, 14, 2004,769-774

[4] T.Bourouina, T.Masuzawa and H.Fujia, Journal of microelectromechanical Systems Vol.13 no.2 April 2004

Coupled-Ring Reflector Laser Diodes Composed of Squared Ring Waveguides

Suhyun Kim[1,2], Doo Gun Kim[3], Nadir Dagli[4], Youngchul Chung[*1], Young Tae Byun[2]

[1]Department of Electronics and Communication Engineering, Kwangwoon University
447-1,Wolgye-Dong, Nowon Gu, Seoul, 139-701, Korea, [*]E-mail: ychung@kw.ac.kr
[2]Intelligent System Research Division, Korea Institute of Science and Technology, Seoul, Korea.
[3]School of Electronic and Electrical Engineering, Chung-Ang University, Seoul, Korea.
[4]Department of Electrical and Computer Engineering, University of California, Santa Barbara, USA.

Abstract- **Coupled-ring reflector laser diodes composed of squared ring waveguides are fabricated and it is demonstrated that the laser diodes exhibit single mode operation and 15 nm tuning range with SMSR exceeding 25 dB**

I. INTRODUCTION

The micro-ring resonators are expected to be used to realize the photonic integrated circuits such as add-drop filters, switches, modulators, and lasers [1]. Recently, a new concept of wavelength selective reflector composed of a straight waveguide coupled with two coupled ring waveguides is introduced [2]. It can be a substitute for a distributed Bragg reflector for a tunable laser due to its reflective property.

In this paper, we fabricated tunable laser diodes composed of semiconductor optical amplifier and squared coupled-ring reflector (CRR) with total internal reflection (TIR) mirrors and demonstrate characteristics of these laser diodes. To realize compact-size ring resonators, squared ring resonators are formed using TIR mirrors and deeply etched multi-mode interference (MMI) coupler [3]. The TIR mirrors and the MMI couplers are fabricated using a self-align process and the offset quantum well techniques are employed to integrate semiconductor optical amplifiers (SOAs) as the gain section and loss compensation in ring resonator [4]. The process and the material used in the fabrication of these devices are compatible with conventional ridge laser diodes.

II. DEVICE DESCRIPTION

Figure 1 is a photograph of a fabricated laser diode. The laser diode is consisted of three SOAs and the squared coupled ring resonator. The SOAs are monolithically integrated in the straight waveguide for gain section and in the ring resonators for loss compensation by offset quantum well technology. The coupling between straight waveguide and ring resonator is accomplished by deeply etched multimode interference coupler and the coupling between rings by directional couplers. The phase control regions are located between SOA1 and the squared CRR and in ring resonators. The refractive index of these regions can be changed by the plasma effect due to current injection. The phase control 1 plays a role of controlling the cavity modes of the laser diode with phase control 2 and 3 to

change the resonance wavelength of each resonator can be adjusted.

For compactness of ring resonators, deeply etched MMI couplers are adopted for coupling between a bus and ring resonators, whose width and length of the MMI are 6 μm and 174.6 μm, respectively. Because the MMI is etched up to the lower cladding layer, it is not influenced by effective index variations due to etch depth variations. Because the widths of rib waveguide and MMI input are 3 μm and 1.5 μm respectively, they are connected by taper waveguide. The TIR mirrors, fabricated by the self-align process, are formed at all the corners of resonators.

The perimeters of each ring resonators are slightly different from each other. The perimeter of the ring 0 is 560 μm and that of ring 1 is 4% longer than that of ring 0. Both facets of the device are cleaved. To prevent interference between reflection wave from CRR and that from cleaved facet, the waveguide near the right facet is tilted by 8 degree. The length of the SOA1 is about 700μm and the total length of the device is about 1.7mm.

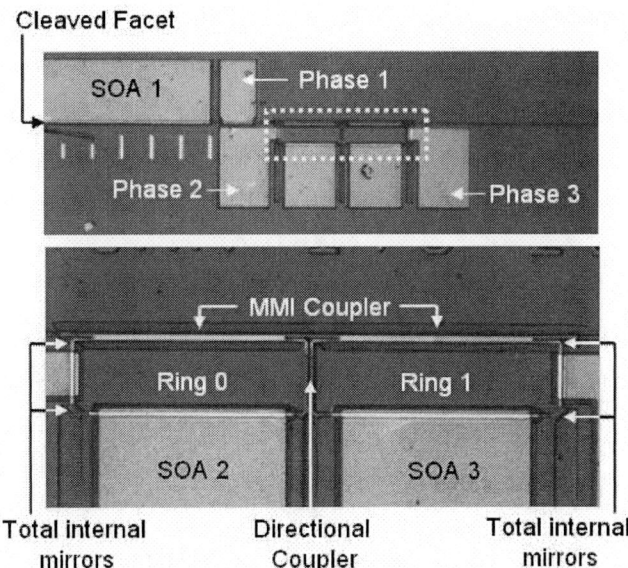

Figure 1. Photograph of a squared coupled ring reflector laser diode

An optical wave created from the gain region (SOA1) is coupled to ring 0, the clockwise wave in the ring 0 is coupled to the count-clockwise wave in ring 1, and then the wave is coupled to the backward wave in the straight waveguide. Because the perimeter of ring resonators is slightly different, the resonance wavelengths of two rings are misaligned except at certain wavelengths. In this case, strong reflection occurs for the wavelength at which both rings are resonant simultaneously. Therefore, amount of resonant wavelength change in one ring result in the reflection wavelength shift of a free spectral range. Using this Vernier effect, wide tuning of the lasing wavelength can be achieved through slight change of the refractive index in the ring resonator. The condition of coupling coefficient for the single reflection peak (flat-top) spectrum of the coupled ring reflector is given by

$$\kappa_1 = \frac{\sqrt{2\kappa_0^2 + 2\sqrt{2}\kappa_0 + 2\kappa_0\sqrt{1-\kappa_0^2}}}{1 + \sqrt{2}\kappa_0} \qquad (1)$$

where κ_1 and κ_0 are field coupling ratios between ring and straight waveguide, and ring waveguides respectively [2]. In this device, the coupling between ring waveguides is accomplished by directional coupler, and the length of the coupler was designed for satisfying (1). The design value of κ_1 and κ_0 are 0.707 and 0.13, respectively.

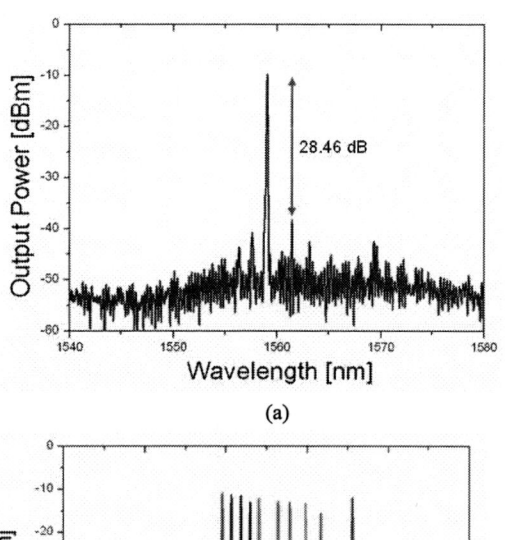

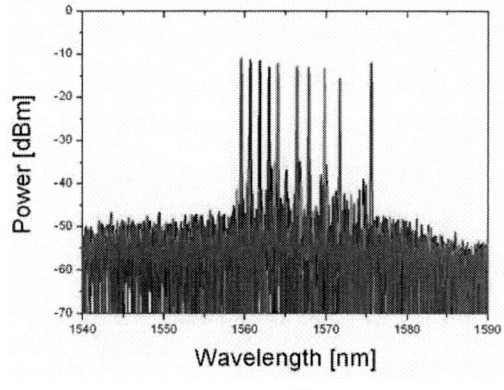

Figure 2. (a) Lasing spectrum of the squared CRR LD,
(b) Superimposed lasing spectra of the LD

III. EXPERIMENTS

The lasing spectrum of the fabricated laser diode is shown in Fig. 2(a). In the experiment, the laser diode is CW-operated at room temperature. The injection current of the gain region (SOA1) and the loss compensation region (SOA2, 3) are 140mA and 10mA respectively, and no injection current is applied in the phase section. The side mode suppression ratio is about 28dB and the output power is about -4dBm. When the injection currents of phase region of ring resonator 1 is increased while that of the other ring resonator is maintained at 0mA, the lasing wavelength is changed from 1559.55nm to 1560.66nm. The wavelength difference is almost equal to the FSR of ring resonator 2. The superimposed lasing spectra of the laser diode are shown in Fig. 2(b). For all lasing wavelengths, we observe a SMSR and the output powers exceeding 25 dB and -5 dBm, respectively.

We suppose that the low output power and narrow tuning range are caused by low reflectivity of coupled ring reflector due to high mirror loss originated from scattering at mirrors. The measured mirror loss was -1.6dB per mirror. These problems could be solved by proper dry etching condition.

IV. CONCLUSIONS

In this paper, we fabricated the coupled-ring reflector laser diode composed of two squared ring resonator with TIR mirrors and straight waveguide integrated with SOA. The laser diode can be tuned as wide as 15nm with SMSR over 25 dB. The high power and widely wavelength tuning operation of this device is expected by proper fabrication condition and device design.

ACKNOWLEDGMENT

This work was supported by grant No. R01-2006-000-10751 from the Basic Research Program of the Korea Science and Engineering Foundation.

REFERENCES

[1] Arkady Kaplan, "Modeling of Ring Resonators with Tunable Couplers", *IEEE J. on Selected Topics in Quan.Elec.*, Vol. 12, No. 1, pp. 86-95, Jan/Feb 2006

[2] Youngchul Chung, Doo Gun Kim, and Nadir Dagli, "Reflection Properties of Coupled Ring Reflectors", *IEEE J. of Lightwave Tech.*, Vol. 24, No. 4, pp. 1865-1874, April 2006

[3] Doo Gun Kim, Jae Hyuk Shin, Cem Ozturk, Jong Chang Yi, Youngchul Chung, and Nadir Dagli, "Total Internal Reflection Mirror-Based InGaAsP Ring Resonators Integrated with Optical Amplifiers", *IEEE Photon. Tech. Lett.*, Vol. 17, No. 9, pp. 1899-1891, September 2005.

[4] Beck Mason, Greg A. Fish, Steven P. DenBaars, and Larry A. Coldren,: "Widely Tunable Sampled Grating DBR Laser with Integrated Electroabsorption Modulator", *IEEE Photon. Tech. Lett.*, Vol. 11, No. 6, pp. 638-641, June 1999.

MEMS-Based Pico Projector Display

Wyatt O. Davis, Randy Sprague, Josh Miller

Microvision, Inc., Redmond WA, USA

Tel 425-402-0826, Fax 425-936-4660, E-mail: wyatt_davis@microvision.com

Abstract

A Pico-projector based on scanned 3-color laser light has been developed. The scanning element is a dual-axis MEMS scanning mirror that produces WVGA display resolution. The laser light sources are red and blue laser diodes and a second harmonic green laser. Use of a MEMS and laser light sources leads to a small volume and a thickness of 7mm, enabling the use of the projector system as a component embedded into portable consumer electronics.

Keywords: MEMS, scanning mirror, projection display

1 INTRODUCTION

In recent years, Microvision, Inc. has used MEMS to create compact scanned beam display and imaging systems such as a head-mounted see-through display [1], a barcode scanner [2], an endoscope [3], a laser printer scan engine [4], and, most recently, a portable projector [5]. The MEMS-based scanned beam approach to image presentation provides a solution to the display bottleneck for increasing the functionality of portable consumer electronics. A dual-axis MEMS scanning mirror combined with 3 laser light sources provides the core of the PicoP™ display engine, providing WVGA display resolution with power consumption and size enabling its use in portable consumer electronics.

2 MEMS SCANNING MIRROR

The MEMS scanning mirror consists of a reflector suspended in a gimbal frame containing a microfabricated electrical coil. Permanent magnets are assembled around the MEMS die to supply a magnetic field. Applying electrical current to the MEMS coil generates a magnetic torque on the gimbal frame with components along both of the desired axes of rotation, as illustrated in Figure 1. One component of the torque is oriented to cause rotation of the gimbal frame about its flexure suspension, as shown in Figure 2. This torque component is used to convert the frequency content of the mirror drive signal below any resonant frequencies of the MEMS to a sawtooth ramp mirror rotation corresponding to the vertical display axis. A second torque component causes rotation of the gimbal frame about an axis normal to its flexure suspension, as shown in Figure 2. This component is used to operate the scanning mirror on a resonant mode of vibration consisting mainly of rotation of the reflector about its flexure suspension to the gimbal frame, and corresponding to the horizontal display axis.

A noteworthy feature of the design is that there is only one drive signal input to MEMS. To create the biaxial motion, the waveforms for both the vertical and horizontal motion are simply superimposed, as illustrated in Figure 3. The frequency response characteristics of the MEMS dynamic system intrinsically filter the composite drive waveform such that a 2-D raster display pattern results. The design

allows for large scan angles with reasonable power consumption operating at atmospheric pressure.

Position feedback for both desired scan angles is implemented using PZR strain sensors microfabricated on both the reflector suspension flexures and the gimbal frame suspension flexures, as shown in Figure 4. Such feedback improves the stability of the projected display over time and environmental conditions.

3 PICO-PROJECTOR DISPLAY SYSTEM

The MEMS scanning mirror is integrated with red and blue laser diodes and a second-harmonic green laser to form a compact color display engine, as shown in Figure 5. The use of a MEMS and compact lasers enables the scanning mirror system including light sources to have an overall volume of 5cc or less, with a height of only 7mm.

The display engine integrated with video and MEMS drive electronics forms the PicoP™ display engine illustrated in Figure 6. The lasers are turned on according to the need to display an individual pixel of a certain color; when one of the three lasers is not needed due to the image content it is left off so that its power consumption is minimized. This system generates a WVGA resolution display at a brightness of 10 Lumens, with a diagonal image size of about 1m at a projection distance of 1.1m. In addition to the low power and small size, one particular advantage of the use of laser light sources is that the image is in focus at any projection distance without any adjustment.

4 SUMMARY

A Pico-projector display engine using a MEMS scanning mirror and laser light sources has a combination of features and performance characteristics that make it suitable for integration into portable electronics.

978-1-4244-1917-3/08/$25.00 ©2008 IEEE

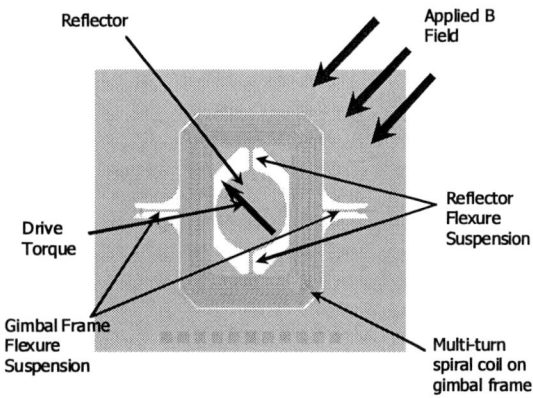

Figure 1 MEMS scanning mirror

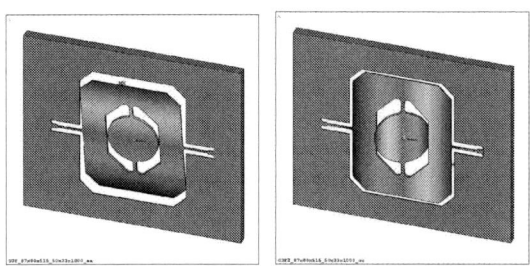

Figure 2 One torque component causes gimbal frame rotation about its flexure suspension (left), while the other component can be used to excite a scanning mirror resonant mode of vibration where the reflector rotates about its flexure suspension to the gimbal frame (right).

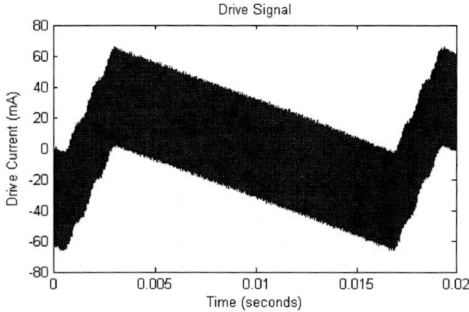

Figure 3 The composite drive waveform contains the superposition of the approximately 60Hz vertical scan sawtooth function and a high frequency sine wave for exciting the horizontal scan resonance.

Figure 4 MEMS scanning mirror die and closeup of PZR strain sensor used for vertical scan position feedback.

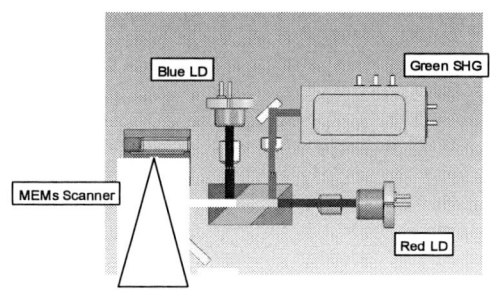

Figure 5 Scan engine subsystem with MEMS scanning mirror and laser light sources.

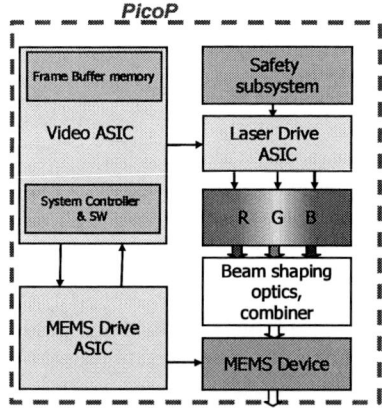

Figure 6 PicoP™ projection display system.

REFERENCES

[1] J. Yan, S. Luanava, and V. Casasanta, "Magnetic Actuation for MEMS Scanners for Retinal Scanning Displays," Proceedings of SPIE Vol. 4985, 2003.

[2] http://www.microvision.com/barcode.

[3] R. James, G. Gibson, F. Metting, W. Davis, C. Drabe, "Update on MEMS-Based Scanned Beam Imager," MOEMS and Miniaturized Systems, Proceedings of SPIE Vol. 6466, 2007.

[4] W. Davis, D. Brown, M. Helsel, R. Sprague, G. Gibson, A. Yalcinkaya, H. Urey, "High Performance Silicon Scanning Mirror for Laser Printing," Proceedings of SPIE Vol. 6466, 2007.

[5] Niesten, M., R. Sprague, J. Miller, "Scanning Laser Beam Displays," Proceedings of SPIE Vol. 7001, 2008.

High-Efficiency MEMS Tuneable Gratings for External Cavity Lasers and Microspectrometers

R. Lockhart, M. Tormen, P. Niedermann, T. Overstolz, A. Hoogerwerf, R.P. Stanley
Swiss Center for Electronics and Microtechnology, Inc. (CSEM)
Neuchâtel, Switzerland

Abstract - Optical characterization of a second generation MEMS tuneable blazed grating with a 1x1mm active area has demonstrated high-efficiency wavelength filtering with reflected outputs >90% and tuning over 25nm in the near-infrared (NIR).

Micro diffraction gratings provide convenient tuning elements for external cavity lasers and microspectrometers. While blazed profiles are typically incorporated in such gratings to improve efficiency, wavelength tuning is accomplished through a macro-rotation which ultimately limits the reflected output of the device as it strays from the Littrow condition [1]. With tuneable MEMS gratings, it is possible to create a stationary tuning element in which the reflected wavelength is shifted by actively modifying the grating period (spacing between the elements) [2, 3]. Li *et al.* achieve this by applying individual control to each grating element. Unfortunately, several drawbacks impair this design. The spacing required for each individual driving mechanism increases the minimum grating period and therefore reduces the achievable free spectral range (FSR). In addition, the number of controls increases with the number of grating elements establishing a tradeoff between ease of control and minimum filter linewidth.

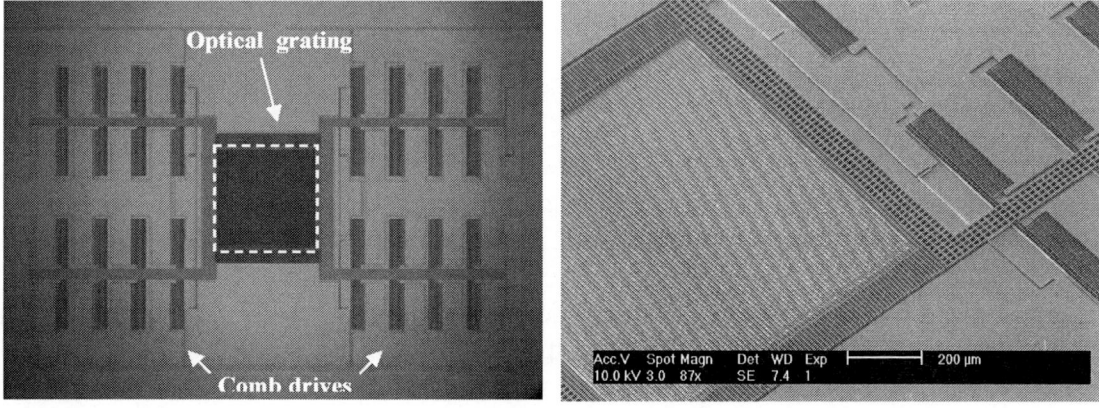

Figure 1: An optical grating membrane is stretched by two sets of opposing comb drives.

In this report, we present the experimental verification of a simple, high-efficiency, wavelength tunable filter based on in-plane stretching of a suspended, blazed grating membrane. In the previous generation of this device, we focused on an easy-to-fabricate, planar design; however, the reflected output of the device was found to be limited [3]. Current devices, fabricated on silicon-on-insulator (SOI) wafers using a combination of deep reactive ion etching (DRIE) and anisotropic KOH etching, consist of a free-standing, 1x1mm optical grating with an initial period of 12µm and two sets of opposing electrostatic combdrives (Figure 1). Blazing of the grating elements accomplished by exploiting the anisotropic nature of KOH for silicon etching produces smooth, angled

978-1-4244-1917-3/08/$25.00 ©2008 IEEE   33

(54.74°), optical surfaces which greatly increase the output efficiency of the device. Manipulation of the grating period is actively controlled through the simple application of a single high and low voltage to the stationary and movable arms of the opposing combdrives. A series of identical, compliant leaf springs interconnecting the individual elements of the grating array allow the membrane to stretch uniformly preventing errors introduced by non-uniform phase shifts occurring between the elements.

The tuneable gratings were measured at wavelengths of 800nm and 1500nm on a setup configured to satisfy the Littrow condition in order to maximize the optical efficiency. The full width at half maximum (FWHM) of the tuneable filter was found to be 0.6nm and 1nm at 800nm and 1500nm, respectively (Figure 2). The linewidth measurements closely resemble theoretical values; however, the small deviations found can be attributed, on the most part, to the small yet finite numerical aperture of the optical setup. Optical efficiencies of gold coated gratings were also measured and found to range between 90-94%.

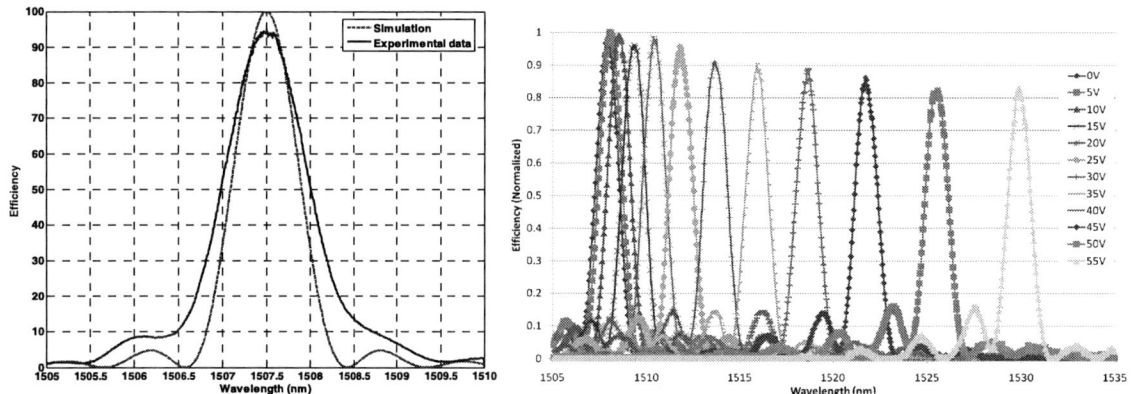

Figure 2: *Measurements demonstrating the optical efficiency, linewidth (left) and tuning response (right) of the MEMS gratings at 1500nm.*

The tuning capability of the gratings has also been demonstrated by applying a potential across the device while recording the optical response. A combdrive displacement of ~15μm produces the observed wavelength shift of 25nm (Figure 2). Maximum tuning thus far has been limited by a weakness in the comb teeth which tend to buckle at voltages above 65V. In addition, a slight decrease in efficiency (~15%) at maximum deflection has also been observed as a result of anti-symmetric forces generated by the opposing combdrives. These minor limitations, however, can be easily overcome in order to achieve high efficiency tuning over an even greater range.

References

[1] A.Q. Liu, X.M. Zhang, "A review of MEMS external-cavity tunable lasers", *J. Micromech. Microeng.*, vol. 17, no.1, pp. R1-R13, 2007.

[2] X. Li, C. Antoine, D. Lee, J.-S. Wang, O. Solgaard, "Tunable blazed gratings," *J. Microelectromech. Syst.*, vol. 15, no. 3, pp. 597-604, 2006.

[3] M. Tormen, Y.-A. Peter, Ph. Niedermann, A. Hoogerwerf, R. Stanley, "Deformable MEMS grating for wide tunability and high operating speed," *Journal of Optics A: Pure and Applied Optics*, vol. 8, no. 7, pp. S337-40, 2006.

Optical Wavelength Selection and Amplification by Silica Microcavities and Erbium Doped Fiber

Sacha Bergeron, Samir Saïdi and Yves-Alain Peter

Ecole Polytechnique de Montreal, Engineering Physics Department

P.O. Box 6079, Station Centre-Ville, Montreal (QC), H3C 3A7, CANADA.

Tel: + 1 514 340 4711, Fax: + 1 514 340 3218, Email: {sacha.bergeron, yves-alain.peter}@polymtl.ca.

Abstract—**In this paper we describe how circular silica micro disk resonators can be coupled to erbium doped fiber to create a desired emission spectrum. This approach could lead to very compact and inexpensive multiple wavelength lasers. We demonstrate here a selective erbium emission at a single wavelength using a silica microcavity.**

I. INTRODUCTION

Very high performance step index optical microcavities have attracted a lot of interest recently. Research on very high quality optical micro resonators started with silica micro spheres [1]. Recently, researchers have modified the spherical micro resonator into a more workable form, the planar micro disks [2]. More suited to conventionnal microfabrication processes and much easier to integrate into complex systems, the micro disk resonators have many advantages over spherical geometries. These resonators offer exceptional performance due to their extremely high quality factor [2]. Very high quality microcavities can be used in a variety of optical systems, such as telecommunication filtering [2], bio-chemical detection [3], [4], and quantum electrodynamics [5]. The cavity's filtering ability can be used to modify signals, and the behavior of active components, such as erbium doped fiber. Coupled to an erbium doped fiber amplifier or laser, the cavity can select the emission spectrum of the doped fiber. With the development of dense wavelength division multiplexing (DWDM), the need for narrow multi wavelength and cost effective laser sources has emerged. In this paper, we propose the basis for a technology that has the potential for a multiple wavelength source and individual optical modulation all in one compact device. We will demonstrate the possibility of selecting the emission spectrum of an erbium doped optical fiber using a micro disk resonator to filter an input broadband signal.

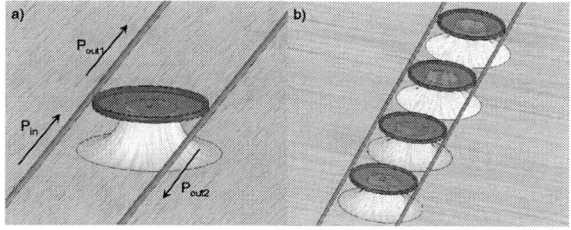

Fig. 1. a) Schematic of the coupled micro disk cavity. b) Schematic of the proposed multiple wavelength system

II. SILICA MICROCAVITIES

Silica disk microcavities use the step in index of refraction between itself and the surrounding media to guide light by total internal reflection. Light trapping in the microcavity is regulated by the geometrical properties of the disk, allowing specific wavelengths to resonate according to specific trajectories or modes. The most useful modes are the whispering gallery modes because they allow a very high level of confinement and thus a very high quality factor (Q). This high Q value translates in very narrow resonant peaks in the resonant spectrum, providing these cavities extremely interesting filtering capabilities.

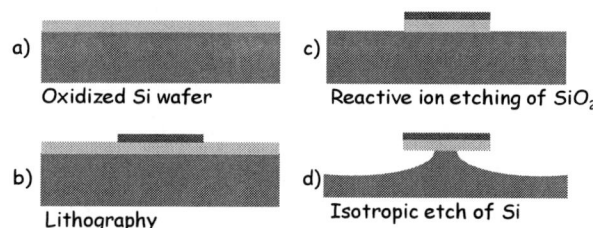

Fig. 2. Schematic of the microfabrication process

We use a simple fabrication process to machine these silica micro disks, as shown in Fig. 2. Using a conventional UV lithography process (Fig. 2b), we transfer a micro disk pattern to an oxidized silicon wafer (Fig. 2a). We then etch the silicon oxide by reactive ion etching (RIE) (Fig. 2c) to ensure good verticality of the side walls. The disk is then released by performing an isotropic etch of the silicon underneath (Fig. 2d). Once the microcavity is fabricated, a tapered optical fiber with a diameter of 1.2 μm is brought to close proximity in order to achieve optical coupling of light with the micro disk. A piezoelectric translation stage is used to optimize the distance between the taper and the edge of the disk. The fabricated circular silica microcavities can be used to select a wavelength in a broad band or multi wavelength signal in order to extract the resonant wavelengths from the input signal [2]. Furthermore, by coupling a second tapered fiber on the other side of the cavity, it is possible to collect these filtered wavelengths as shown in Fig. 1a [6]. The coupling of resonant wavelengths into a second fiber was achieved experimentally and Fig. 3 presents the power spectrum collected by both input

978-1-4244-1917-3/08/$25.00 ©2008 IEEE

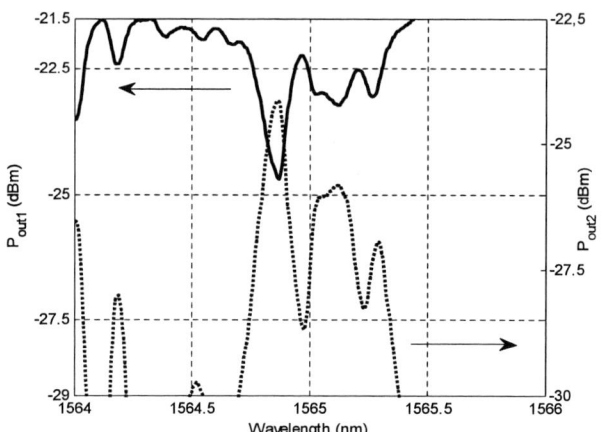

Fig. 3. Output power spectrum of both tapered fibers P_{out1} and P_{out2} as seen in Fig. 1a.

and output tapers. Note that the powers are offset du to losses in the tapers. This demonstrates clearly that the optical signal collected from the second taper does indeed correspond to the cavity's resonant spectrum. Although it is important to note that the addition of a second taper causes additional loss to the cavity, lowering the Q factor and broadening the resonant peaks. Also, the presence of secondary peaks suggests that the cavity used in this experiment is not single mode.

III. ERBIUM EMISSION SELECTED BY MICROCAVITY FILTERING

The very high quality factor of these cavities implies that the light collected out of it is highly coherent. This coherent signal can then be amplified using an Erbium doped fiber amplifier (EDFA) to generate a stimulated emission with narrow linewidth centered on the resonant wavelength of the microcavity. Figure 4 presents a schematic of the optical setup used to filter and amplify a broadband ASE (amplified spontaneous emission) source into a narrow coherent emission. To amplify the filtered signal, we use a five meter Erbium doped fiber optically pumped by a 980 nm laser diode.

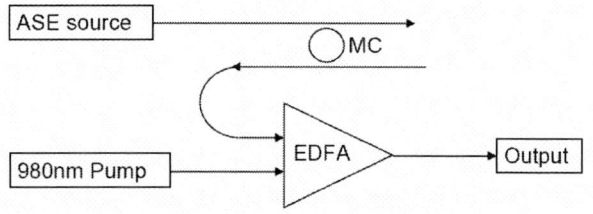

Fig. 4. Schematic of the optical filtration and amplification setup

Figure 5 presents the filtered output previously discussed, before and after its amplification. This shows maximum gain on the highest and thinnest peak of the filtered signal. The emission peak is perfectly aligned with the maximum of the filtered spectrum at 1564.85 nm. This emission has a linewidth

(FWHM) of less than 0.06 nm. 8 dB amplification is observed between the filtered and amplified peaks. But, looking at the peak at 1565.29 nm, we note 11 dB of attenuation. As a consequence, the total effective amplification is more than 20dB. The large attenuation observed is due to a loss in the amplification setup. By reinjecting this amplified signal into the input, it would be possible to generate a fiber laser which would increase the output power and emission selectivity. Furthermore, the linewidth of the emission would be much narrower. Finally, by introducing multiple cavities in such a system it will be possible to create a multiple wavelength source.

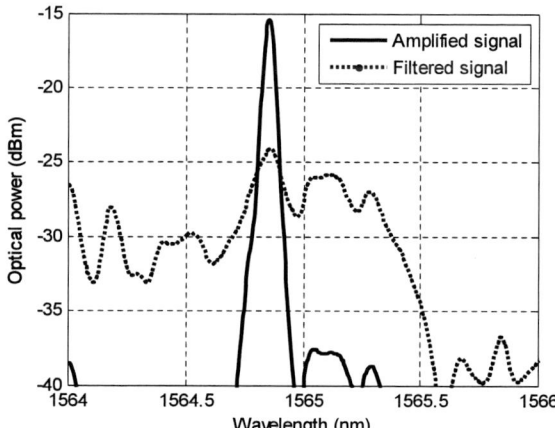

Fig. 5. Emission spectrum of the filtered signal before and after amplification

IV. CONCLUSION

We have shown that it is possible to control the emission spectrum of an erbium doped fiber amplifier by using the filtering ability of a silica micro disk resonator. We have obtained an emission wavelength of 1564.85 nm which corresponds to one of the resonant modes of the microcavity that was used in this experiment. This emission peak has a linewidth smaller than 0.06 nm. This approach could lead to very compact and inexpensive multiple wavelength lasers.

REFERENCES

[1] M. L. Gorodetsky, A. A. Savchenkov and V. S. Ilchenko, "Ultimate Q of optical microsphere resonators", *Opt. Lett.*, Vol. 21, no. 7, pp. 453-455, 1996.
[2] T. J. Kippenberg, S. M. Spillane, D. K. Armani and K. J. Vahala, "Fabrication and coupling to planar high-Q silica disk microcavities", *Appl. Phys. Lett.*, Vol. 83, no. 4, pp. 797-799, 2003.
[3] F. Vollmer, D. Braun, A. Libchaber, M. Khoshsima, I. Teraoka and S. Arnold, "Protein detection by optical shift of a resonant microcavity F.", *Appl. Phys. Lett.*, Vol. 80, no. 21, pp. 4057-4059, 2002.
[4] A. M. Armani, R. P. Kulkarni, S. E. Fraser, R. C. Flagan and K. J. Vahala, "Label-Free, Single-Molecule Detection with Optical Microcavities", *Science*, Vol. 317, pp. 783-787, 2002.
[5] V. Lefkvre-Seguin, S. Haroche, "Towards cavity-QED experiments with silica microspheres", *Mater. Sci. Eng.*, Vol. 48, pp. 53-58, 1997.
[6] H. Rokhsari and K. J. Vahala, "Ultralow Loss, High Q, Four Port Resonant Couplers for Quantum Optics and Photonics", *Phys. Rev. Lett.*, Vol. 92, no. 25, p. 253905, 2004.

Micropatterned Complex Optical Surface for Wide Angle Illumination

Sun-ki Chae, Hyukjin Jung, Ki-Hun Jeong*
*Corresponding author : kjeong@kaist.ac.kr
Department of Bio and Brain engineering, KAIST, Daejeon, Korea

Abstract: A direct-type LED based Back-Light-Unit (BLU) in LCD displays requires wide angle illumination with high uniformity. As a single lens surface, a micropatterned complex optical surface can provide wide angle LED illumination angle over 150° by designing the micropattern arrays on the hemispherical surface.

This work presents a novel single optical component for wide angular illumination. Unlike a conventional single lens surface, a micropatterned complex optical surface (μ-COS) based single lens helps light efficiently scattered by total internal reflection or diffraction. The angular illuminations differ in angles and intensity following with the shape, size and aspect ratio of micropattern arrays on a lens curvature as well as the lens curvature and refractive index. The micropattern arrays were designed with ASAP™ simulation to differentiate the angle of light illumination (Fig. 1). A direct-type LED (light-emitting diode) based Back-Light-Unit (BLU) in LCD displays requires wide illumination angle with high uniformity. The μ-COS can be used as an innovative light-scattering component in non-imaging optical systems especially including a direct-type LED based BLU with the small numbers that requires the protection against heat.

The microfabrication of μ-COS is done by a replica molding technique of pressurized flexible elastomer membrane with micropatterns (Fig. 2). The prepatterned microarrays by photolithography is replicated with thin polydimethylsiloxane (PDMS) elastomer and the reconfigurable microtemplate is then made by transferring the thin elastomer membrane to another elastomer with a cavity (2.5mm in diameter) that is connected with a microchannel. Under negative pressure, the membrane is deformed to a hemispherical shape with negative pattern arrays. Finally, the curved surface with micropattern arrays is recasted by a high index optical polymer resin (Fig. 3). The μ-COSs with miscellaneous shapes, aspect ratios, lens's materials, depending on the controllability of illumination angle were designed and microfabricated to control the angular illumination.

The angular illumination through μ-COSs with micropattern variation was measured in the near-field with incoherent and collimated uniform light source, impinged onto the bottom flat surface of μ-COS (Fig.4). The diffraction due to the micropatterns and total internal reflection inside μ-COS remarkably contribute to the wide illumination angle more than 60 degrees. The experiment shows the illumination angle substantially increases with narrower gap size and smaller patterns, compatible to wavelength.

The micropatterned complex optical surface is a single optical component, however, it can provide wider illumination angle over 150° under the LED illumination with angular distribution of 90°. The uniformity can also be improved by varying the local packing density of the micropattern arrays on a dome lens. This novel device will provide an innovative pathway to wide angle illumination in non-imaging optical applications such as direct-type LED based LCD Back-Light-Units (BLU) or small illumination systems.

Reference
1. Y. Xia, Science Vol. 273. no. 5273, pp.347-349 (1996)
2. K. Jeong, Science Vol. 312. no. 5773, pp. 557-561 (2006)
3. Tasso R. M. Sales, Proc. of SPIE Vol. 5530 (2004)

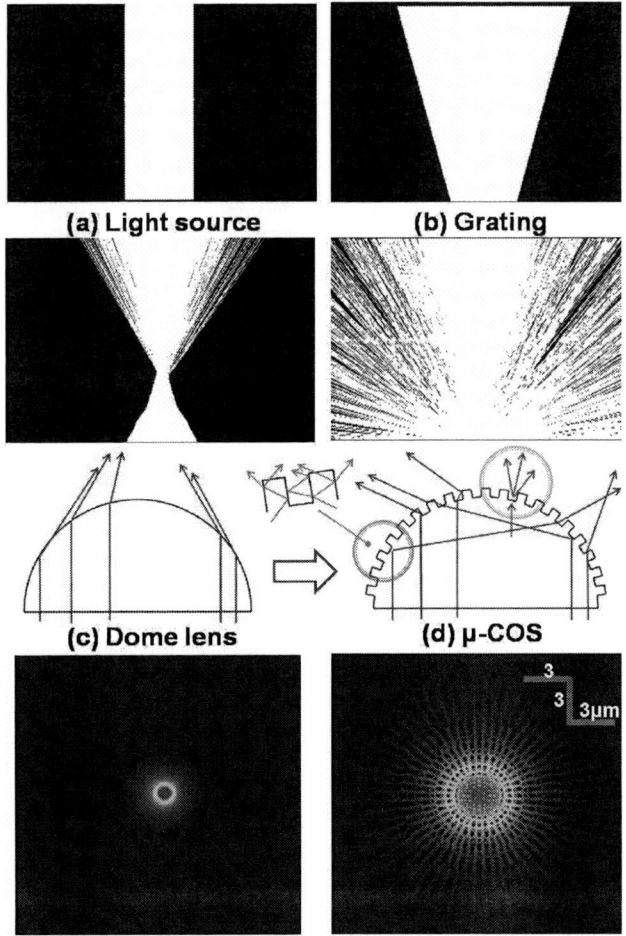

(a) Light source   **(b) Grating**

(c) Dome lens   **(d) μ-COS**

Fig.1 The comparison of angular distributions of (a) uniform collimated light source by (b) a transmission grating, (c) a dome lens and (d), micropatterned complex optical surface on a dome lens, using ASAP™ optical simulation.

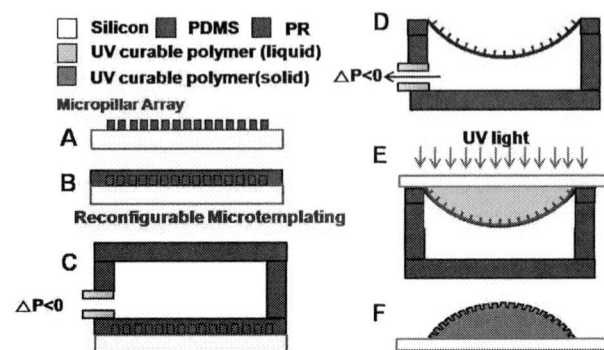

Fig.2 Microfabrication procedures of μ-COS involving reconfigurable microtemplating.

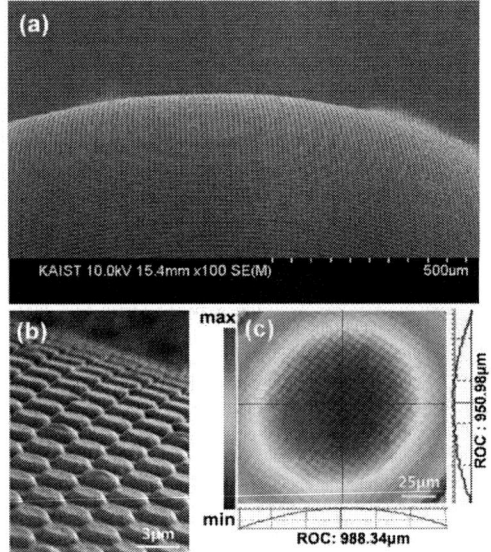

Fig. 3 SEM images of μ-COS and micropatterned cylinder arrays on a hemispherical dome(a,b). Optical profile measurements of μ-COS with a white light interferometer for the measurement of the uniformity and radius of curvature of μ-COS(c).

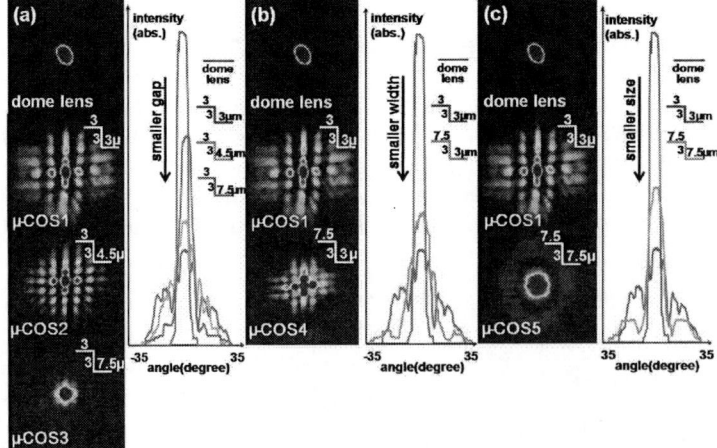

Fig. 4 the near-field measurement of the angular distribution of micropatterned complex optical surface lens (μ-COS)s, depending on (a) the gap, (b) the width, and (c) the dimension of each micropattern on a hemispherical dome surface. Total internal reflection and diffraction through a micropatterned complex optical surface substantially increase the illumination angle, compared to a dome lens.

Large linear micromirror array for UV femtosecond laser pulse shaping

Severin Waldis*, Stefan M. Weber[†], Wilfried Noell*, Jérôme Extermann[†], Denis Kiselev[†], Luigi Bonacina[†], Jean-Pierre Wolf[†], and Nico F. de Rooij*

Tel: +41 32 720 5520, Fax: +41 32 720 5711, email: stefan.weber@physics.unige.ch
*Institute of Microtechnology, Université de Neuchâtel, Rue Jaquet-Droz 1, CH-2002 Neuchâtel
[†]GAP - Biophotonics, Université de Genève, Rue de l'École-de-Médecine 20, CH-1211 Genève 4

Abstract—**We are fabricating a bulk-micromachined micromirror device for laser pulse shaping applications on femtosecond time scales. An array of micromirrors is used to individually retard or diminish certain laser frequencies spanning from the UV to the near-infrared. The individual mirrors are fixed by two springs on either side and can be tilted (amplitude modulation) or moved out-of plane (phase modulation) using asymmetrical or symmetrical vertical comb drives, respectively.**

Keywords: micromirror, vertical comb drive, SOI, timed DRIE, UV-NIR femtosecond laser pulse shaping, biomolecules

I. INTRODUCTION

Controlling coherent reactions in molecules using modulated femtosecond laser pulses has recently lead to a revolution in photochemistry [1]. A lack of pulse shaping devices in the UV and below has, however, excluded fundamental compounds of organic chemistry and biochemistry such as aromatic rings, which prefer optical excitation in the 250-400 nm region. The technology employed to modulate pulses in the UV was mainly indirect, by mixing IR-shaped pulses in nonlinear crystals using liquid crystal displays or acousto-optical modulators [2]. A more direct approach of pulse shaping in the UV is the utilization of micromirror, MEMS based devices [3], which can be reflective over larger bandwidths when coated accordingly.

II. CONCEPT

Our device is designed to perform continuous phase and binary amplitude modulation, therefore two independent degrees of freedom are needed for the mirrors. In the pulse shaping setup (Fig. 1), the incoming femtosecond beam is first dispersed by a grating and then, by a combination of lenses or mirrors, focused in a Fourier plane where the MEMS device will be placed [4]. This way, each pixel of the device corresponds to a well-defined frequency band which can be either retarded by performing an out-of-plane motion of the mirror (phase modulation), or deflected on a different outgoing beam path by tilting (amplitude modulation). The device must achieve 1° of tilt and 2 μm piston in static mode, including a calibration reserve. The mirror tilt is achieved by a set of asymmetrical, vertical comb drives on both sides (Fig. 2 - top shows the right part) and the piston by similar, symmetrical comb drives. The mirror shape is rectangular (80 x 1000

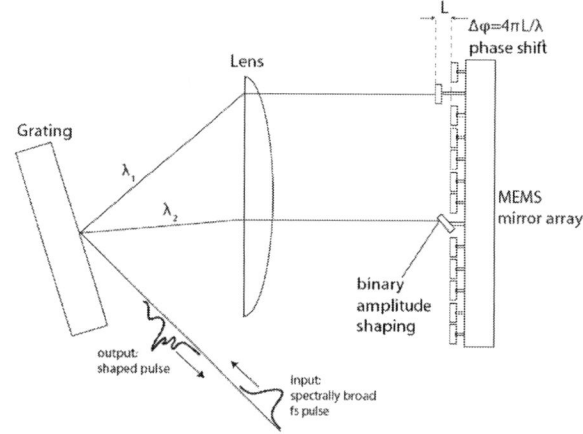

Fig. 1. Optical concept of the pulse shaping setup.

μm) to match the laser profile in the Fourier plane when cylindrical mirrors are used [4]. Broader mirrors would cause an inferior spectral resolution, ultimately leading to limited temporal pulse shaping capabilities. After the spring follows a suspension structure compensating for potential stresses and strains in the surrounding chip and package.

III. FABRICATION AND CHARACTERIZATION

The micromirror device is fabricated using silicon-on-insulator (SOI) substrates. We fabricated single mirrors, and arrays of seven and 100 micromirrors (see Fig. 2). The vertical electrodes are realized utilizing a self-aligned, delay-mask deep reactive ion etching (DRIE) process [5].

A close-up of the piston actuator (Fig. 3) shows the resulting step heights which were 8 μm for the lower level and 20 μm for the upper part of the device layer. The fabricated devices were then packaged in ceramic substrates and wire-bonded.

The mirror surface quality was measured with a Veeco/Wyko NT1100 Optical profiler. The peak-to-valley (PTV) deformation of unpackaged and uncoated mirrors was only 8-15 nm over a length of 1 mm; which corresponds to $\lambda/20$ at $\lambda=400$ nm (Fig. 4a), allowing the full surface to be used as reflective area. The aluminum coating for UV-operation only slightly increased the PTV to 20-25 nm,

978-1-4244-1917-3/08/$25.00 ©2008 IEEE

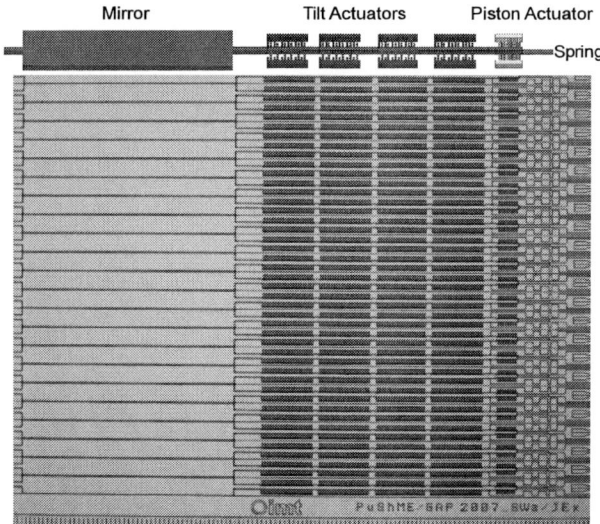

Fig. 2. Top: Sketch of the actuation concept with vertical comb drives. Center: Detail of the fabricated 100-micromirror device.

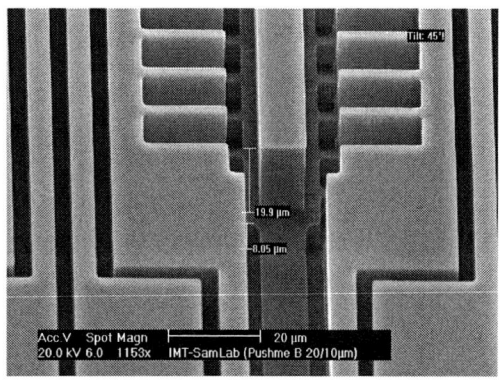

Fig. 3. SEM image of the piston actuator, fabricated using a self-aligned delay-mask DRIE process.

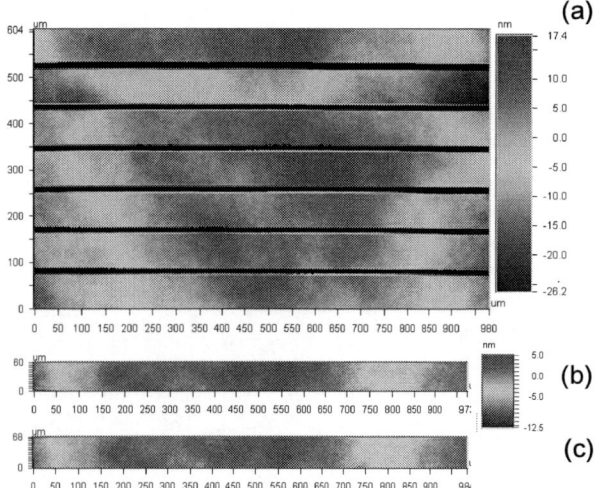

Fig. 4. Surface flatness measurements of a) a seven-mirror array b) before and c) after packaging, exhibiting an excellent surface quality over 1 mm.

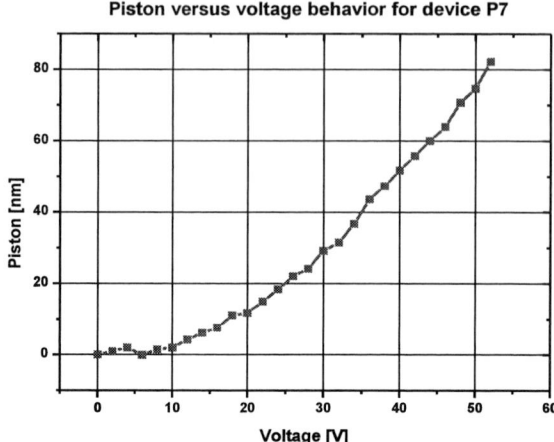

Fig. 5. Pull-in limits the achievable piston to 80 nm.

whereas the packaging did not influence the surface quality due to the decoupling of mirror and surrounding device (see Figs. 4b and 4c). The maximum achievable piston and tilt were 80 nm and 0.2°, respectively, thus not yet within the design specifications. The reason for this limited range of the first generation device was a lateral pull-in of the vertical comb-drives (Fig. 5).

Currently, we are working on a re-design featuring a new comb design with a reduced orthogonal force component to minimize lateral pull-in, and a displaced tilt actuator with increased torque for larger tilts at lower voltages.

IV. CONCLUSION

We showed the first prototype of a linear micromirror array to be used for femtosecond pulse shaping. A suitable, first application for the device could be for example Optical Dynamic Discrimination (ODD) between bacteria and traffic related organic particles [6] in the UV, utilizing the different temporal behavior of quasi-identical biomolecules.

ACKNOWLEDGMENT

The authors would like to thank the SAMLAB and the service for micro- and nanoscopy at the IMT.

REFERENCES

[1] R. S. Judson and H. Rabitz, "Teaching lasers to control molecules," *Phys. Rev. Lett.*, vol. 68, pp. 1500–1503, 1992.
[2] P. Nuernberger, G. Vogt, R. Selle, S. Fechner, T. Brixner, and G. Gerber, "Generation of shaped ultraviolet pulses at the third harmonic of titanium-sapphire femtosecond laser radiation," *Applied Physics B: Lasers and Optics*, vol. 88, no. 4, pp. 519–526, 2007.
[3] M. Hacker, G. Stobrawa, R. Sauerbrey, T. Buckup, M. Motzkus, M. Wildenhain, and A. Gehner, "Micromirror SLM for femtosecond pulse shaping in the ultraviolet," *Applied Physics B: Lasers and Optics*, vol. 76, no. 6, pp. 711–714, 2003.
[4] A. M. Weiner, "Femtosecond pulse shaping using spatial light modulators," *Rev. Sci. Instrum.*, vol. 71, no. 5, p. 1929, 2000.
[5] D. Hah, C. Choi, C. Kim, and C. Jun, "A self-aligned vertical comb-drive actuator on an SOI wafer for a 2 D scanning micromirror," *Journal of Micromechanics and Microengineering*, vol. 14, pp. 1148–1156, 2004.
[6] F. Courvoisier, V. Boutou, V. Wood, A. Bartelt, M.Roth, H. Rabitz, and J.-P. Wolf, "Femtosecond laser pulses distinguish bacteria from background urban aerosols," *Applied Physics Letters*, vol. 87, no. 6, p. 063901, 2005.

Cooling and Amplifying Micro-Mechanical Motion with Light

Kerry Vahala

California Institute of Technology, Pasadena, California

vahala@caltech.edu, www.vahala.caltech.edu

Recent years have witnessed a series of developments at the intersection of two, previously distinct subjects. Optical (micro-) cavities [1] and micro (nano-) mechanical resonators [2], each a subject in their own right with a rich scientific and technological history, have, in a sense, become entangled experimentally by the underlying mechanism of optical forces. These forces and their related physics have been of major interest in the field of atomic physics for over 5 decades [3-5], and the emerging opto-mechanical context for these forces has many parallels with this field. There is also a rich theoretical history that considers the implications of optical forces in this new context [6-9]. Despite this theoretical promise, the manifestations of these forces on micro-mechanical objects have only recently become an experimental reality [10]. We will review recent demonstrations of both mechanical amplification and cooling by radiation pressure forces in micron-scale toroidal resonators and also in silicon cantilever-based resonators. These devices contain high-Q optical modes in coexistence with high-Q mechanical modes. Resonantly enhanced optical forces couple these mechanical and optical degrees of freedom, creating two distinct dynamical regimes. In the first, mechanical amplification can overcome intrinsic loss to induce regenerative mechanical oscillation up to microwave rates [12]. In the second, mechanical cooling to low temperatures is possible (sub Kelvin cooling from room temperature has been demonstrated). After discussing progress directed to ground-state cooling of a macroscale mechanical oscillator [13,14], the possible future directions of this emerging field of *cavity opto-mechanics* will be considered.

[1] K. Vahala, "*Optical Microcavities*," Nature, vol. 424, No. 6950, August 2003.
[2] K. Schwab and M L Roukes, "*Putting Mechanics into Quantum Mechanics*," Physics Today, July 2005.
[3] A. Ashkin, *Physical Review Letters* **24**, 156 (1970).
[4] T. W. Hänsch, A. L. Schawlow, Optics Communications 13, 68 (1975).
[5] D. Wineland, H. Dehmelt, *Bulletin of the American Physical Society* **20**, 637 (1975).
[6] V. B. Braginsky, *Measurement of Weak Forces in Physics Experiments* (University of Chicago Press, Chicago, 1977), pp.
[7] C. M. Caves, *Physical Review D* **23**, 1693 (1981).
[8] V. B. Braginsky, S. P. Vyatchanin, Phys. Lett. A, **293**, 228 (2002).
[9] S. Mancini et. al., Phys. Rev Lett., **88,** no. 12, 120401-1 (2002); Marshall, W. et. al. Phys. Rev. Lett., **91**, 130401 (2003).
[10] T. Kippenberg and K. Vahala, "*Cavity Optomechanics*," Optics Express Review, Dec 10, 2007.
[11] D. K. Armani, et. al. Nature, **421**, pp. 925-929, 27 February (2003).
[12] T. J. Kippenberg, et. al. Phys. Rev. Lett. **95**, 033901, 2005. ; T. Carmon, et. al., Phys. Rev. Lett., **94**, 223902, June 2005.; H. Rokhsari, et. al. Optics Express, **13**, No. 14, July 2005.
[13] A. Schliesser, et. al. Phys. Rev. Lett., 97, 243905, Dec 15, (2006); S. Gigan, H.R. et. al., Nature (London) **444**, 67 (2006); O. Arcizet, et. al., Nature (London) **444**, 71 (2006).
[14] Kleckner, D.& Bouwmeester, D., *Nature* 432, 75-78 (2006).

MEMS based Dual-Axes Confocal Clinical Endoscope
for Real Time *in vivo* Imaging

W. Piyawattanametha[1,2,3], M. J. Mandella[1,3], H. Ra[1], J. T. C. Liu[1,3], E. Garai[1,3], G. S. Kino[1], O. Solgaard[1], and C. H. Contag[3]

[1]*Edward L. Ginzton Laboratory, Stanford University, Stanford, CA 94305, USA*
[2]*NECTEC/NANOTEC, Pathumthani, Thailand 12120,* [3]*James H. Clark Center for Biomedical Engineering & Sciences*
Stanford University, Stanford, CA 94305, USA. E-mail:wibool@gmail.com

ABSTRACT

We demonstrate a dual-axes confocal endoscope in a 5.5 mm diameter package for clinical use. Miniaturization is achieved by using a barbell-shaped, gimbaled, two-dimensional MEMS scanner that is actuated by self-aligned, vertical-comb actuators. The maximum DC optical scan angles are ±2.6° on the inner axis and ±0.8° on the outer axis, and the corresponding resonance frequencies are 3.1 kHz and 1.1 KHz. The maximum imaging rate is 10 frames/second, and the microscope achieves full-width-half-maximum transverse and axial resolutions of 5μm and 7 μm, respectively when operated in the near infrared wavelength (785 nm).

INTRODUCTION

Confocal microscopy offers opportunities for three-dimensional (3-D) imaging due to its optical sectioning capabilities, and has had a tremendous impact on the study of cells in culture and small transparent organisms. However, transition to a miniature *in vivo* confocal microscope has been limited by high numerical aperture (NA) optics and the need for a scalable beam-scanning mechanism. Advances will be required to further develop the field of intravital microscopy and clinical imaging. A conventional single-axis confocal (SAC) microscope uses a large NA lens while a dual-axes confocal (DAC) microscope utilizes two overlapping low NA beams from two smaller lenses. The transverse and axial resolutions of the DAC are both proportional to 1/NA, while in the SAC they are proportional to 1/NA and 1/NA^2, respectively. For 3-D imaging, the axial resolution is most important for achieving optical sectioning. Performance tradeoffs among axial resolution, field of view (FOV), and objective lens size of both systems can be further discussed as follows: By assuming the SAC and DAC to have the same FOV (implying the same focal length, *f*) as shown in Fig. 1a, lens diameter, *D*, is defined from the geometry as $D_{DAC} \sim f[(\tan \alpha+\theta) - (\tan \theta-\alpha)]$ and $D_{SAC} \sim 2f \times \tan \rho$. The normalized axial resolution [1], given by δ, is defined as $\delta_{SAC} = \lambda/\Delta z_{SAC} = 1.1n(1-\cos \rho)$ and $\delta_{DAC} = \lambda/\Delta z_{DAC} = 2.7n\alpha\sin \theta$. Figure 1b shows a plot of *D* versus δ for both SAC and DAC systems. It reveals that for a SAC system to achieve the same axial resolution and FOV, it requires ~4x the lens

diameter of a DAC system, thus requiring ~8x the scan-mirror area. DACs can therefore be implemented in much smaller packages than SACs. Our current DAC design parameters, θ=23° and α= 6° (δ_{DAC}=0.15) have an optimum ratio α/θ=0.26 and enables miniaturization down to a 5.5. mm diameter package.

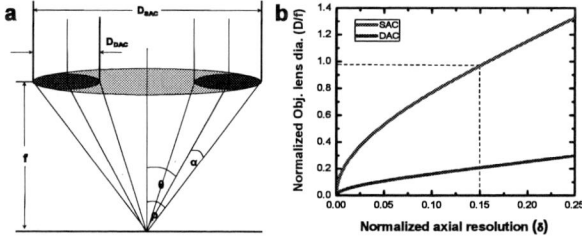

Fig. 1: a) Basic architecture of SAC versus DAC. b) Objective lens diameter vs. axial resolution for SACs and DACs.

Another advantage of the DAC design is improved image contrast of optical sections in a scattering medium (tissue) because the light scattered along the input illumination path before the focal volume (scattered photon noise) has very low probability of coupling back to the output collection path [1].

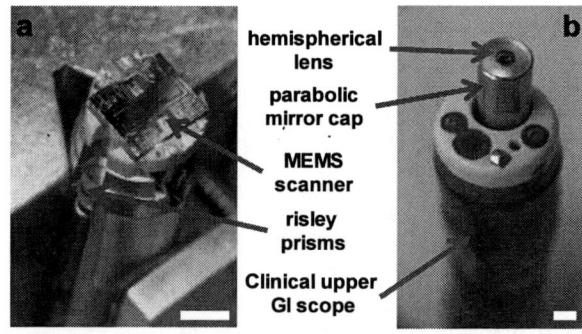

Figure 2: 5-mm diameter DAC endoscope: a) Photograph with its top open. b) Photograph while in a biopsy channel of a therapeutic upper GI endoscope. The scale bars are 2 mm.

Previously, MEMS scanner based DAC microscopes have been demonstrated in 10 mm diameter packages [2-4]. In this abstract, we present an endoscopic MEMS DAC with a 5.5 mm diameter package capable of *in vivo* real time imaging. A photograph of the DAC endoscope with its top open showing internal parts and

978-1-4244-1917-3/08/$25.00 ©2008 IEEE

the MEMS scanner is shown in Fig. 2a. When in use, the 5 mm DAC endoscope is sealed inside a 5.5 mm diameter package and is inserted through the biopsy channel of a therapeutic gastrointestinal (GI) scope as shown in Fig. 2b.

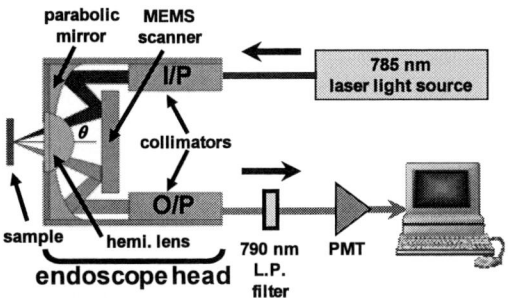

Fig. 3: Schematic of the DAC imaging setup.

The MEMS gimbal two-dimensional (2-D) scanners are batch fabricated on silicon-on-insulator wafers that have two bonded single-crystalline silicon device layers (each 30 μm thick) [5]. The scanner has a barbell shape to accommodate both illumination and collection beams and to ensure overlapping of two foci throughout the scan. A schematic drawing of the endoscope head is shown in Figure 3. The detail of the optical design has been described in ref. [3]. The endoscope is capable of imaging in both reflectance and fluorescence imaging modes. A 790-nm long pass filter is used for fluorescence imaging mode. The NA of the input and output beams is 0.15. The output laser power on the sample is 2.5 mW. An electric micromotor can be used to enable vertical translation (depth interrogation) for volumetric imaging. The endoscope is capable of imaging in both reflectance and fluorescence imaging modes.

RESULTS & DISCUSSIONS

Table 1: MEMS scanner performance

	Inner	Outer
Mirror radius of curvature	> 0.6 m	
Resonant freq.	3110 Hz	1100 Hz
Max. DC scan angle (Optical)	± 2.6°	± 0.8°
Voltage at Max. DC angle	130 V_{dc}	160 V_{dc}

Images are acquired by the MEMS scanner performing a 2-D raster scan. The inner axis (slow-axis) has its opposing comb actuator banks driven 180° out of phase to linearize the angular deflection [1] while the outer axis (fast-axis) is driven by either one of the comb actuator banks. The fast-axis is driven at

resonance by a unipolar sinusoidal waveform (V_{pp} = 110 V). The slow-axis is driven in a DC driving mode (1-10 Hz depending on the required imaging frame rate) with smooth-turn-around sawtooth waveforms (V_{pp} = 130 V) to avoid mirror ringing. Small image distortion can be observed at the edges of the images due to deploying uncorrected biasing waveforms to linearize the scan when actuating the MEMS scanner. Table 1 summarizes the MEMS scanner performance.

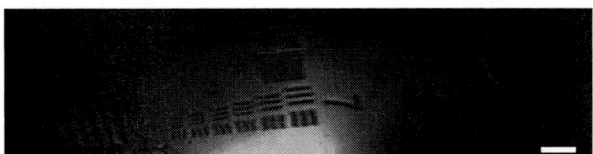

Fig. 4: Reflectance image of USAF resolution target. The scale bar is 25 μm.

Figure 4 shows a reflectance image of a US Air Force (USAF) resolution target (group 7) acquired from the DAC endoscope. The image is a single frame collected at 10 frames per second. Dark areas on the image are from imaging a flat USAF resolution target with an arc-shaped focal plane. The FOV is 475 × 200 μm^2 with 400 pixels × 110 pixels. The scale bar is 25 μm.

CONCLUSION

We have successfully demonstrated the first DAC endoscope in a 5.5 mm diameter package with a 2-D MEMS scanner. Reflectance images with large FOV have been acquired at 10 Hz. We anticipate a broad set of future *in vivo* imaging applications for this endoscope for early cancer detection in sub-mucosal tissue layers in the GI tract and for intravital imaging.

REFERENCES

[1] "Improved rejection of multiply scattered photons in confocal microscopy using dual-axes architecture," L. K. Wong, M. J. Mandella, G. S. Kino, and T. D. Wang, Optics Letters. 32, 1674-1676 (2007).

[2] "Three-dimensional *in vivo* Real Time Imaging by a Miniature Dual-axes Confocal Microscope based on a Two-dimensional MEMS Scanner,"W. Piyawattanametha, H. Ra, M. J. Mandella, J. T. C. Liu, L. K. Wong, C. B. Du, T. D. Wang, C. H. Contag, G. S. Kino, and O. Solgaard, the 14th International conference on Solid-State Sensors, Actuators, and Microsystems (Transducers 2007 & Eurosensors XXI), Lyon, France, June 10-14, 2007, pp. 439-442.

[3] "MEMS Based Dual-axes Confocal Reflectance Handheld Microscope for *in vivo* Imaging," W. Piyawattanametha, J. T. C. Liu, M. J. Mandella, H. Ra, L. K. Wong, P. Hsiung, T. D. Wang, G. S. Kino, and O. Solgaard, IEEE/LEOS International Conference on Optical MEMS, Montana, USA, August 21-24, 2006.

[4] "A miniature near-infrared dual-axes confocal microscope utilizing a two-dimensional MEMS scanner," J. T. C. Liu, M. J. Mandella, H. Ra, L. K. Wong, P. Hsiung, T. D. Wang, G. S. Kino, W. Piyawattanametha, and O. Solgaard, Optics Letters: Vol. 32 (2006), Issue 3, pp. 256-258.

[5] "Two-dimensional MEMS scanner for dual-axes confocal microscopy," H. Ra, W. Piyawattanametha, Y. Taguchi, D. Lee, M. J. Mandella, G. S. Kino, C. H. Contag, and O. Solgaard, IEEE Journal of Micro Electromechanical Systems (JMEMS), August 2007, pp. 969-976.

Tunable Multi-micro-lens System for High Lateral Resolution Endoscopic Optical Coherence Tomography

Khaled Aljasem, Andreas Seifert, and Hans Zappe

Laboratory for Micro-optics, Department of Microsystems Engineering – IMTEK, University of Freiburg
Georges-Köhler-Allee 102, 79110 Freiburg, Germany
Khaled.Aljasem@imtek.uni-freiburg.de

Abstract—A pneumatically actuated tunable liquid micro-lens is developed for integration in an endoscopic optical coherence tomography (OCT) system allowing dynamic focusing along the scan depth. An evaluation of the optical performance, particularly transverse resolution, is presented and its utility in OCT discussed. A transverse resolution of about 15 μm which does not vary with a scan depth of 9 mm scan depth has been obtained, performance which cannot be obtained with a fixed focal length lens.

Index Terms—Endoscopic optical coherence tomography, micro-lens, polydimethylsiloxane (PDMS), optical optimization

I. INTRODUCTION

Optical coherence tomography (OCT) has emerged as an essential tool for in-vivo imaging of internal human tissues [1]. High transverse resolution (TR) imaging is often required, particularly for optical coherence microscopy (OCM); lateral resolution on the same order as the axial resolution (1–15 μm) is usually desirable. Maintaining a high TR during the axial scan is desirable but difficult to achieve.

Tunable membrane micro-lenses provide a means to tune the focal length along the axial scan of OCT, but the variation of TR during the scan has not yet been studied.

We present here resolution measurements of pneumatically actuated tunable membrane-lenses covering the required range for OCT axial scan length (typically 3–4 mm). The results will show that single membrane lenses are not practicable for tunable endoscopic applications due to the low TR obtained as the lens is tuned. We present a convenient solution to improve and maintain a high TR along the axial scan of OCT by combining the micro-lens with a fixed-focal length lens. Optical characteristics of the single micro-lens and the multi-micro-lens system are presented. An endoscopic design including the improved system is illustrated.

II. DESIGN AND CHARACTERIZATION

A. Design of the micro-lens

The micro-lens illustrated in Fig. 1 consists of silicon bonded onto a transparent substrate (borosilicate glass [BK7] or Pyrex) [2]. Two openings and a 200 μm wide channel connecting the two openings were defined on the backside of the Si substrate using deep-reactive ion-etch processes. A polydimethylsiloxane (PDMS) membrane is spun on the front

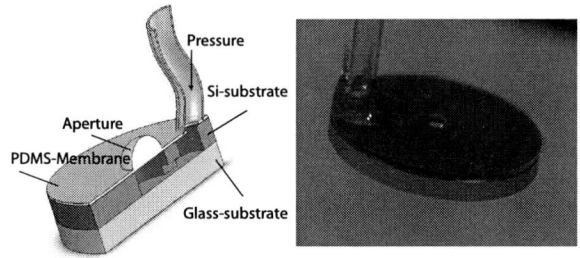

Fig. 1: Cross sectional view and a photo of the lens design

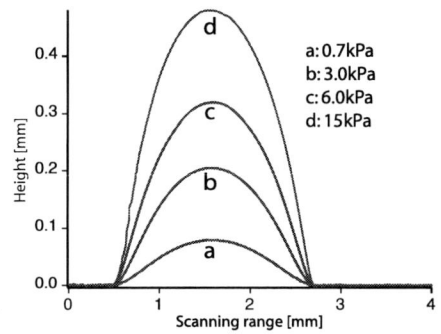

Fig. 2: Surface profiles of the lens as a function of applied pressure.

side of the Si substrate. The cavity is filled with water to give a higher refractive index than the surroundings (air). By applying a pneumatic pressure on the lens, a change of the curvature will occur, and hence a tuning of the focal length. The lens has an aperture of 2 mm.

B. Characterization of the micro-lens

Modulation transfer function (MTF) analysis is applied to estimate the transverse resolution of the micro lens and the improved system. The TR of the lens is mainly dependent on the NA and limited by the diffraction limit and spherical aberration. To obtain the surface shape of the lens as a function of applied pressure, a surface profilometer was used to measure the deformation of the membrane.

Fig. 2 illustrates the measured curvature of the lens for different pressure values. The data shown were fit by 4th order polynomials. The fitted polynomial coefficients were used as

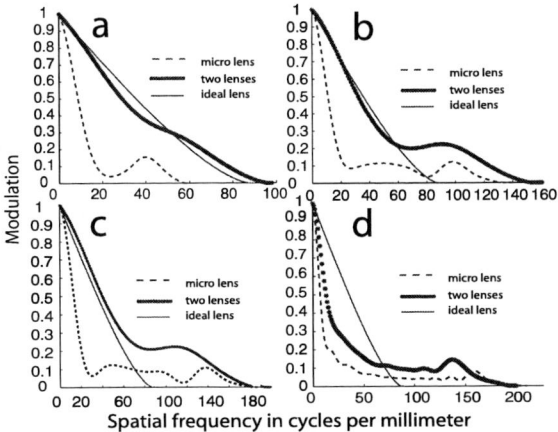

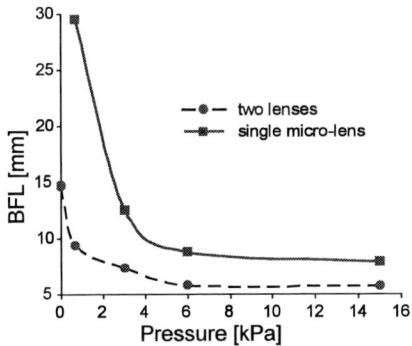

Fig. 4: Back focal length as a function of pressure for the single lens and the combined lens-system.

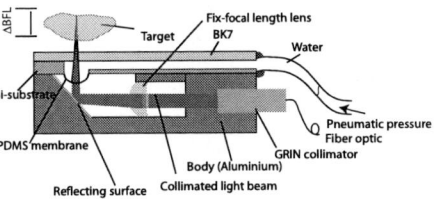

Fig. 5: OCT-probe design with a tunable micro-lens and fixed-focus lens.

Fig. 3: MTF diagrams of the single micro-lens, the entire optical system (combination of two lenses), and a 15 µm TR ideal lens as a reference. a, b, c, and d correspond to the pressure values given in Fig. 2.

input for optical simulation of the lens using ZEMAX to obtain the calculated MTF diagrams of the lens.

III. SIMULATIONS AND EXPERIMENTAL RESULTS

A. Single micro-lens

Fig. 3 illustrates the MTF of the lens at various pressures and thus focal lengths. An ideal diffraction limited lens with 15 µm TR was used as a reference to estimate the TR of the realistic lenses proposed for tunable probe integration.

The dashed curves in Fig. 3 present the MTF of the micro-lens obtained by using the fitted polynomial coefficients of the measured curves from Fig. 2. It is clearly seen that the modulation drops rapidly for higher spatial frequencies, indicating a low TR of the lens in comparison to the ideal lens. The MTF diagrams clearly show that spherical aberration is the primary cause for degradation of MTF.

The TR can be improved using one of two approaches: either by optimizing the surface profile of the lens which is very complicated technologically, or by reducing the aberration through a combination of the micro-lens with a second fixed-focus lens. We pursue the second option here.

B. Advanced micro-lens system

The improved micro-lens system consists of the tunable lens combined with a fixed-focus lens, which leads to a change of the focal tuning range and the TR. The structure is shown in Fig. 5.

Fig. 4 shows the back focal length of the single micro-lens and the lens combination as a function of the applied pressure. It is clearly seen that the tuning range of the combined lens system is about 10 mm, starting at zero pressure.

The MTF curves of the multi-lens system are shown in Fig. 3; these curves show a significant improvement of the TR which is approximately maintained at 15 µm for a wide pressure range (0 – 6 kPa). The obtained TR has been improved significantly and is on the same order as the axial resolution of most conventional OCT systems.

Comparing Figs. 3 and Fig. 4, one can see that an invariant TR along 9 mm scan depth has been achieved. The advanced optical system offers a fine-tuning in sub-millimeter steps, an additional preferable and practicable feature for such an application. It is important to point out that a higher TR can still be achieved, but the tuning range will be smaller.

C. Design of the probe

Fig. 5 illustrates the final design of an OCT-probe with two lenses. The 1 mm wide collimated light beam is focused using the fixed-focus lens. After passing a 45°-deflection mirror, the beam is fine-tuned by the micro-lens. The probe has a 5 mm diameter and 10 mm length.

IV. CONCLUSION

In this paper, a simple, but efficient method is used to enhance the optical resolution of a tunable endoscopic OCT probe including a membrane based tunable liquid micro-lens. The method is based on the integration of an additional objective lens with fixed focus into the optical path. An invariant TR of about 15 µm along 9 mm in the axial scan has been obtained, which promises a significant enhancement of biological imaging using endoscopic OCT.

REFERENCES

[1] Z. Yaqoob, J. Wu, E. J. McDowell, X. Heng, and C. Yang, "Methods and application areas of endoscopic optical coherence tomography," *Journal of Biomedical Optics*, vol. 11, no. 6, p. 063001, 2006.

[2] A. Werber and H. Zappe, "Tunable microfluidic microlenses," *Applied Optics*, vol. 44, pp. 3238–3245, Jun. 2005.

Resonant Cantilever Bio Sensor with Integrated Grating Readout

H. İlker Ocaklı, Alibey Öztürk, Natali Özber, Halil Kavaklı, Erdem Alaca, and Hakan Urey

Koç University, College of Engineering, Istanbul, TURKEY

Phone: +90-212-338-1474, E-mail: hurey@ku.edu.tr

Abstract— Microcantilever based biosensor is fabricated and tested. The main features are the simple single mask fabrication, grating based sensitive optical readout, magnetic thin-film actuation, and suitability for parallel array operation. Sub-picogram mass detection sensitivity demonstrated.

Keywords: cantilever sensor, magnetic actuation, mass sensor

Introduction

MEMS (Microelectromechanical Systems) based resonant mass sensors have been widely utilized in diverse fields of research with a number of applications including detection of chemical analytes, viruses, cells, bacteria or DNA molecules. [1-7] The change in the resonant behavior of the sensors can be monitored via optical, electrical or mechanical readout sensors. Most of the recent studies showed that smaller structures lead to better sensitivity since they provide higher resonance frequencies and higher quality factors. [8]

In this study, a microcantilever sensor array is fabricated and tested. Magnetic actuation using thin film Nickel structural layer and the integrated grating optical readout results are discussed. Bio-molecule detection with high SNR is demonstrated.

Fabrication and Characterization

Since the resonance frequency of a cantilever depends on the geometry, cantilevers with varying sizes are designed and fabricated. The fabrication sequence is illustrated in Figure 1a. For patterning, UV lithography with a single mask is used. Main structure is formed by electroplating nickel film on silicon to enable magnetic actuation. In order to overcome the adhesion problem between Ni and Si interfaces and to increase the conductivity, thin chromium and gold layers are sputtered on the Si wafer. Finally cantilevers are released by gold, chromium and anisotropic KOH etching. Gap between the cantilevers and the silicon substrate is about 8um, which makes the release process straight-forward and avoids squeezed-film effects.

The cantilevers are tested in ambient pressure with magnetic actuation by applying an AC signal through an external electromagnetic coil, which can easily be micromachined underneath or on the back side of the wafer. The resonance frequencies of the cantilevers are measured with a photodetector which utilizes the grating interferometry principle as shown in Figure 1b. Q-factor ranges 10-950, which is comparable with the literature for ambient operation. The silicon substrate serves as the stationary reference surface. The displacement of the cantilevers can be monitored by measuring the sinusoidal intensity modulation in the zeroth and first order diffracted light that is caused by the dynamic deflection of the cantilevers. [9] All the scattered light from the silicon and nickel surface contribute to a bias signal mainly at the 0th grating order, a high SNR signal can easily be extracted at the 1st grating order due to the dynamic operation mode.

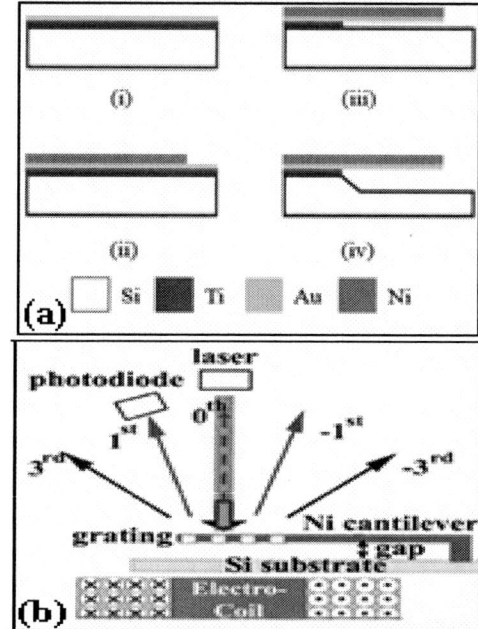

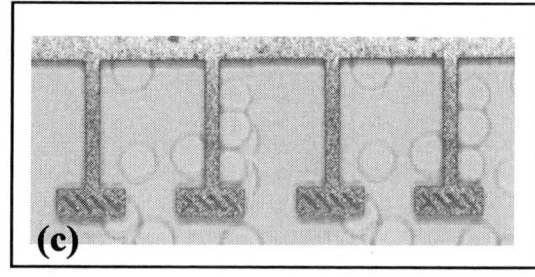

Fig.1. (a) Fabrication flow (b) Optical readout and magnetic actuation mechanism. A PD setup is used to monitor the modulation of 0th and 1st orders of the diffracted light beam. (c) Microscope picture of the fabricated cantilevers with diffraction grating.

Figure 2 illustrates the optical readout signal. Since the photodetector output is cyclic with the cantilever deflection, different output curves are observed for

978-1-4244-1917-3/08/$25.00 ©2008 IEEE

large and small deflection cases. Main interest is not the deflection amplitude but rather the frequency of operation and the gap can be tuned with a DC magnetic field to provide maximum output signal around the operating point. Photodetector signal output signal can be used to keep the system at resonance and the shift in the resonance in response to the change in the cantilever mass is detected.

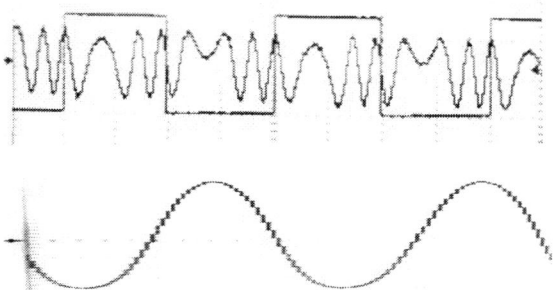

Figure 2: Top: square-wave excitation signal and the large deflection (>λ/4) optical readout signal. Bottom: Small deflection case (<λ/4) optical readout signal.

Mass and Bio Detection Results

By applying inorganic and organic loads on the cantilever, the sensitivity is tested. In the first method known amounts of Au particles are deposited on a 6 μm x 60 μm x 1.4 μm (width x length x thickness) cantilevers in two steps using a fully calibrated sputtering system. The change in the resonance frequency is monitored in each step and illustrated in Figure 2. The results show that the cantilever has a mass responsivity of 0.32 pg/Hz, which leads to a minimum detectable mass of 0.5pg using 1.5Hz minimum detectable frequency at 72KHz calculated using RMS phase noise analysis around the resonance.

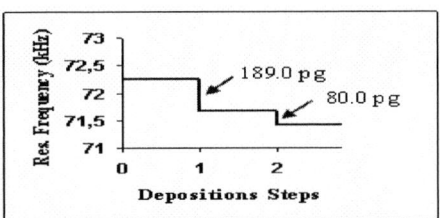

Fig 2. Estimation of the mass responsivity.

In the second demonstration, a purified human kappa opioid receptor is immobilized on the gold surface of a 7 μm x 70 μm x 1.3 μm cantilever via thiol groups of DSP linker through His tag. [10] The shift in the resonance frequency is measured as 180Hz and illustrated in Figure 3. The corresponding mass loading is calculated from FEM analysis as 85pg at 55KHz. The theoretical estimation for monolayer of receptors is 40pg.

Conclusions

A high-sensitivity mass sensor and bio-functionalization with human opioid receptors were demonstrated. A diffraction grating interferometer is integrated with the cantilevers that are actuated with magnetic forces. The electromagnetic actuation makes the system suitable for liquid operation. Dynamic operation provides better SNR compared to static deflection based cantilever sensors. Last but not the least, the integrated grating readout makes the system suitable for parallel array readout with a single photodetector, which is an important advantage compared to other cantilever sensors. Individual cantilever signals can be seperated by Fourier filtering at the output.

In summary, electromagnetic actuation, grating based optical readout, parallel operation, and one mask simple microfabrication are the main important advantages compared to other cantilever sensors in the literature and allows for low-cost, portable, and disposable bio and chemical sensor system development.

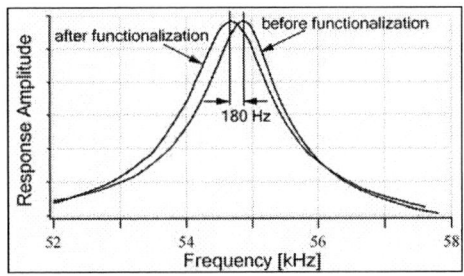

Fig.3. A shift of the resonance frequency is caused by loading human opioid receptors.

References

[1] M. K. Baller, et al., "A cantilever array-based artificial nose", *Ultramicroscopy*, vol.82, pp 1-9, 2000

[2] A. Gupta, et al., "Single virus particle mass detection using microresonators with nanoscale thickness", *App. Phys. Lett.*, vol.84, 1976-1978, 2004

[3] B. Ilic, et al., "Single cell detection with micromechanical oscillators", *J. Vac. Sci. Technol. B*, vol.19, 2825-2828, 2001

[4] D. Ramos, et al., "Origin of the response of nanomechanical resonators to bacteria adsorption", *J. Appl. Phys.*, vol.100, 106105-1-106105-3, 2006

[5] B. Ilic, et al., "Enumeration of DNA Molecules Bound to a Nanomechanical Oscillator", *Nano Lett.*, vol.5, 925-929, 2005

[6] K. K. Park, et al., "Capacitive micromachined ultrasonic trasnducers for chemical detection in nitrogen, " Appl. Phys. Lett., 91, 094102, 1-3, 2007

[7] F. M. Battison, et al., "A chemical sensor based on microfabricared cantilever array with simultaneous resonance-frequency and bending readout", *Sensors and Actuators B*, vol.77, 122-131, 2001

[8] Y. T. Yang, et al., "Zeptogram-Scale Nanomechanical Mass Sensing", *Nano Lett.*, vol.6, 583-586, 2006

[9] N. A. Hall and F. L. Degertekin, "An Integrated Optical Interferometric Detection Method for Micromachined Capacitive Acoustic Transducers", *Appl. Phys. Lett.*, vol.80, 3859-3861,2002

[10] S. A. Trammell, et al., "Oriented binding of photosynthetic reaction centers on gold using Ni---NTA self-assembled monolayers", *Biosensors and Bioelectronics*, vol.19, 1649-1655, 2004

Design and Fabrication of Parylene-hinged Slow-scan Optical Scanner for OCT Endoscope Application

M. Nakada[1], K. Takahashi[1], A. Higo[2], H. Fujita[1], and H. Toshiyoshi[1]

[1]Institute of Industrial Science, The University of Tokyo, 4-6-1 komaba Meguro-ku, Tokyo 153-8505, Japan

Phone: +81-3-5452-6277 / FAX: +81-3-5452-6250 / Email: nakada@iis.u-tokyo.ac.jp

[2]Research Center for Advanced Science and Technology, The University of Tokyo

ABSTRACT

We report a new design and fabrication of parylene-hinged electrostatic optical scanner made by the SOI-MEMS process technology. Parylene is a CVD-processed organic material of small elastic constant and high chemical stability, and it was found to be suitable to make a low-voltage MEMS scanner of low resonance frequency.

Keywords: scanner, parylene, low resonance frequency, endoscope

1. INTRODUCTION

High speed MEMS optical scanners are usually preferred for as laser displays and optical communication systems but an intentional low speed is needed for a scanner used in a medical endoscope that acquires faint signal of scattered photons. Previously on our works, MEMS optical scanner for OCT (optical coherence tomography) endoscope has been developed [1, 2]. However, making thin and soft silicon springs lead to a processing difficulty. Silicon has the shear modulus of rigidity of 60~70 GPa[3], and hence it is difficult to make a MEMS scanner of low resonance frequency such as 30 Hz or 60Hz of video frame rate. To resolve this problem, we newly used CVD (chemical vapor deposition) processed parylene-C; it has elastic constant of 2.76 GPa [4], which is 50~60 fold lower than that of silicon.

Torsion bars of small rigidity could also be made by using thin film polymer or silicon nitride [5, 6] but such geometrical shapes are occasionally associated with undesired oscillation mode. In this paper, we study optical scanners with torsion bars of thick but soft polymer film for the sake of low resonance frequency. Parylene-based process and the MEMS scanner characterization are reported.

2. FABRICATION

Vertical-comb type electrostatic actuation is used in the scanner. Fig. 1 (a) shows an SEM (scanning electron microscope) image of the developed scanner; the chip footprint was 3.0 mm by 2.4 mm, and the mirror area was 1.5 mm x 1.5 mm square. Fig. 1 (b) is a close-up view of the torsion bar of 300 μm long, 20 μm wide, and 13 μm thick. Fig. 1 (c) shows a close up view of the comb fingers of 130 μm long, 5 μm wide, and with an 8 μm air gap. Parameter are summarized in Table I. The torsion bar thickness (parylene) could be designed to be different from that of the SOI layer.

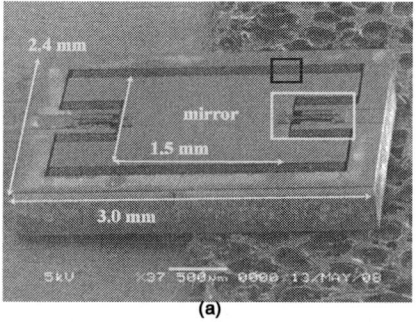
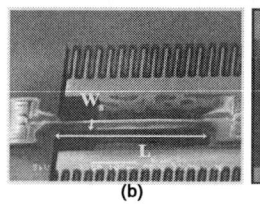
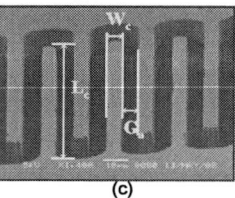

Fig.1 SEM image of parylene-hinged MEMS scanner. (a) Entire image, (b) close up view of the parylene torsion bar, and (c) close up view of the comb actuators.

Table I   Parameters of fabricated MEMS scanner

Parameter	Symbol	Design
Mirror area	S	1.5 mm x 1.5 mm
Torsion bar length	Ls	300 μm
Torsion bar width	Ws	20 μm
Torsion bar thickness	Ts	13 μm
Comb length	Lc	130 μm
Comb width	Wc	5 μm
Air gap	Ga	8 μm

Fig. 2 shows the fabrication process used an SOI wafer (10-μm-thick silicon, 0.5-μm-thick oxide, and 450-μm-thick substrate). (1) The SOI was trench-etched by DRIE (deep reactive ion etching) for the parylene torsion bar, and the trench was filled the with a 13-μm-thick parylene film. (2) Protecting the torsion bar with aluminum mask, we removed the parylene in unwanted area by using oxygen ashing. (3-4) The top and bottom sides of the wafer were processed by DRIE to make the shape of the scanner. (5) The buried oxide was selectively removed in a hydrofluoric acid and finally aluminum was deposited on the top for electrical conductivity. It should be noted that the parylene survived throughout the MEMS processes.

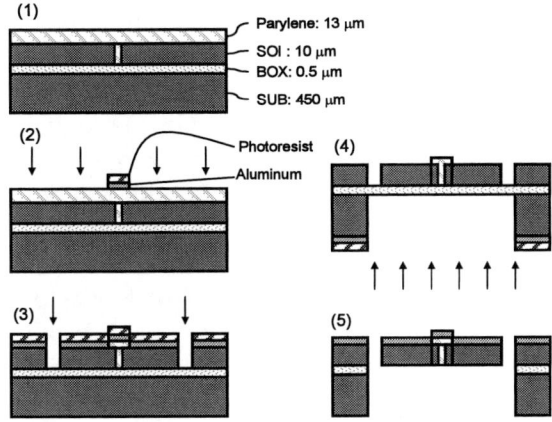

Fig.2 Fabrication process for parylene-hinged MEMS scanner.

3. EXPERIMENTAL RESULTS

Fig. 3 shows the frequency response of the developed scanner, operated with a 0-10V sinusoidal voltage; the frequency was set to double the mechanical resonance. It exhibited a hysteresis behavior depending on the frequency-sweep direction. The peak mechanical angle in Fig. 3 is 4.3° at a mechanical frequency of 305Hz, observed on the way the excitation frequency was lowered from 1 kHz. Fig. 4 plots the scan angle at resonance; the maximum was 15.6 degrees at 43.4V.

4. DISCUSSION AND CONCLUSION

We have estimated the impact of parylene in designing the torsion hinges for a given resonant frequency. Fig. 5 plots the resonant frequency for a structure with parameter in Table I but with different suspension width. Parylene torsion bars could reach a low resonance of 60 Hz at a suspension width of 4.7-μm-width. On the other hand, silicon beams should be made as thin as 1.2-μm-width, which was difficult to make with negligible processing deviation. In conclusion, CVD-processed parylene was found to be a reliable material from both mechanical and chemical points of view to make optical MEMS scanners of low resonance.

ACKNOWLEDGEMENTS

The photomasks used in this work were made by using the EB pattern generator within the VDEC, the University of Tokyo. We also thank Dr. K. Kuribayashi and Prof. Takeuchi for evaporating parylene.

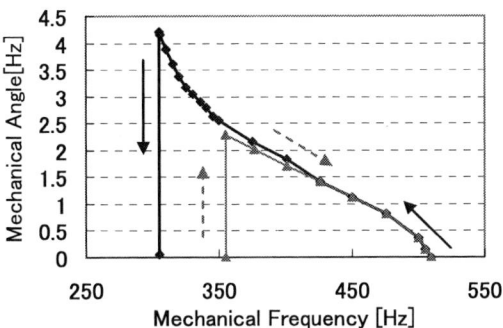

Fig.3 Frequency response of fabricated MEMS scanner.

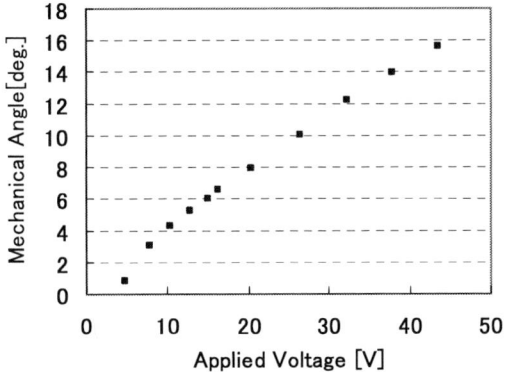

Fig.4 Mechanical angle at resonance as a function of applied voltage.

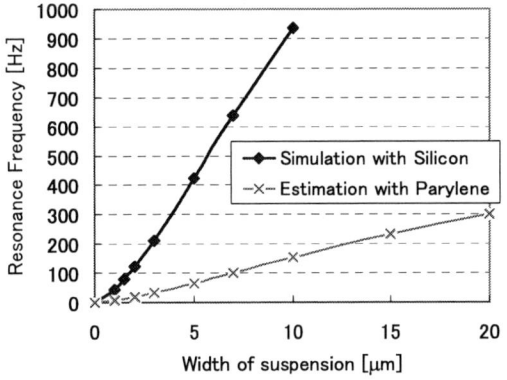

Fig.5 Simulation result of resonant frequency as a function of suspension width for given geometrical dimensions of scanner.

REFERENCES

[1] M. Nakada, et al., Optical MEMS '06, pp. 34-35.

[2] M. Nakada, et al., APCOT '08 (to be presented)

[3] http://www.memsnet.org/

[4] Japan parylene Co.: http://www.parylene.co.jp/

[5] A. Debray, et al., Journal of Institute of Industrial Science, the University of Tokyo, vol. 56, pp. 116-120, 2004.

[6] M. Sasaki, et al., IEEE Journal of Selected Topics in Quantum Electronics, vol. 13, pp. 290-296, 2007.

Photonic Crystal Biosensors and Tunable Resonant Optical Devices

Prof. Brian T. Cunningham
University of Illinois at Urbana-Champaign
Department of Electrical and Computer Engineering

Photonic crystals represent a diverse class of devices that have found important applications in optical sources, waveguides, filters, and detectors. While the first demonstrated photonic crystals were static, researchers soon realized that their properties could be modulated through dynamic changes in their structure and materials. Through visualization of electromagnetic near-fields using precise measurements and computer simulation tools, it became possible to engineer photonic crystal structures with easily-measured far-field properties that function as sensors and optical actuators. This talk will summarize the development of tunable photonic crystal structures for a variety of applications within the Nano Sensors Group at the University of Illinois. The talk will focus on the design, fabrication, and application of large, plastic-based photonic crystal slabs for label-free biosensing, fluorescence enhancement, mechanical strain sensors, and tunable reflectors for laser modulation.

While the sub-wavelength features of photonic crystal structures fabricated by methods such as electron beam lithography offer tremendous precision, many applications require large-area (up to 100 cm^2) inexpensive (single-use disposable) fabrication. These cost and size requirements have led our group to the development of polymer-based 1D and 2D photonic crystal structures that can be produced by nanoreplica molding, such as shown in Figure 1. The basic structure, also sometimes referred to as a Guided Mode Resonant Filter (GMRF) incorporates a low refractive index periodic surface structure that is coated with a high refractive index material, which is covered by a low refractive index superstrate. Through proper selection of the period, materials, grating depth, and thicknesses, such structures are capable of producing narrow bandwidth ($\Delta\lambda$ ~1-5 nm) optical resonances for wavelengths spanning the ultraviolet to infrared. In the far-field, the resonance manifests itself as a narrow band of wavelengths that are reflected with near perfect efficiency, while in the near-field, optical standing waves are produced resulting in enhanced electromagnetic fields in the vicinity of the grating structure, shown in.

The structure shown in Figure 1 is tunable by several means. If the superstrate of the device is water, the displacement of water within the evanescent field region by biomolecules with higher dielectric permittivity will result in a modification of the effective refractive index of the optical resonance, causing a shift of the resonance to longer wavelengths. Due to the narrow bandwidth of the resonance, small changes in the reflected wavelength can be resolved, resulting in the ability to detect biomolecules, virus particles, and cells with high sensitivity. Because resonances are optical standing waves without the capability for lateral propagation, spatially localized attachment of biomaterial results in spatially localized changes in reflected wavelength, resulting in the ability to perform label-free imaging of single-cell attachment to the photonic crystal surface. Biosensors, detection instruments, and applications of these concepts will be shown. Likewise, fabrication of the structure of Figure 1 using materials with refractive index that can be manipulated by application of an electric potential or high intensity optical illumination will result in a tunable resonance. For example, we have demonstrated structures in which the superstrate is comprised of a liquid crystal for electrical tuning (Figure 2), or an azobenzene-containing solution for optical tuning. We have also demonstrated sensitive photonic crystal strain gauges in which the device is fabricated upon a deformable substrate, such as PDMS, and the resonant wavelength is modulated by mechanically stretching (Figure 3). Finally, we demonstrate photonic crystals with resonances that are tuned to the excitation and emission wavelengths of fluorescent dye molecules commonly used for gene expression microarrays, protein biomarker diagnostics, and cell imaging. We use the resonance-enhance near fields to excite fluorophores to emit with higher intensity than on an optically inactive surface, and we extract fluorescent emission toward a sensor with greater collection efficiency.

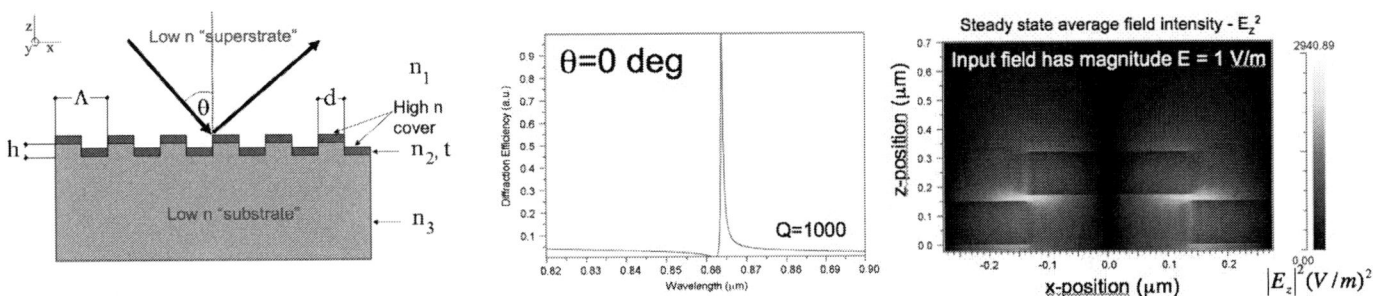

Figure 1. (Left) Photonic crystal cross section structure, (center) far-field narrowband reflectance, and (right) near-field electric field profile at the resonant wavelength.

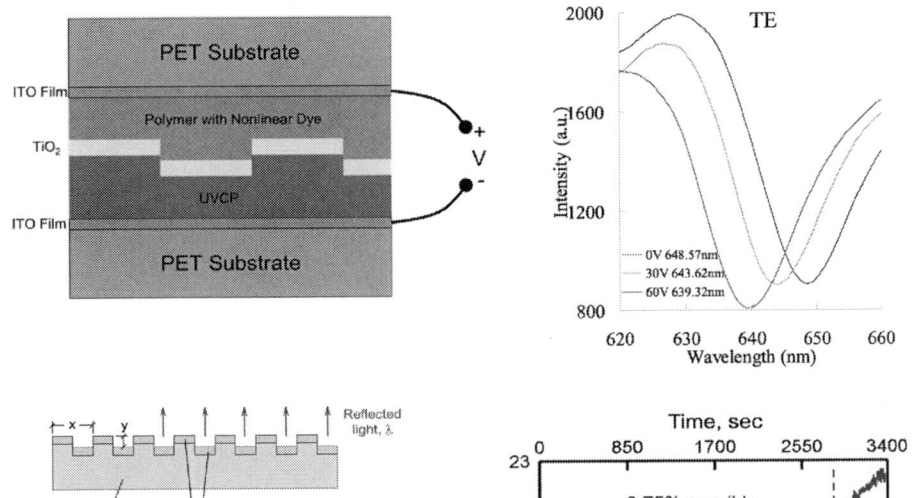

Figure 2. Electrically tunable photonic crystal structure incorporating a liquid crystal superstrate and tuned reflectance behavior of the TE mode.

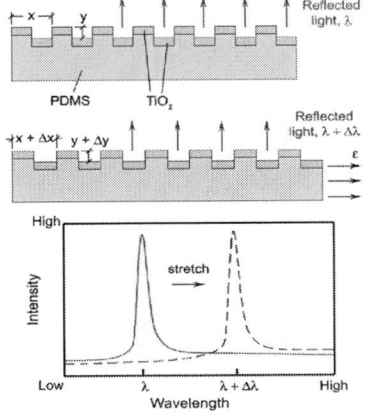

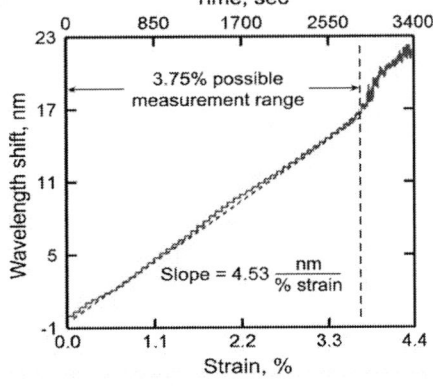

Figure 3. Mechanically tuned photonic crystal structure and relationship between reflected wavelengh and applied strain

Ultrahigh Sensitivity Slot-Waveguide Biosensor on a Highly Integrated Chip for Simultaneous Diagnosis of Multiple Diseases

Daniel Hill, Universidad Politécnica de Valencia, Spain, et. al.

SABIO is a multidisciplinary project involving the emerging fields of micro-nano technology, photonics, fluidics and bio-chemistry, targeting a contribution to the development of intelligent diagnostic equipment through the demonstration of a compact polymer based and silicon-based CMOS-compatible micro-nano system. It integrates optical biosensors for label-free biomolecular recognition based on a novel photonic structure named slot-waveguide with immobilised biomolecular receptors on its surface. The slot-waveguides provide high optical intensity in a subwavelength-size low refractive index region (slot-region) sandwiched between two high refractive index strips (rails) [1] leading to an enhanced interaction between the optical probe and biomolecular complexes (antibody-antigen). As such a biosensor is predicted to exhibit a surface concentration detection-limit lower than 1 pg/mm^2, state-of-the-art in label-free integrated optical biosensors, as well as the possibility of multiplexed assay, which, together with reduced reaction volumes, leads to the ability to perform rapid multi-analyte sensing and comprehensive tests. This offers the further advantageous possibility of assaying several parameters simultaneously and consequently, statistical analysis of these results can potentially increase the reliability and reduce the measurement uncertainty of a diagnostic over single-parameter assays. In addition, the SABIO micro-nano system device applied to its novel protein-based diagnostic technology has the potential to be fast and easy to use, making routine screening or monitoring of diseases more cost-effective.

The flexibility of label-free biomolecule optical sensing technologies permits simplification of assays and time–resolved kinetics measurement for biomolecular interactions. Furthermore the use of integrated photonic devices permits high sensitivity, small size and high scale integration. A recently reported [2] integrated photonic sensor based on slot-waveguide sensor exhibited a bulk ambient sensitivity as high as 212.13 nm/RIU (refractive index unit), which is more than twice as large as that exhibited by other ring resonator optical sensors based on conventional strip waveguides. More recently the detection of label-free molecular binding reactions on the surface of a slot-waveguide ring resonator has been demonstrated [3] through the use of Bobine Serum Albumin (BSA) protein and antiBSA. The device consists of a 70-µm-radius slot-waveguide ring resonator made of Si_3N_4 on SiO_2 with the Si_3N_4 rails of the slot-waveguide ring separated by 200 nm (wslot) and their widths are 400 nm and 550 nm for the outer and inner rails, respectively[2]. Sensing of antiBSA-glutraldehyde and BSA-antiBSA affinity reactions were then studied on the sensing chip through optical characterization. The estimated limits of detection of the device for antiBSA and BSA molecules are 1.66 ng/ml and 52.6 ng/ml, limited by wavelength resolution, comparing favourably with the performance of state-of-the-art integrated photonic biosensors based on conventional strip waveguides.

For coupling light into integrated optics structures surface grating couplers are ideal due their high efficiency and enablement of light injection anywhere on the wafer. The refractive index contrast of silicon nitride is about 0.5, high enough in photonic structures to guarantee a reasonably strong light confinement, while allowing a sensitivity reduction to fabrication imperfections (such as interface roughness) that is problematic for higher confinement levels such as that for silicon-on-insulator structures. Experimental coupling efficiencies >60% for totally and partially etched silicon nitride grating couplers have been seen [4] and more recently angular and wavelength -3 dB tolerances reaching 4° and 50 nm, respectively, have been determined. Alignment tolerances, i.e. the variations of the coupling efficiency when the incident beam waist position relative to the grating is changed in the 3 space dimensions were also determined. The alignment tolerance parallel to the grating is slightly greater than the grating width itself and depends on the beam waist, and along the optical propagation axis it can reach several 100 μm. Perpendicularly to the grating grooves, the coupling efficiency profile appears to be particularly influenced by the duty ratio of the grating.

Summary

1. V. Almeida, Q. Xu, C.A. Barrios, and M. Lipson, "Guiding and confining light in void nanostructure," Opt. Lett. **29**, 1209-1211 (2004).

2. C.A. Barrios, K.B. Gylfason, B. Sanchez, A. Griol, H. Sohlström, M. Holgado and R. Casquel, "Slot-waveguide biochemical sensor," Opt. Lett. **32**, 3080-3082 (2007).

3. C. A. Barrios, M. J. Bañuls, V. Gonzalez-Pedro, K.B. Gylfason, B. Sanchez, A. Griol, A. Maquieira, H. Sohlström, M. Holgado, and Rafael Casquel, "Label-free optical biosensing with slot-waveguides", Opt. Lett. **33**, 7, 708-710 (2008)

4. G. Maire, L. Vivien, G. Sattler, D. Marris-Morini, E. Cassan, S. Laval, A. Kazmierczak, D. Giannone, B. Sanchez, A. Griol, D. Hill, K.B. Gylfason, H. Sohlström, Opt. Exp. **16**, 1, 328-333 (2008)

Nanostructured effective-index micro-optical devices based on blazed 2-D sub-wavelength gratings with uniform features on a variable-pitch

David L. Dickensheets[a], Hans Peter Herzig[b], Wataru Nakagawa[a], Kaspar Suter[b], Urs Staufer[c]

[a]*Electrical and Computer Engineering Department, Montana State University, Bozeman, Montana 59717 USA*
406 994-7147 (ph) 406 994-5958 (fax) davidd@ee.montana.edu
[b]*Institute of Microtechnology (IMT), University of Neuchâtel, Neuchâtel, Switzerland*
[c]*Micro and Nano Engineering Lab, Mechanical, Maritime and Materials Engineering, Delft University of Technology, The Netherlands*

Abstract

Sub-wavelength two-dimensional gratings are formed by etching holes into silicon using DRIE followed by thermal oxidation. Smoothly graded effective index blazing is possible using uniform holes on a variable grating pitch, exhibiting minimal etch lag.

Introduction

Diffractive microlenses and gratings are desirable for many applications because of the possibility for achieving precise phase control, large refractive power and the ability to combine focusing with wavelength dispersion in a single element. Achieving high diffraction efficiency for a diffractive optical element (DOE) requires blazed phase zones. One approach is to create a surface relief grating with smoothly varying grooves using gray scale lithography or a multiple-mask binary optics technique. Another approach is to create a grating with smoothly varying index of refraction within a single phase zone. This paper describes a method for creating blazed gratings by controlling the effective index of refraction within the phase zone using sub-wavelength gratings operating in the zero order of diffraction.

Several investigators have reported blazed gratings based on zero-order diffraction from sub-wavelength gratings (SWG)[1-3]. To create the blaze profile, the fill factor of the grating is varied continuously. The underlying grating may be one-dimensional or two dimensional, and effective medium theory is often used as a starting point for analysis of the resulting nanostructured material[4]. Figure 1 shows a typical device with variable size holes (or columns) defined on a fixed sub-wavelength period. A challenge in creating these structures is etch lag and microloading associated with anisotropic reactive ion etching of grating features with variable size. Fabrication errors resulting from these problems limit diffraction efficiency to be significantly less than theory would predict.

Hypothesis

We propose here a method to create blazed phase structures based on a SWG with uniform etch features on a variable pitch, potentially minimizing problems stemming from feature-size dependent etch lag. The two-dimensional SWG consists of holes on a rectangular grid, with the grating pitch varied continuously in both x and y as a linear function of position x. We anticipate achieving high diffraction efficiency and low polarization dependence over a broad range of incidence angles and grating period, with gratings fabricated using a single-mask process. This variable index material based on blazed SWG structures should be useful to create efficient DOEs for many applications.

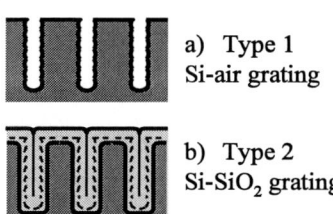

a) Type 1
Si-air grating

b) Type 2
Si-SiO₂ grating

Figure 2 Grating structures studied are of two types: a) Si-air gratings based on DRIE etched holes and b) Si-SiO₂ gratings created with post-etch thermal oxidation to completely back-fill DRIE holes with oxide, and also create an antireflection oxide layer on the grating surface.

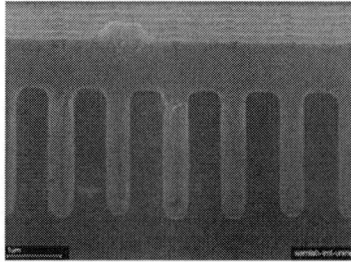

Figure 3 Scanning electron micrograph of a type-2 grating with post etch oxidation. Scale bar is 1 μm.

Methods

We are constructing our gratings from silicon (n=3.4), and silicon dioxide ($1.3<n<1.5$), with the illuminating wave in air. We have designed gratings for use at wavelengths of either λ=1.5 μm or λ=5.2 μm. The basic structure is illustrated in Figure 2. To determine specific parameters for

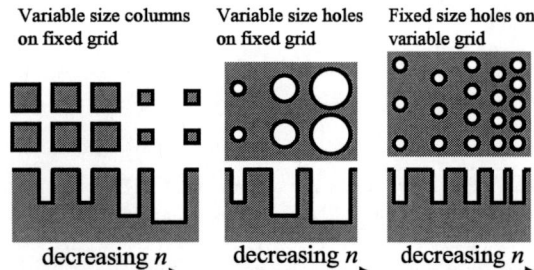

Variable size columns on fixed grid   Variable size holes on fixed grid   Fixed size holes on variable grid

decreasing n decreasing n decreasing n

Figure 1 Blazed binary SWG based on variable size features located on a fixed grid experience pronounced etch lag; fixed size holes on variable grid experience less.

978-1-4244-1917-3/08/$25.00 ©2008 IEEE

the SWG (hole size, hole depth, period in x and y) we have analyzed the local SWG as a uniform grating using rigorous coupled wave analysis (RCWA), which predicts diffraction efficiency and phase delay into the 0th diffraction order. Adjusting the grating parameters changes the phase delay, corresponding to a change in the local effective index of refraction for the structured material. Using sufficiently high aspect ratio holes and a variable period that remains less than $\lambda/2$ everywhere, we can achieve a full 2π phase variation in the 0th diffraction order. We evaluated both Si-air gratings (Figure 2a) and Si-SiO$_2$ gratings (Figure 2b). The latter structures showed the most promise, with simulations predicting better than 85% diffraction efficiency over a broad range of incidence angles for both linear polarization states (it may be noted that Fresnel transmission for a simple air-Si interface is only 70%). Figure 4 shows a typical result, showing diffraction efficiency and phase variation as a function of grating period for a fixed hole size and depth.

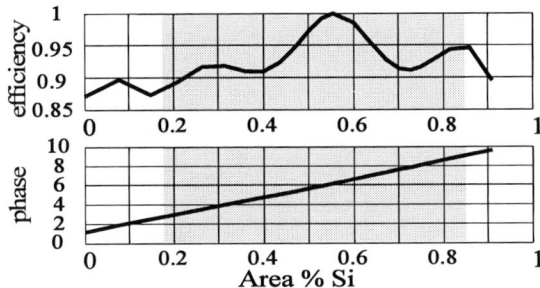

Figure 4 RCWA simulation of local SWG grating showing 0th order diffraction efficiency and phase delay as a function of grating period (scaled as area % of silicon). Parameters: design wavelength 5.2 μm, substrate material Si (n=3.42), hole material SiO$_2$ (n=1.35), hole width 0.12λ, hole depth 0.7λ, top layer SiO$_2$ thickness 0.18λ. Calculated for normal incidence linearly polarized plane wave.

Test gratings were fabricated at IMT of the University of Neuchâtel, Switzerland. Electron beam lithography was used to pattern a PMMA layer, followed by RIE etching to transfer the pattern into a thin mask oxide. Holes were etched into the silicon layer with high aspect ratio of approximately 10:1 using DRIE. The wafers were cleaned in O$_2$ plasma and the mask oxide was removed in BHF. Then a post-etch thermal oxidation was used to completely fill the holes and simultaneously form a top-surface silicon-dioxide layer that serves as an index matching layer which improves transmission efficiency of our gratings. An SEM of the completed SWG structure is shown in Figure 3.

Results and Conclusions

Test gratings were made, each having several phase zones composed of continuously variable SWG structures as shown in the upper image of Figure 5. Other structures such as the Fresnel lens in the lower image in Figure 5 were also fabricated. A primary design feature was use of uniform hole sizes to minimize etch lag across a phase zone. As seen in Figure 6, the etch depth variation is less than 10%, which is an encouraging result. We performed

initial optical measurements of a type-2 22° grating designed for use at λ=1.5 μm, which showed overall diffraction efficiency of approximately 60% (including Fresnel losses), significantly poorer than the simple RCWA simulations predict. This is at least partially due to fabrication tolerances and the departure of our physical devices from the design target dimensions. Measurements are in progress to further characterize these gratings as well as the DOEs designed for 5.2 μm.

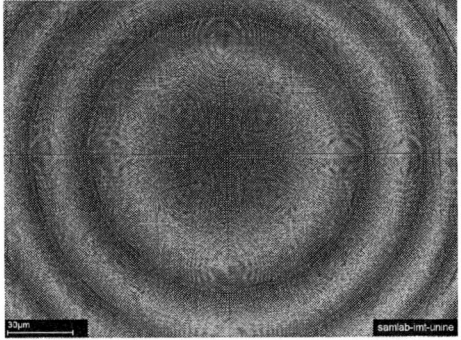

Figure 5 Continuous-phase grating (upper) and Fresnel lens (lower) created using blazed SWG structures.

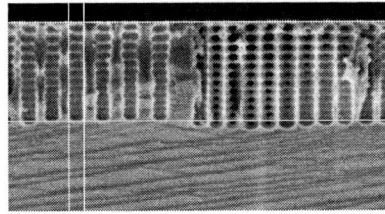

Figure 6 SEM image of one phase zone, with zone boundary near the center of the image. Target devices have more complete separation between etched holes and exhibit even less depth variation of holes across the phase zone. The hole sidewalls show scalloping which is typical for DRIE using the Bosch process.

Based on our analysis and preliminary experimental devices, we believe the use of blazed SWG structures based on uniform sizes of etched features holds promise as a practical approach to create DOEs within the limitations of current DRIE and nanolithography technologies. Hole depth variation across a phase zone was minimal and can be compensated with iterative design techniques.

References

[1] Chen *et al.*, Opt. Letters **20**(2), 121 (1995).
[2] Lalanne *et al.*, J. Opt. Soc. Am. A **16**(5), 1143 (1999).
[3] Sauvan, *et al.*, Opt. Letters **29**(14), 1593 (2004).
[4] Mait *et al.*, J. Opt. Soc. Am. A **16**(5), 1157 (1999).

Characterization of Silicon-on-Insulator Waveguide Chirped Grating for Coupling to a Vertical Optical Fiber

X. Chen, C. Li and H. K. Tsang

Dept. of Electronic Engineering, The Chinese University of Hong Kong, N.T., Hong Kong, P. R. China

Abstract: We compare experimentally the performance of a linearly chirped waveguide grating with a uniform grating for coupling light to a vertical fiber. The measurements show the chirped grating reduces reflection and coupling loss by 2dB.

Keywords: silicon photonics, waveguide, grating coupler, chirped grating

Introduction

Submicron-sized silicon waveguides are the basic building blocks of photonic integrated circuits. However, coupling of light from optical fibers to such waveguides can suffer large losses because of the large mode mismatch. Waveguide grating couplers [1, 2] and adiabatic tapers with polished facets [3] have been proposed to tackle this problem. The grating coupler has some key advantages: they could be placed anywhere on a chip (they need not be located at the edge of a die), they do not need facet polishing, and they are compatible with wafer-scale optical testing. Previously an off-vertical tilt [2, 6] of the optical fiber was used to improve the coupling efficiency by reducing the back reflection into waveguide. However the precise angled alignment of an optical fiber to the grating may not be well suited for low cost optical packaging and grating couplers for perfectly vertical fiber coupling have been proposed [4, 5]. However, these designs either used a slanted grating which requiring a non-standard fabrication process [4] or include additional polishing and etching steps [5] to reduce the back reflection from the grating.

Here we investigate a grating coupler that may be fabricated simply in a single dry etch process and which is based on a section of linearly chirped grating to eliminate the back reflection into waveguide. The performance of the chirped grating is characterized experimentally and compared to the performance of a conventional vertical grating coupler with uniform grating.

2. Chirped grating coupler

The grating coupler was fabricated using deep UV lithography at IMEC on a silicon-on-insulator (SOI) wafer with 220nm top silicon layer and 2μm buried oxide. The fundamental mode of submicron-sized may be expanded by adiabatic taper to a width of 12μm and coupled out vertically using a 70nm shallow etched grating structure. The lateral TE mode profile of the 12μm-widthed waveguide is well-matched to conventional single mode optical fibers and allows theoretical coupling efficiency as high as 97%. Thus, two-dimensional (2-D) simulations may be used for such grating structures [2]. The period needed for vertical out-of-plane coupling is $\Lambda = \lambda/n_{eff}$ [6], where n_{eff}

is the average effective index of the waveguide in the grating range. The proposed design was optimized for TE mode in the waveguide.

The device design is shown in Figure 1. The 22 periods of the grating were divided into two sections. The total length is about 13μm. The grating in the front section was linearly chirped. Grating period Λ varies linearly with p (p=0,1,2,3,4,5,6,7,8) according to: $\Lambda=\Lambda_0+(p/8)\Delta$, where $\Delta=-120nm$ is the max grating period deviation, $\Lambda_0=640nm$ is the period of first grating at the front end. The average period is 580nm with a duty cycle of 53%. A uniform grating with 66% duty cycle was used for the rear section. The period is 590nm. This rear section has a larger back reflection to enhance the coupling efficiency.

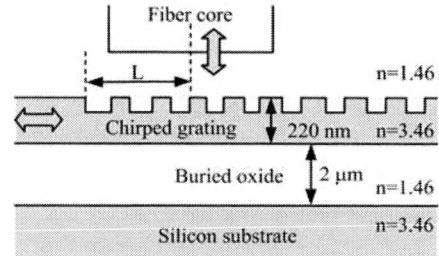

Figure 1: Cross-section schematic of proposed grating coupler.

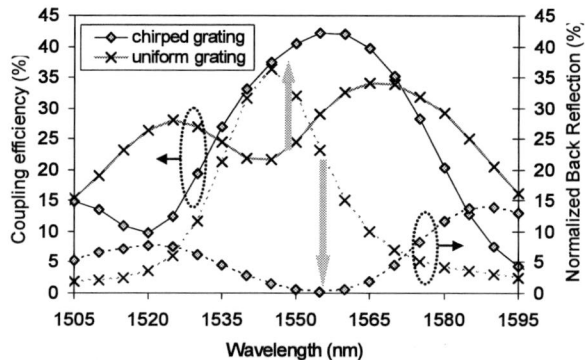

Figure 2: Comparison on coupling efficiency (solid) and back reflection into waveguide (dashed) between our proposed design and an optimized uniform grating coupler

We use 2D FDTD software to calculate the performance of the grating couplers. A Gaussian waveform with 1/e width of 10.4μm was employed to represent the fiber mode. Coupling efficiency and the back reflection into the waveguide are shown in Figure 2. By employing a section of linearly chirped grating, back reflection into waveguide was dramatically suppressed by the chirped grating and the

978-1-4244-1917-3/08/$25.00 ©2008 IEEE   56

maximum coupling efficiency is increased to about 42% with a 3dB bandwidth of 48nm. Simulation results of an optimized grating coupler with uniform grating are also presented for comparison. The period is 580nm with a duty cycle of 53%.

3. Experimental result

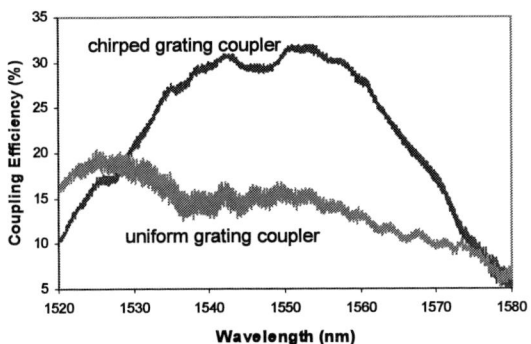

Figure 3: Measured coupling efficiency versus wavelength.

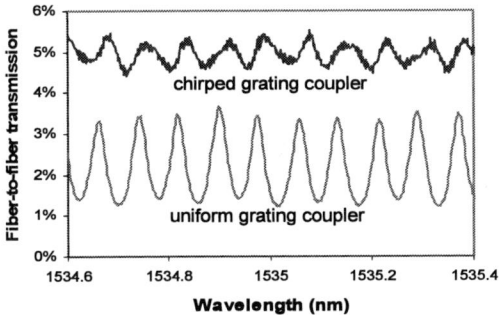

Figure 4: Transmission spectrum of the waveguide with two grating coupler. Strong Fabry-Perot interference was found for the couplers with uniform granting.

The devices which were experimentally characterized consisted of a section of 500nm-widthwaveguide with the adiabatic taper and grating couplers described above for input and output coupling. The measured coupling efficiency is shown in figure 3, which was calculated assumed that the waveguide and taper are lossless and that the input and output grating couplers have the same coupling loss. The maximum coupling efficiency is around 32% with a 3dB bandwidth of 45nm. The actual coupling efficiency of the grating is higher if one takes into account the propagation losses of the waveguide and the 4% reflection loss of the fiber facets are included. The back reflection of the grating coupler could be estimated by the Fabry-Perot interference of the waveguide. As the length between two grating couplers is 4mm including adiabatic tapers and the section of 500nm-width waveguide, the period of the transmission spectrum is calculated to be 0.1nm, assuming the average index of the fundamental mode is 2.9. This agreed well with the experimental result shown in figure 4. The Fabry-Perot interference oscillation of the waveguide with uniform grating couplers was much stronger than the one with chirped grating coupler because of their larger reflection. The uniform grating has a

reflection of about 22% while the chirped grating has about 4% reflection. The alignment tolerance to fiber positioning was studied by scanning the fiber position along x and y axis. The result shown in figure 5 indicates good tolerance, with less than 1dB additional loss from misalignment by 1μm.

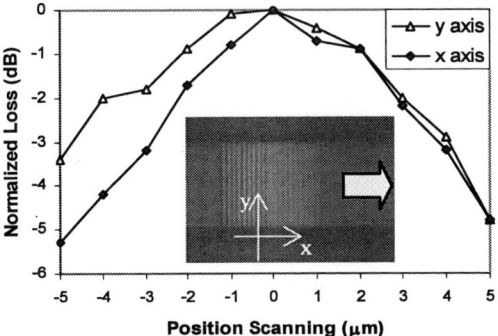

Figure 5: Coupling loss versus fiber position along x and y axis. A photograph of the grating coupler is shown in the inset.

4. Conclusion

In summary, the performance of a linearly chirped waveguide grating with a uniform grating for coupling light between submicron-sized silicon waveguide and a perfectly vertical optical fiber was studied experimentally. The measurements show the chirped grating had reduced reflection and improved the coupling efficiency by 2dB. We obtained 32% coupling efficiency with a 3dB bandwidth of 45nm.

Acknowledgements: This work was funded fully by RGC Earmarked Grant 415905.

5. References

[1] D. Taillaert, et al.: "An out-of-plane grating coupler for efficient butt-coupling between compact planar waveguides and single-mode fibers," IEEE J. Quantum Electron. **38**, 949-955 (2002).

[2] D. Taillaert, et al: "Compact efficient broadband grating coupler for silicon-on-insulator waveguides," Opt. Lett. **29**, 2749-2751 (2004).

[3] T. Shoji, et al: "Low loss mode size converter from 0.3μm² Si wire waveguides to singlemode fibers," Electron. Lett. **38**, 1669-1670 (2002).

[4] B. Wang, et al.: "Embedded slanted grating for vertical coupling between fibers and silicon-on-insulator planar waveguides," IEEE Photon. Technol. Lett. **17**, 1884-1886 (2005).

[5] G. Roelkens, et al.: "High efficiency grating coupler between silicon-on-insulator waveguides and perfectly vertical optical fibers," Opt. Lett. **32**, 1495-1497 (2007).

[6] D. Taillaert, et al: "Grating couplers for coupling between optical fibers and nanophotonic waveguides," Jap. J. Appl. Phys. **45**, 6071-6077 (2006).

Microspectrometer: From Ideas to Product

Hans-Peter Herzig, Toralf Scharf, Omar Manzardo
University of Neuchâtel
Institute of Microtechnology (IMT)
Rue Breguet 2
CH-2000 Neuchâtel, Switzerland

Abstract- **Spectroscopy with miniaturized systems is one of the fastest developing fields and enters now industrial applications. We will discuss near infrared sensing systems based on MEMS spectrometers starting from concepts, explain prototypes and show final product.**

I. INTRODUCTION

Portable and nomadic devices will gain more and more importance but sensing quality and power consumption are main issues. To increase stability of chemometrics recognition the extension of the visible wavelengths range to the near infrared (NIR) and mid infrared (MIR) spectral region is necessary. The main drawbacks for using grating spectrometers in the NIR are the price of detector arrays, the cooling and its power consumption and the size given by the optical path that has to be realized when dispersion is used for wavelengths separation. The upcoming mechanical optical electrical microsystems (MOEMS) attracted a lot of interest and created the hope to enable small reliable high performance sensing systems for NIR and MIR. The technical innovation comes mainly due to the possibility to incorporate scanning elements into the spectrometer. There are several operational principles that have lead to first prototypes or products based on MEMS.

In this paper we will go through the development steps of a high performance scanning microspectrometer developed at IMT-UniNE and commercialized with ARCOPTIX (Neuchâtel). Our discussion will be limited to systems with in-plane actuation. The scanning system moves in the plane of the wafer and all optical functions have to be implemented in such a way to effectively couple and detect light. In the beginning of the development classical spectrometer designs like a Michelson interferometer were copied. Performance of a functional model will be discussed. Miniaturization leads to major performance problems because of the limited size of optical elements or not appropriate material characteristics. The invention of the lamellar principle applied to MEMS brought the breakthrough for MEMS scanning interferometers, which lead to several prototypes and a commercialized product.

II. FOURIER TRANSFORM SPECTROMETERS FUCNTIONAL MODELS

Fourier transform (FT) spectroscopy is a well-known technique to measure the spectra of a weak and extended light source whereas it offers a higher signal-to-noise ratio than other methods [1]. Commonly used FT spectrometers (FTS) require a mirror scanning mechanism with very high precision, resulting in a large device size at high costs. Low-cost miniature spectrometers, however, are key components that permit the fabrication of small-size, portable sensor solutions for applications such as photometers in quality management and spectrometers in analysis tools.

The most common way of fabricating an FTS is utilizing a Michelson interferometer configuration with scanning mirrors. The output signal of the interferometer is the variation of the light intensity I_R as a function of the optical path length δ. The relation of I_R to δ is referred to as the interferogram. The wavelength power spectrum and the recorded interferogram $I_R(\delta)$ are related by a Fourier transformation. Therefore, a spectrum of a light source can be obtained by recording $I_R(\delta)$ and a subsequent FT.

Most microfabricated FTS utilized a micromechanical concept to build a Michelson interferometer but they don't apply MEMS devices for different reasons [2]. Such set-ups, however, offer only limited performance such as measurement speed and stability. Therefore, a new fabrication concept was pursued that includes a compact MEMS scanning mirror with a small footprint of only 75 μm x 500 μm shown in Fig. 1 [3]. The light from the source is divided by a standard macroscopic beam splitter to the movable MEMS mirror and the photodetector. The other mirror opposing the photodetector (PD) is fixed. In order to achieve the obligatory linear relation between the driving voltage and the mirror displacement, a push-pull type driving voltage configuration was chosen by setting the inner combs to one and the same electrical potential and the other two combs on opposing voltages. This way the displacement Δx is proportional to the driving voltage.

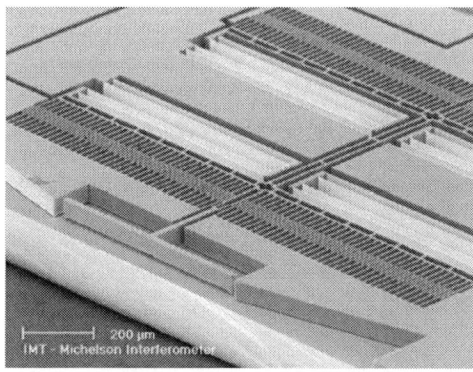

Fig. 1 Scanning electron microscopy photograph of the electrostatic actuator, showing the 75 μm x 500 μm movable mirror, the combs, and the springs.

Note that the interferometer was realized by miniaturizing the scanner and using a conventional optical interface for signal detection. To test the set-up, a He-Ne laser was focused onto the two mirrors via a beam splitter. The driving voltage for the movable mirror was either ±5 V for half the travel distance or ±10 V for the maximum achievable travel distance of $\Delta x = 38.5 \, \mu m$ which corresponds to an optical path difference of $\delta = 77 \, \mu m$. While the small voltage range yielded an almost linear mirror displacement response, the full range showed a slight non-linearity as well as a small hysteresis. Both effects, however, remain constant and can be eliminated by calibrating the voltage response of the mirror displacement. Interferometric calibration allowed measurements of the He-Ne laser linewidth with a precision of about 6 nm, whereas the peak position was repeatedly accurate to about 1 nm. New designs with integrated collimators were tried but lead to unsatisfying results. A different concept was needed to get better performance and that was found with a tunable laminar grating [4].

III. LAMELLAR GRATING FOURIER TRANSFORM SPECTROMETERS PROTOTYPES

A lamellar grating interferometer is a binary grating with a variable depth, which operates in the zero order of the diffraction pattern. This type of apparatus was invented by Strong and Vanasse in 1960 [5]. A lamellar grating interferometer is used as a Fourier spectrometer, but, in contrast to a Michelson interferometer, which splits wave amplitudes at the beam splitter, a lamellar grating interferometer divides the wave front. At the grating the wave front is separated such that one part of the beam is reflected by the front facets (fixed mirrors) and the other part by the back facets (mobile mirrors). The distance between the two series of mirrors determines the optical path difference (OPD). In general, this type of spectrometer is used for wavelengths larger than 100 μm; below that level the tolerances are too tight for most machine shops. We identified that silicon micromachining is the ideal technology to overcome these limitations and realized prototypes for shorter wavelengths.

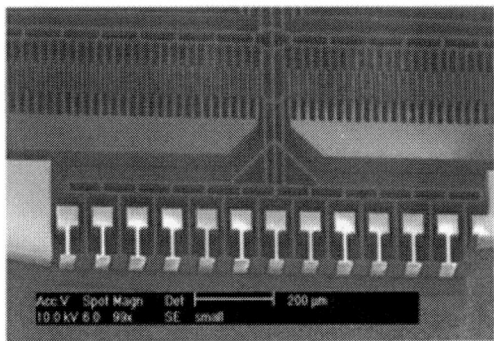

Fig. 2. First lamellar grating chip prototype. The number of mirrors is very limited giving a small active surface and low efficiency.

The detailed theoretical concept of the micro-fabricated lamellar grating interferometer is described in Ref. [6]. Experimentally, the variable depth of the grating is obtained by realizing a series of fixed and mobile mirrors as shown in Fig 2. The intensity of the zero order of the diffraction pattern is recorded in function of the depth of the grating and then Fourier transformed in order to recover the spectrum. First devices were capable to record visible light spectra by using a scanning range of 180 μm. The height of the mirrors was 75 μm, the number of illuminated periods of the grating was 12 and the grating period was 90 μm. The MEMS assembly had a total dimension of 5 mm x 3,5 mm x 1 mm. The quality of the surface of the mirrors is a critical issue for visible light applications and was ensured to have a surface roughness lower than 36 nm rms. The position accuracy of the emission peaks of a mercury vapor lamp were better than 0.5 nm and the measured resolution was 1.6 nm at λ=400 nm and 5.5 nm at λ=800 nm. The wavelength range extends from 380 to 1100 nm. The device provided already a wide wavelength range and faire resolution. It became evident that higher resolution and better stability has to be achieved.

IV. CONTINUOUS SCANNING PRODUCTS

A continuous scanning lamellar MEMS chip as used today is photographed in Fig. 3. The scanner is driven in resonance to move with best repeatability and high stability. Interfacing is still based on conventional optical components like fiber optics and lead to the smallest operational interferometer based spectrometer on the market. The resolution is comparable with conventional miniaturized grating spectrometers but no cooling is required and the system has low power requirements. The stability is several orders of magnitude better than for grating systems and the signal to noise ratio is limited by the detector noise only. Averaging allows to bring up the signal to noise ration to values as high as 20'000. Si-InGaAs detectors enable a large wavelength range from 500 to 2600 nm.

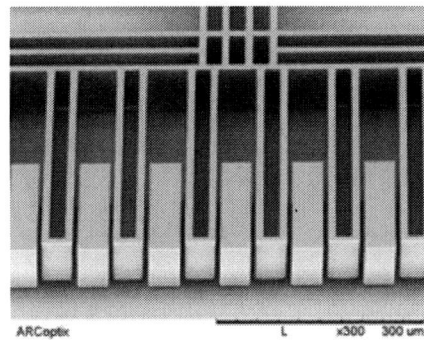

Fig. 3 Lamellar grating as used in actual products from ARCOPTIX (Neuchâtel). Such gratings allow for path differences of several hundreds micrometers and give high reproducibility of the MEMS movement.

V. OUTLOOK

Current developments going to devices for the mid-infrared (3 – 15 μm) wavelengths range with improved resolution. Conceptually, the only limitation of the lamellar grating spectrometer is the ratio between the grating period and the wavelength. At longer wavelength, the crucial point is the height of the mirrors that will cause diffraction and losses.

ACKNOWLEDGMENT

We thank the CTI for funding parts of the project and IMT-Samlab headed by Prof. N. de Rooij for the MEMS fabrication.

REFERENCES

[1] R. J. Bell, "Introductory Fourier Transform Spectroscopy", *Academic*, New York, 1972.
[2] R. F. Wolffenbuttel, "MEMS-based optical mini- and microspectrometers for the visible and infrared spectral range," J. Micromech. Microeng. **15** (2005) S145–S152.
[3] O. Manzardo, H.-P. Herzig, C.R. Marxer, N.F. de Rooij, "Miniaturized time-scanning Fourier Transform Spectrometer based on silicon technology", *Opt. Lett.* **24**, 1705-1707, 1999
[4] O. Manzardo, B. Guldimann, C. Marxer, N.F. de Rooij, H.-P. Herzig, "Optics and Actuators for Miniaturized Spectrometers", Digest IEEE/LEOS Intern. Conf. on Optical MEMS, Hawaii, August 5-9, 2000, pp. 23-24.
[5] J. Strong and G. A. Vanasse, "Lamellar grating far-infrared interferometer," J. Opt. Soc. Am. **50**, 113-118 (1960).
[6] O. Manzardo, R. Michaely, F. Schädelin, W. Noell, T. Overstolz, N. De Rooij, and H. P. Herzig, "Miniature lamellar grating interferometer based on silicon technology," Opt. Lett. **29**, 1437-1439 (2004).

Fabrication of aberration-corrected tunable micro-lenses

Daniel Mader, Andreas Seifert, and Hans Zappe

Laboratory for Micro-optics, Department of Microsystems Engineering – IMTEK, University of Freiburg
Georges-Köhler-Allee 102, 79110 Freiburg, Germany
daniel.mader@imtek.uni-freiburg.de

Abstract—**Pressure-actuated liquid-filled micro-lenses may be tuned in focal length from several millimeters to infinity and are thus suitable for use in tunable lens systems. A novel fabrication method for pneumatic multi-chamber lens systems which correct imaging errors such as chromatic and spherical aberrations is presented. The approach is a highly flexible, accurate and inexpensive way to implement lens combinations on the micro-scale which are similar to those in conventional macroscopic systems. A three-chamber lens system will demonstrate the potential of the process.**

Index Terms—**silicone, PDMS, fabrication, assembly, tunable, micro-lens, multi-chamber, stack, aberration correction.**

I. INTRODUCTION

SINCE the correction of higher order imaging errors requires more degrees of freedom than are available in a single micro-lens, a stacked fluidic-based multi-lens system has been proposed for the realization of aberration-corrected micro-optical imaging [1]. In this system, the refractive solid materials used in conventional lenses are replaced by specialized liquids with known optical properties [2]. The wide range of available optical liquids offers a choice in refractive indices and dispersion in the same range as those provided by the manufacturers of optical glasses.

In the multi-chamber micro-lens, the fluids are separated by thin and flexible membranes into different cavities, as shown in Figure 1. The pressure in each chamber (and thus the curvature of the membranes) is individually controllable and is thus used to tune the refractive properties of each lens element.

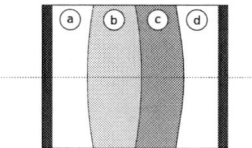

Fig. 1: Schematic of a multi-chamber membrane lens system. The chambers are filled with different fluids, and the curvature of the membranes can be adjusted individually by pressure.

This type of multi-chamber micro-fluidic optical systems requires new strategies for cheap, reproducible and highly accurate fabrication and assembly. Based on previous work on single-cavity liquid-filled lenses [3], [4] a suitable process technology has been developed and is presented here.

II. FABRICATION & PROCESS TECHNOLOGY

The optical performance of the system is defined solely by the optical characteristics of the fluids; the thin separating membranes should only be transparent. As shown in Figure 2, a membrane material with good transmission properties at visible wavelengths is available and is used below.

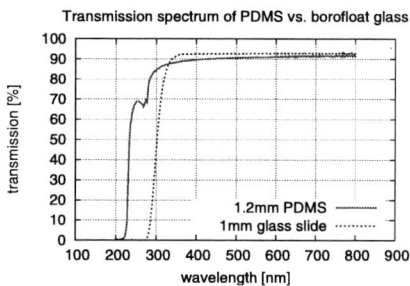

Fig. 2: Measured transmission spectrum of a PDMS membrane elastomer compared to that of a glass microscope slide.

The multi-lens stack requires the fabrication and assembly of individual fluidic lens cavities with the desired aperture and optical properties. We have developed two fabrication approaches for these individual lens components, one using Si micro-machining and the other PDMS molding.

A. Silicon micro-machining

The fluidic channels and the lens cavity itself are defined by two-step silicon micro-machining. Figure 3 shows both RIE-only and combined KOH-RIE processes. Though technically feasible, spin-coating of PDMS membranes on the substrate and using these as stop layers for the etch-through of the apertures [3] has been shown to be inadvisable since the selected polymers are very sensitive to plasma damage.

The PDMS membrane thus has to be bonded to the silicon lens frame after completion of the etch process by means of plasma-activated bonding. Proper plasma treatment parameters for the chosen silicone have been developed so that strong covalent Si–O–Si bonds can be formed. Figure 4a shows resulting silicon frames used for single lens cavities.

The membrane itself is formed in a separate spin-coating process on a substrate which allows for easy membrane liftoff; Teflon-coated glass and polypropylene substrates have been shown to be suitable. Soaking in a mild organic solvent leads to an easy detachment of the membranes due to the swelling of the polymer.

978-1-4244-1917-3/08/$25.00 ©2008 IEEE

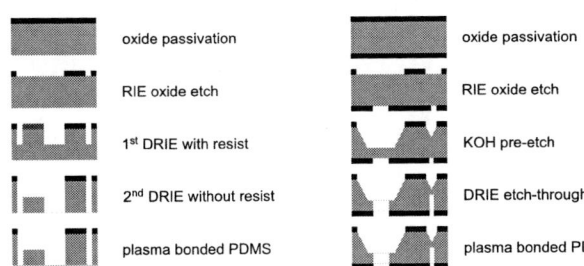

Fig. 3: Schematic of the fabrication process based on deep reactive ion etching (DRIE) and KOH-etching.

B. Molding technique

As an alternative, molding offers a cheap means for mass-production of single lens cavities. Again, the critical feature is the thin membrane. It can be formed between the high-quality surfaces of a glass substrate and the flat end of a GRIN-lens. The latter is precisely fixed in a milled mold to create a membrane of the desired thickness and is removed after molding. Figure 4b shows such a successfully utilized mold.

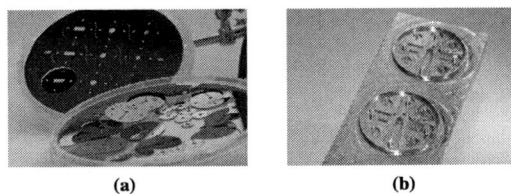

(a)   **(b)**

Fig. 4: (a) Micro-machined silicon lens frames. (b) The casting mold (aluminum) used for lenses with an aperture of 2 mm.

C. Stack assembly

After fabrication using one of the processes above, the individual single cavity lenses are stacked to form the lens system with up to four individually-controllable fluidic cavities. Oxygen-activated plasma-bonding was used to bond the cavities together.

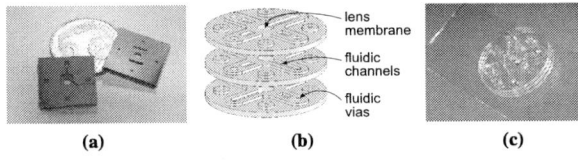

(a)   **(b)**   **(c)**

Fig. 5: (a) Assembled prototypes of the combined KOH-DRIE process. (b) Expanded view of a multi-chamber system based on molded individual lens cavities. (c) Photo of a bonded all-silicone stack.

Successfully assembled stacks of the micro-machined type are shown in Figure 5a, and Figures 5b and 5c show a schematic of the stack from the molding process type and a finished three-chamber lens, respectively.

Since the plasma-activated surfaces will bond within very short time when placed in contact, accurate alignment of the lens apertures is not possible. Therefore, a thin film

of methanol is pipetted onto the PDMS layer prior to the contact [5] allowing for precise alignment of the parts before bonding.

III. CONCLUSION AND OUTLOOK

An oxygen plasma bonding technique has been applied successfully for the fabrication and assembly of liquid multi-chamber micro-lenses. No adhesive or glue is required, the bond is irreversible, and the cavities can be accurately aligned prior to the bonding. The resulting devices withstand fluid pressures up to 10^5 Pa before any delamination occurs.

The approach allows a flexible combination of optical parts such as aperture stops together with lenses in a single device. It also enables the integration of actuation principles such as piezo displacement elements or thermo-pneumatics [6] within the encapsulated assembly.

The fabricated lens stacks are currently being tested and their optical performance (Figure 6) is compared with the results of numerical simulations.

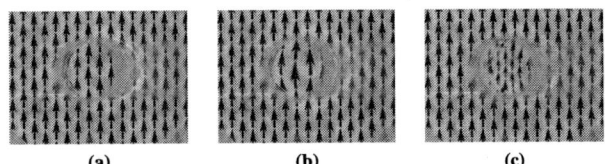

(a)   **(b)**   **(c)**

Fig. 6: First experiments to evaluate the characteristics of the molded lenses.

ACKNOWLEDGMENT

The authors would like to thank Bernd Aatz and the clean room service team of the IMTEK for their invaluable assistance and support.

REFERENCES

[1] S. Reichelt and H. Zappe, "Design of spherically corrected, achromatic variable-focus liquid lenses," *Optics Express*, vol. 15, no. 21, pp. 14 146–14 154, 2007.

[2] Cargille Laboratories – Precision calibrated laboratory products, *Specifications of Cargille Optical Liquids (personal communication 2007)*.

[3] A. Werber and H. Zappe, "Tunable microfluidic microlenses," *Applied Optics*, vol. 44, no. 16, pp. 3238–3245, June 2005.

[4] M. Agarwal, R. A. Gunasekaran, P. Coane, and K. Varahramyan, "Polymer-based variable focal length microlens system," *Journal of Micromechanics and Microengineering*, vol. 14, pp. 1665–1673, Dec. 2004.

[5] B.-H. Jo, L. M. V. Lerberghe, K. M. Motsegood, and D. J. Beebe, "Three-Dimensional Micro-Channel Fabrication in Polydimethylsiloxane (PDMS) Elastomer," *Journal of Microelectromechanical Systems*, vol. 9, no. 1, pp. 76–81, March 2000.

[6] A. Werber and H. Zappe, "Thermo-pneumatically actuated, membrane-based micro-mirror devices," *Journal of Micromechanics and Microengineering*, vol. 16, no. 12, pp. 2524–2531, 2006.

Three-Dimensional Integration of Optical Multi-Chips Using Surface-Activated Bonding for High-Density Microsystems Packaging

Eiji Higurashi[1], Daisuke Chino[2], Tadatomo Suga[2], and Renshi Sawada[3]

[1] Research Center for Advanced Science and Technology, The University of Tokyo

4-6-1, Komaba, Meguro-ku, Tokyo 153-8904, Japan

Tel +81-3-5452-5180, Fax +81-3-5452-5184, E-mail eiji@su.t.u-tokyo.ac.jp

[2] School of Engineering, The University of Tokyo, 7-3-1 Hongo, Bunkyo-ku, Tokyo 113-8656, Japan

[3] Department of Intelligent Machinery and Systems, Kyushu University, Motooka, Nishi-ku, Fukuoka 819-0395, Japan

Abstract

The hybrid integration of multiple chips in three dimensions is an important technology for realizing highly functional, compact optoelectronic microsystems. In this study, three-dimensional integration of optical chips was successfully performed using Au-Au surface-activated bonding (bonding temperature: 150 °C). The fabricated microsystem consists of two elements: i) a micromachined silicon optical bench incorporating a bonded laser diode, micromirrors, and through-hole electrodes; and ii) a glass substrate incorporating bonded photodiodes, electrical interconnections, and refractive microlenses. The silicon optical bench and the glass substrate were vertically stacked by surface-activated bonding. The size of the fully integrated microsystem is only 2.6 mm × 2.6 mm × 1 mm thick. The design and fabrication process allow chip-scale and wafer-level packaging. In addition, application for laser Doppler velocimeter sensor is discussed.

Keywords: Three-dimensional integration, Surface activated bonding, Hybrid integration, Compact packaging.

1 INTRODUCTION

The optical hybrid integration of optical chips on silicon (Si) substrate has opened many possibilities for constructing small, reliable, and highly functional optical devices, such as communication modules (transmitters and receivers) and microsensors [1, 2]. To fabricate such highly functional microsystems, it is essential to develop a high density multi-chip integration method using passive alignment techniques. In the bonding process, eutectic AuSn is commonly used because AuSn has very high yield strength and is typically free of thermal fatigue and creep movement.

However, the high thermal conductivity of Si substrate and a multiple reflow process make solder formed on the substrate tend to be damaged during several repeated bonding steps. If high bonding accuracy is required, remelting of solder should be avoided. Furthermore, optoelectronic chips often do not withstand long exposure time at high temperature. Therefore, a low-temperature and solderless bonding process is desirable for three-dimensional integration of multiple chips.

This paper reports on low-temperature direct flip-chip bonding of multiple chips for high-density microsystems packaging. This method uses surface activation by radio frequency (RF) plasma irradiation.

2 STRUCTURE AND MULTI-CHIP BONDING

Figure 1 shows the schematic configuration and photograph of the fabricated microsystem. The size of the fully integrated microsystems is 2.6 mm × 2.6 mm × 1 mm thick. It consists of two elements: i) a micromachined Si optical bench incorporating a bonded laser diode (LD, λ = 1310 nm), micromirrors, and through-hole electrodes (Fig. 2); and ii) a pyrex glass substrate incorporating bonded photodiodes (PDs), electrical interconnections, and refractive microlenses (Fig. 3). The top half of the microsystem package is a pyrex glass cap which serve as an optical window with microlenses and a protective cover. The Si optical bench with a cavity defined by {111} planes was used for free–space micro-optical bench. The path followed by the optical beam is illustrated in Fig. 1(a). The laser beam is reflected by <111> face of the cavity wall and then encounters the refractive microlens. After passing through the lens, the beam is collimated.

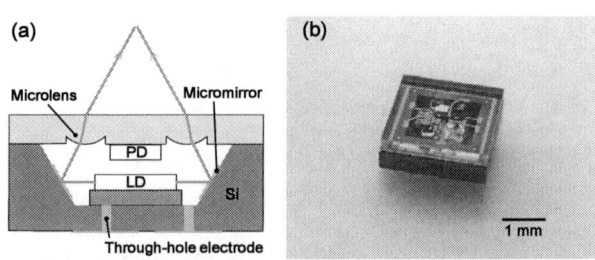

Fig. 1 Integrated microsystem. (a) Cross-sectional configuration, (b) Photograph showing the fabricated microsystem.

Fig. 2 (a) Photograph showing the microfabricated Si optical bench, (b) SEM photograph showing the integrated LD on the Si optical bench using surface activated bonding (bonding temperature: 150 °C).

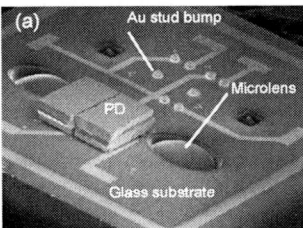 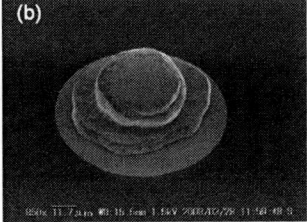

Fig. 3 (a) SEM photograph showing the integrated PDs on the Pyrex glass substrate, (b) SEM photograph showing Au stud bump for surface activated bonding of PDs

To perform multi-chip integration, we adopted Au-Au surface activated bonding [3]. Surface-activated bonding is a direct bonding method that joins two clean surfaces using the adhesive force of surface atoms at room temperature or low temperature. The clean active surfaces are prepared by irradiation of argon RF plasma. After surface activation, the bonding is carried out by contact in ambient air. Static pressure is applied to the chip during heating, which is kept at a peak temperature (150 °C) for 30 s. Not only optical chips such as LDs and PDs, but also the glass substrates were bonded on the silicon optical benches by surface-activated bonding. Figure 4 shows typical L-I-V characteristics as measured by the L-I-V test system, before (unbonded bare chip) and after surface-activated bonding. Compared with the unbonded LD, the flip-chip-bonded LD does not show any changes in electrical and optical data. However, when the AuSn bonding was used instead of surface-activated bonding, it was observed that the degradation of LD during repeated bonding steps due to the thermal damage.

Currently, the two elements of the Si optical bench and the glass substrate are assembled after dicing the pieces individually. However, the design and fabrication process allow chip-scale and wafer-level packaging.

3 APPLICATION

The fabricated microsystem has been successfully applied to laser Doppler velocimeter (LDV) sensor. Figure 5 shows the beat frequency as a function of object velocity. The object (grating) was moved by a piezoelectric actuator. Good linearity between the relative velocity and the Doppler

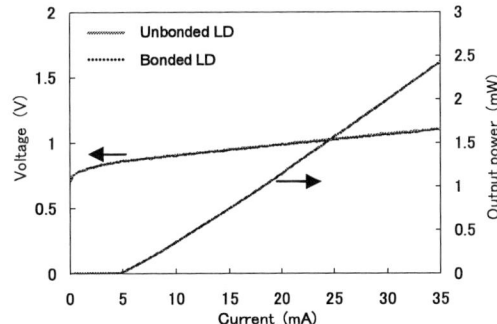

Fig. 4 Comparison of the L-I-V characteristics of the LD chip before and after bonding (bonding temperature: 150 °C).

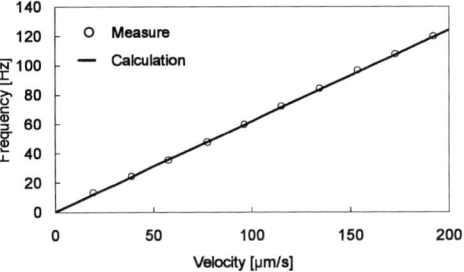

Fig. 5 Beat frequency versus velocity

beat signal frequency was obtained. The experimental results were very close to the calculated output.

4 SUMMARY

Multiple flip-chip assembly using surface-activated bonding was demonstrated for high-density microsystems packaging. Bonding was carried out in air at low temperature (150 °C). The feasibility of measuring velocity was demonstrated using prototype microsensors. The results show that this technique is very promising for producing highly functional compact optical microsystems.

ACKNOWLEDGMENT

The authors would like to thank T. Oguchi (NSK Ltd) for technical support in fabricating the sensor chip and Oki Electric Industry Co., Ltd for supplying us with PDs. This work was partly supported by the Toray Science Foundation.

REFERENCES

[1] E. Higurashi, R. Sawada, and T. Ito, J. Lightwave Tech., **21**, 591–595 (2003).
[2] R. Sawada, E. Higurashi, and Y. Jin, J. Lightwave Tech., **21**, 815- 820 (2003).
[3] E. Higurashi, T. Imamura, T. Suga, and R. Sawada, Photon. Tech. Lett., **19**, 1994-1996 (2007).

PULSED THERMAL EXCITATION OF LUMINESCENT MICROPARTICLES FOR RADIATION DOSIMETRY

M. E. Manfred[1], N. T. Gabriel[1], M. L. Mah[1], E. G. Yukihara[2], and J. J. Talghader[1]

[1]University of Minnesota, Minneapolis, MN 55455
[2]Oklahoma State University, Stillwater, OK 74078

Abstract

A technique using temperature pulses was used to measure thermoluminescence in single aluminum oxide microparticles. This method can excite a large fraction of trapped carriers to luminesce simultaneously, which increases the maximum intensity and makes it ideal for probing the luminescence of micro- and nano-particles. Using 50ms and 10ms pulses, the technique resulted in curves similar to standard thermoluminescence curves with peaks near 240°C and 300°C respectfully. Increasing the temperature of a single pulse resulted in an increase in intensity.

Background

A traditional means of monitoring radiation dose uses a plastic badge filled with ceramic particles, such as alpha-phase carbon-doped aluminum oxide (α-Al$_2$O$_3$:C)[1]. The traps in these particles are initially empty, but exposure to ionizing radiation excites electrons from the valence level to the conduction band, from where they subsequently drop to fill deep traps in the oxide[2]. Eventually, the badge is sent to a central laboratory. The radiation dose is then determined by heating the α-Al$_2$O$_3$:C with a macroscopic heater. The light intensity is measured which is due to trap emptying and subsequent radiative decay. This results in a thermoluminescence (TL) peak in a graph of intensity verses temperature. The peak's height is related to the number of filled traps for which the absorbed dose can be determined.

This technique could be vastly improved using pulsed thermoluminescence (PTL) to assess radiation exposure in quasi-real time. A microheater and ceramic particle system could easily be integrated on a badge with a small battery and avalanche photodiode. It could take a measurement every few seconds to minutes to determine if a "radiation event" had taken place, and then warn the user if they had been exposed to a dangerous amount of radiation. In addition, the ability to excite a large fraction of the carriers with a single pulse can make the instantaneous intensity very high,

Traditional TL measurements are taken with a photomultiplier tube. To overcome the noise inherent in an avalanche photodiode, an increased light intensity during thermoluminescence would be desirable. This can be accomplished by using a single pulse in temperature rather than the traditional temperature ramp. The pulse allows for a majority of the filled traps to empty at the temperature where light intensity is the greatest. This also makes the technique particularly appropriate for measuring small volumes of material, i.e. micro- and nano- particles. Of course, the temperature may also impact the rate of trap emptying and luminescent efficiency.

PTL is of significant scientific interest as well: since the microheaters have thermal constants shorter than that of the F-center emission decay, (35ms for α-Al$_2$O$_3$:C) the majority of the luminescence can occur after the particle has cooled toward room temperature. Competing theories on quenching luminescent response [3,4] in certain ceramic particles propose either increased non-radiative recombination at high temperatures or deep-level trap filling. PTL may be used to distinguish these.

Experimental Details

In our experiment, we used microheaters which consisted of suspended microfabricated plates. The plates had a resistor on them to heat the structure. The TL material was α-Al$_2$O$_3$:C formed in particles approximately 50μm in size. This material is well known for its high luminescence response to radiation. The microheaters have thermal time constants ranging from 100μs to 50ms depending on size and etch release. A series of 10ms heat pulses with gradually increasing voltage were applied, producing instantaneous powers varying from zero to 6W. The resulting TL curves have similar shapes to those produced using standard technology (linear DC heating with macroscopic heaters). The temperature was calculated by measuring the change in resistance of the heating element. All temperatures stated are of the resistor.

Figure 1: Placement of an Al$_2$O$_3$ particle on a first generation microheater.

Figure 1 shows an α-Al$_2$O$_3$:C particle on a fabricated microheater composed of silicon nitride structural layers with polysilicon resistors. The thermal time constants of the devices are approximately 50ms. Figure 2 shows an unreleased device with platinum resistors. These "second generation" devices have particularly well-characterized temperature-resistance relationships and a thermal response of less than 10ms.

Before testing, the particles are radiated with a deep-UV lamp with peak emission at λ=205nm, which

978-1-4244-1917-3/08/$25.00 ©2008 IEEE

fills the traps. Figure 3 shows a traditional TL curve taken with the microheaters using a linear heating ramp of 2°C/sec. The two curves demonstrate how luminescence differences that can occur simply by placing a particle over a resistor or over a dielectric area of a first-generation heater. The peaks can be seen to shift based on the particles location, indicating a temperature gradient across the heater. The second-generation heaters were modified to minimize this effect.

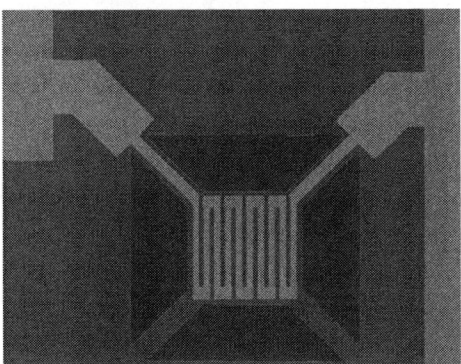

Figure 2: SEM of an unreleased second generation microheater. The resistor evenly covers the surface to provide uniform heating.

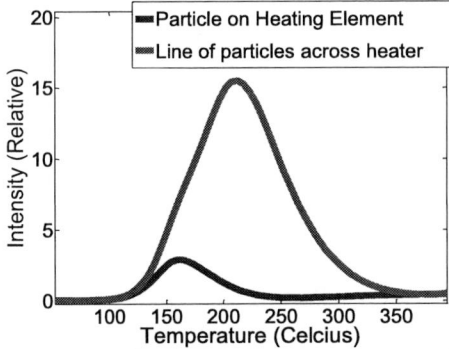

Figure 3: A comparison of two TL curves taken on a first generation microheater. In one curve, there was a line of particles across the heater's plate. In the other, there was one particle located on the resistor.

Figure 4 shows a PTL curve where the temperature was increased with each pulse (i.e. increasing pulse voltage amplitude). It can be seen that the shape of these curves are similar to those of standard TL curves. The peak is seen to shift higher temperatures for shorter pulse lengths. This can be explained by imperfect thermal contact between the particle and heater causing decreasing heat transfer for shorter pulses, as well as shorter pulses causing the peak temperature to hold for shorter times, resulting in fewer traps emptying.

Figure 5 shows PTL with pulses of constant power. The heater was pulsed repeatedly, and the total photon count from each pulse was recorded. At low temperatures, the intensity was near constant over a large number of pulses. This is significant because one can set the readout low enough in certain cases that the overall trap population is only slightly affected by read-out. The intensity is seen to increase with increasing temperature. There appears to be significant trap emptying at higher temperatures due to a decrease in the intensity over the pulse number.

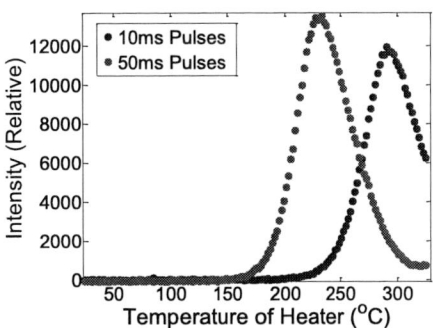

Figure 4: PTL on a second generation microheater. The heaters were heated by a brief voltage pulse. The light emitted due to the pulse was recorded with a PMT. The temperature of each pulse to was increased by 2 °C per pulse.

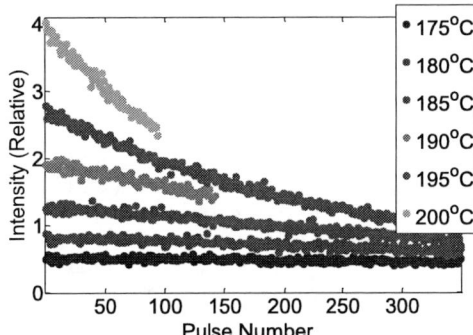

Figure 5: PTL on a second generation microheater. The temperate of each pulse was kept constant.

Conclusion

A technique of PTL was demonstrated. It was able to recreate results similar to standard TL curves. However, it also has the advantage of being able to increase the light intensity during a TL scan for use with micro- and nano-particles.

Acknowledgements

This work was supported by the Defense Threat Reduction Agency (DTRA) under grant HDTRA1-07-1-0016.

References

[1] E.G. Yukihara and S.W.S. McKeever, *Radiation Protection Dosimetry*, vol. 119, no 1-4, 2006, pp. 206-217.

[2] S. W. S. McKeever, *Thermoluminescence of Solids*, Cambridge University Press: Cambridge, 1985.

[3] V.S. Kortov, et. al., *Radiation Protection Domimetry*, 2006 vol. 119, no. 1-4, pp. 41-44.

[4] M.S. Akselrod, et. al., *Journal of Applied Physics*, 1998, Vol. 84, No. 6, pp. 3364-3373

Widely Tunable Fabry-Perot Optical Filter Using Fixed-Fixed Beam Actuators

J. Milne[1], J. Dell, A. Keating, L. Schuler, L. Faraone

Microelectronics Research Group, School of EE & CE, University of Western Australia, Perth, Western Australia
[1] jsmilne@ee.uwa.edu.au

Abstract—**Wide spectral tuning from 1620 nm to 2425 nm has been achieved in a MEMS-based tunable Fabry-Pérot filter. Fixed-fixed beam actuators were used to extend the tuning beyond the inherent limit of a parallel plate actuator.**

Index Terms—**Electrostatic actuation, Micro-electro-mechanical systems (MEMS), Fabry-Pérot etalon, tunable filter**

I. INTRODUCTION

Combining a MEMS-actuated Fabry-Pérot optical filter with a broadband infrared sensor creates a voltage-tunable narrowband infrared sensor, with the detected wavelength controlled by an electrical signal applied to the MEMS actuator. Parallel-plate actuators are commonly employed in electrostatic MEMS-based tunable Fabry-Pérot filters [1], and these actuators exhibit pull-in when the actuator has traveled 33% of the initial distance between electrodes; 33% of d_0 [2]. When a Fabry-Pérot cavity is designed to operate in first-order mode, pull-in at 33% of d_0 limits the tunable wavelength range of the cavity [1].

Our application demands a wavelength tuning range of 1600 nm – 2500 nm, which is only possible in first-order operation [3]. The optical modeling shown in Figure 1 demonstrates that if pull-in were to occur at 33% of the gap, the wavelength tuning range would be limited to 1890 nm – 2500 nm, whereas the desired wavelength range requires travel of 50% of d_0.

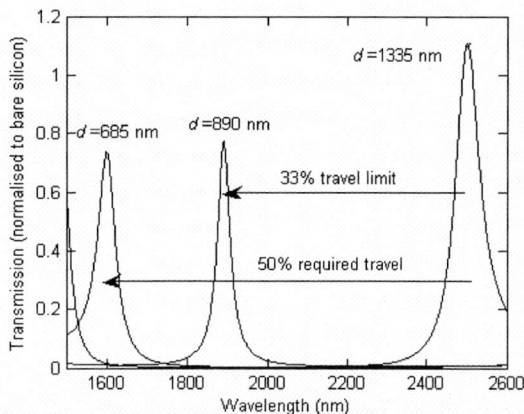

Figure 1: Modeled transmission spectra for a Fabry-Pérot cavity with Ge-SiO-Ge mirrors, normalized to the transmission through a bare silicon wafer. The mirror separation, d, used in the model is shown for each of the three spectra, demonstrating that wavelength tuning from 2500 nm to 1600 nm requires an actuator that can move the mirror further than 33% of d_0. The roll-off in peak transmission is due to absorption in the Ge mirror layers.

It has been reported that fixed-fixed beam actuators can travel over a wider range than parallel plate actuators; from 40% up to 67% of d_0 depending on the geometry and material parameters of the actuator [2]. Using a combination of finite element modeling and optical modeling, it has also been demonstrated that fixed-fixed beam actuators can be used to achieve a tuning range of 1600 nm – 2500 nm in a Fabry-Pérot filter [2]. This work reports on the design, fabrication and optical measurement of a tunable Fabry-Pérot optical filter using fixed-fixed beam actuators. The measured wavelength tuning range of 1620 nm – 2425 nm is, in terms of percentage travel, the widest reported to date for an electrostatically actuated Fabry-Pérot cavity, yielding a travel range of 45% of d_0.

II. DESIGN AND FABRICATION

The layout of the device, including the actuator and mirror placement, is shown in Figure 2. Devices were fabricated with mirror diameters of 100 μm to 300 μm.

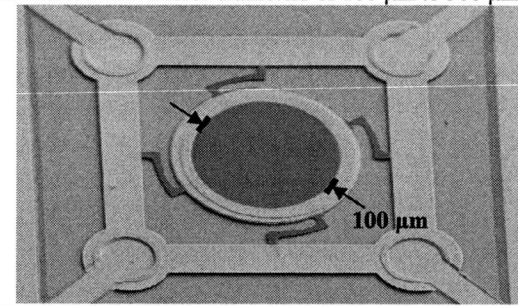

Figure 2: Scanning electron microscope image of a fabricated MEMS-actuated Fabry-Pérot filter with fixed-fixed beam actuators. The mirror diameter is 100 μm.

The fabricated device uses four fixed-fixed beam actuators in a square arrangement, supporting a Ge-SiO-Ge Bragg mirror. The mirror is attached to the centre of the arms using 'S' shaped silicon nitride tethers, as depicted in Figure 2.

The fixed-fixed beam actuators are composed of a 20 nm thick thermally evaporated gold electrode deposited on a 200 nm thick silicon nitride structural layer. The tuning range of fixed-fixed beams increases with decreasing beam stress [2], so the silicon nitride was deposited using PECVD to achieve a nominally zero-stress (± 5 MPa) condition.

Stress gradients in the Ge-SiO-Ge mirror lead to mirror curvature, which causes linewidth broadening in Fabry-Pérot filters [3]. Stress gradients in the mirror structure were compensated using a compressively stressed silicon nitride layer deposited before the mirror, with a thickness of

978-1-4244-1917-3/08/$25.00 ©2008 IEEE

50 nm chosen to achieve a nominally zero net stress gradient across the mirror structure.

III. RESULTS

Transmission spectra were obtained using a tunable monochromatic source and an InGaAs detector. Light was prevented from leaking around the mirror by the bottom metal electrode, which covers the entire silicon supporting substrate except for a circular opening below each movable mirror.

An actuation voltage was applied to the device under test, and the transmission from 1300 nm – 2600 nm was measured in 10 nm steps with the actuation voltage held constant. Spectra were obtained for a range of actuation voltages up to the estimated pull-in voltage of each device.

Once the transmission spectra were obtained, a reference spectrum for the silicon substrate was obtained from a circular opening in the bottom electrode, having the same diameter as the opening in the bottom metal electrode below the measured device. Dividing each measured device spectrum by the reference spectrum yields a transmission spectrum normalized to bare silicon.

A range of filter sizes were measured, with the smallest device (100 μm mirror diameter) yielding the optimal travel range and linewidth. Selected spectra from this device are presented in Figure 3. The difference between the measured zero-voltage wavelength of 2425 nm and the desired wavelength of 2500 nm was likely due to the sacrificial layer in the fabrication process being slightly thinner than the designed thickness of 1335 nm.

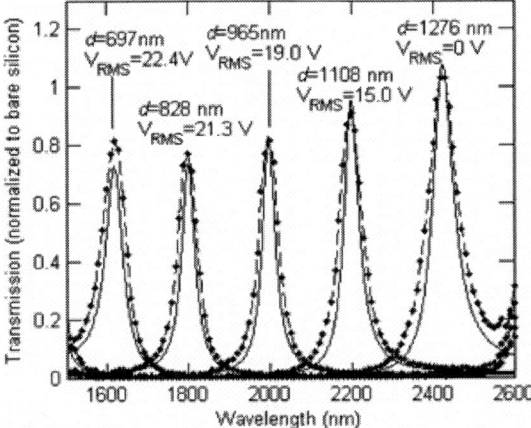

Figure 3: Measured (closed circles) and modeled (solid lines) transmission spectra of a device with a mirror diameter of 100 μm. The modeled mirror separation and the applied voltage during the measurement is given for each spectrum.

Voltage-displacement curves of selected devices were also obtained, using an optical profilometer (Zygo NewView 6300) to generate a height map of the device whilst the actuation voltage was held constant. The voltage-displacement data at the centre of the mirror for the 100 μm diameter device is presented in Figure 4, along with a voltage-displacement curve obtained from an analytical model of a single fixed-fixed beam actuator [2], fitted to the

voltage-displacement data. The displacements obtained from optical modeling of the transmission spectra presented in Figure 3 are also plotted against the actuation voltage of each transmission spectrum.

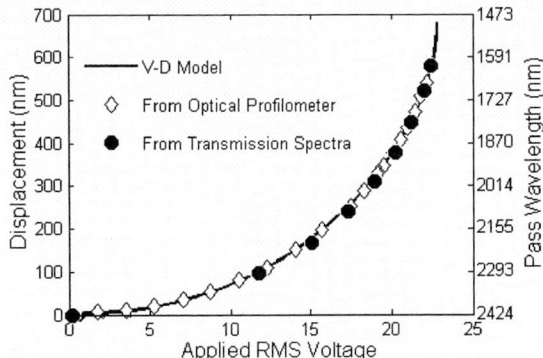

Figure 4: Voltage-displacement data for the 100 μm diameter device. The solid line is a voltage-displacement model fitted to the optical profilometer data (open diamonds). The closed circles are obtained from optical modeling of measured transmission spectra; every alternate circle corresponds to one of the spectra from Figure 3.

The voltage-displacement model for the fixed-fixed beam suggests that the maximum travel range of this device should have been 682 nm, or 53% of d_0. The measured tuning range was 579 nm or 45% of d_0. The device may have tuned over the full 53% of d_0, however, further tuning was not attempted because of the risk of pull-in and consequential device failure.

IV. CONCLUSIONS

Tunable spectral filters based on MEMS-actuated Fabry-Pérot cavities offer the possibility of low-cost, portable and potentially robust instruments for short-wave infrared spectral data acquisition. Large tuning ranges are required for some applications, but the actuator type usually used with this type of device does not permit tuning beyond 33% of the zero-voltage electrode separation, d_0. Fixed-fixed beams can theoretically provide tuning of up to 66% of d_0, making them a candidate for providing the required travel 50% of d_0, which corresponds to a wavelength tuning of 1600 nm – 2500 nm. Devices using fixed-fixed beam actuators were fabricated, and the 1620 nm – 2425 nm measured wavelength tuning of one device is, in terms of percentage travel, the widest tuning reported for a MEMS-actuated Fabry-Pérot cavity.

REFERENCES

[1] R. F. Wolffenbuttel, "State-of-the-art in integrated optical microspectrometers," *IEEE Transactions on Instrumentation and Measurement*, vol. 53, pp. 197-202, 2004.

[2] J. S. Milne, A. J. Keating, J. Antoszewski, J. M. Dell, C. A. Musca, and L. Faraone, "Extending the tuning range of SWIR microspectrometers," presented at Infrared Technology and Applications XXXIII, Orlando, FL, United States, 2007.

[3] A. J. Keating, K. K. M. B. D. Silva, J. M. Dell, C. A. Musca, and L. Faraone, "Optical characterization of Fabry-Pérot MEMS filters integrated on tunable short-wave IR detectors," *IEEE Photonics Technology Letters*, vol. 18, pp. 1079-81, 2006.

Impact of High Optical Power on Optical MEMS

Olga Blum Spahn, Leslie M. Phinney, C. Channy
Wong, William D. Cowan, David P. Adams
Sandia National Laboratories, Albuquerque, NM 87185
oblum@sandia.gov

Grant D. Grossetete
L&M Technologies, Albuquerque, NM 87185

Abstract— a number of applications require optical MEMS devices to handle high optical power. We present theoretical and experimental results for SUMMiT ™ fabricated mirrors and other structures illustrating performance limitations and several mitigation strategies.

Keywords-MEMS mirrors, deformation ,optical power handling

BACKGROUND

In this work, we focus on the effects of exposure of optical MEMS such as mirrors, shutters and other elements, to high optical power. The primary consequence of exposure to high optical power is heating, resulting in surface deformation. Devices described here are specifically those fabricated by SUMMiT™ technology [1], consisting of thin (~2.5 μm) layers of polysilicon in close proximity (~10 μm) to the Si substrate. In most cases, polysilicon surfaces have been metallized by a thin (~100 nm) layer of Au or AlCu.

EXPERIMENT

A. Incidence of High Optical Power

Incident laser power heats the mirror surface through the residual absorption in the mirror coating. Due to mismatch of the coefficient of thermal expansion between the metal coating and the polysilicon, stress develops and deforms the surface. This is a problem for a mirror, such as that used for adaptive optics applications [2], where the phase of the reflected light is being manipulated and thus flatness to fractions of a wavelength is required. Figure 1 shows the interferometric image of a hexagonal polysilicon mirror coated with 100 nm of AlCu before, during and after exposure to an 808 nm laser. It worth noting that the curvature does not return to its initial value upon removal of laser power (Figure 1c). This is caused by an inelastic deformation of the metallized layer [3]. As shown in Figure 2, the curvature of a metallized mirror decreases as the incident laser power is increased. As in the case of ambient temperature increase, increase in the optical power flattens the mirror and eventually causes it to flip from a concave to a convex shape. Less obvious, but still present, is the increasing curvature of the mirror after laser exposure (the red curve has a small slope). We will discuss various means of mitigating this effect. These interferometric measurements of MEMS during laser exposure are obtained using the setup illustrated in Figure 3. If the mirror coating is not present or fails during the laser incidence, catastrophic failure can occur (Figure 4), where laser heating can locally re-crystallize polysilicon and drive the coating material into the grain boundaries. Local temperature measurements obtained via micro-Raman and thermal imaging confirm existence of high temperature capable of modification of the mirror material.

B. Mitigation strategies

There are several strategies available to minimize mirror deformation occurring during the laser exposure. One is to increase the mirror stiffness, most simply by making it thicker. This has the disadvantage of also making the mirror heavier and thus slower. The other option is to improve the reflective coating, such that less incident laser power is absorbed (and more reflected) and converted into heat. Figure 5 shows the curvature of a MEMS mirror coated with metal only and the same mirror coated with metal and a dielectric "helper" stack during the 808 nm laser exposure. It can be clearly seen that the mirror with a metal and a dielectric layer coating deforms considerably less upon exposure to large laser power.

THEORY

We modeled and experimentally verified behavior of simplified, bare polysilicon cantilevers. A coupled physics analysis code, Calagio [4], was used to simulate the thermo-mechanical response of the cantilevers under laser radiation. Figure 6 shows interferometric images of laser irradiated cantilevers and a plot of vertical displacement obtained experimentally and theoretically.

CONCLUSION

We quantify response of optical MEMS to high optical powers; discuss causes of this phenomenon and present experimental strategies for its mitigation.

REFERENCES

[1] Sniegowski, J. J., and de Boer, M. P, "IC-Compatible Polysilicon Surface Micromachining," *Annual Review of Material Science*, vol. 30, pp. 299-333, 2000

[2] Dagel, D.J., Cowan, W. D., Spahn O.B., Grossetete G.D., Grine A.J., Shaw, M.J., Resnick P.J, Jokiel B., "Large-stroke MEMS deformable mirrors for adaptive optics", *Journal of Microelectromechanical Systems*; vol.15, no.3, p.572-83, 2006

[3] Gall K., West N., Spark K., Dunn M. L., Finch D. S., "Creep of thin film Au on bimaterial Au/Si microcantilevers", *Acta Materialia* vol. 52, pp. 2133–2146, 2004

[4] Phinney, L.M, Spahn, O.B, Wong, C.C, "Experimental and computational study on laser heating of surface micromachined cantilevers", *Proceedings of SPIE* v. 6111, pp.11108-11108, 2006

Sandia is a multiprogram laboratory operated by Sandia Corporation, a Lockheed Martin Company, for the United States Department of Energy's National Nuclear Security Administration under contract DE-AC04-94AL85000.

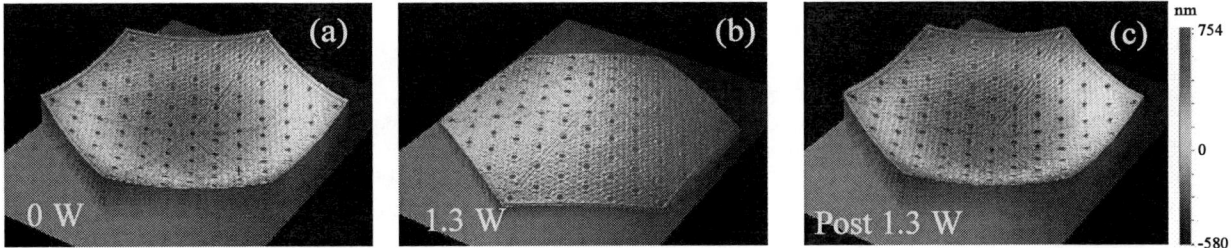

Figure 1 Interferometric images of mirror surfaces exposed to 808 nm laser a) before exposure c) during 1.3 W exposure c) after 1.3 W exposure. Hexagonal 500 μm mirrors consist of metallized (100 nm AlCu) 2.5 μm thick polysilicon layer supported by a 2.5 μm patterned polysilicon layer

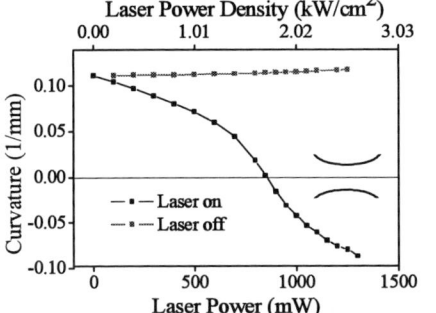

Figure 2 Curvature of a MEMS mirror as a function of laser power. Inset shows mirror shape.

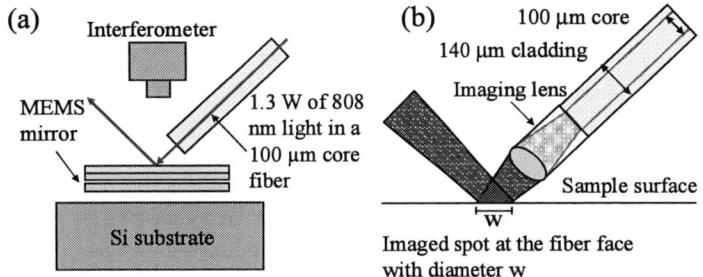

Figure 3 a) Experimental setup for interferometric imaging of the mirror surface during laser exposure b) detail of the setup of the fiber tip

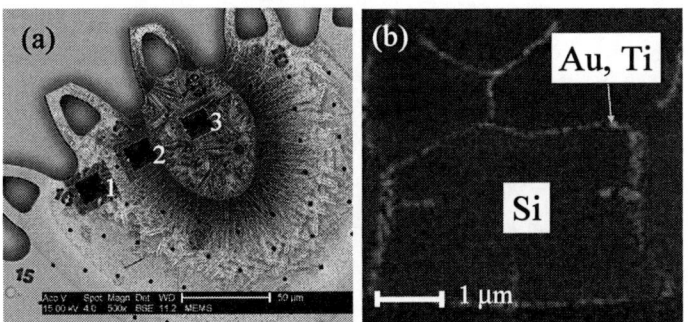

Figure 4 a) SEM view of a laser burn on a MEMS shutter b) False color image of the TEM cross section taken from region 3 shows an enlarged Si grain with Au and Ti from the coating resident in the grain boundary.

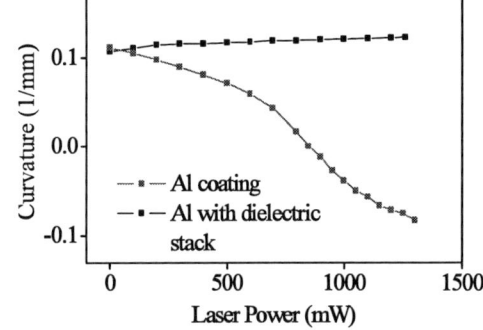

Figure 5 Curvature of a MEMS mirror coated with metal only (red curve) and metal with dielectric stack (black curve) during 808 nm laser exposure

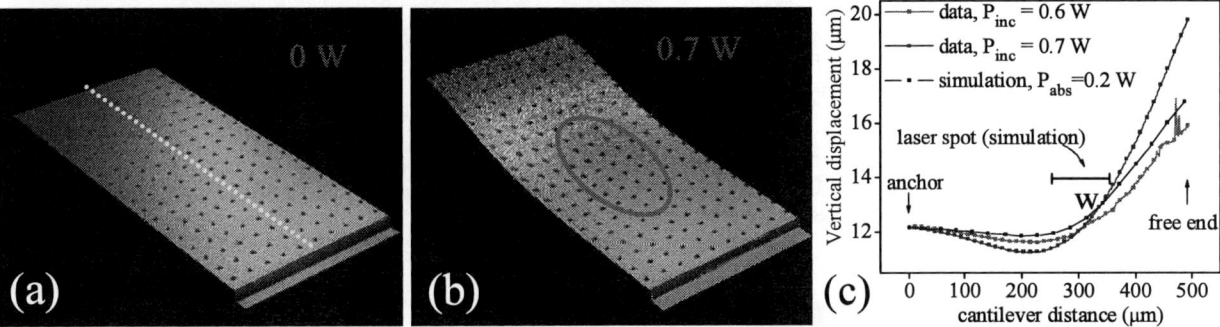

Figure 5 Interferometric images of a polysilicon cantilever illuminated with a 808 nm laser a) before illumination b) during illumination with 0.7 W – pink ellipse shows approximate spot size/location c) experimental vertical displacement along a dashed yellow line in a) of a cantilever illuminated with 0.6 W (red line), 0.7 W (blue line) and simulated vertical displacement of a cantilever which absorbed 0.2 W of illumination (~0.6 W incident, assuming 30% absorption) (black line).

A High-Power Handling MEMS Optical Scanner for Display Applications

Yasutaka Ohira[1,2], Aleksandr Checkovskiy[2], Toshio Yamanoi[3], Takashi Endo[3], Hiroyuki Fujita[1], and Hiroshi Toshiyoshi[1,2]

1 Institute of Industrial Science, The University of Tokyo,
4-6-1 Komaba, Meguro-ku, Tokyo 153-8505, Japan
Phone: +81-3-5452-6277 / Fax: +81-3-5452-6250 / E-mail: ohira@iis.u-tokyo.ac.jp
2 Also with Kanagawa Academy of Science & Technology, Japan, 3 Koshin Kogaku Co. Ltd., Japan

ABSTRACT

This paper reports a design and fabrication technique of high-power handling MEMS (Micro Electro Mechanical Systems) optical scanner for laser 3-D image display application. The MEMS scanner is designed to control the reflection of high-power YAG-laser beam of 5W (0.5mW/pulse) by the optomechanical combination of a dielectric-film coated mirror cube and an electrostatic MEMS scanner platform.

Keywords: 3-D Display, YAG-laser, laser induced breakdown, hybrid MEMS

1 INTRODUCTION

In our recent publication [1], we proposed a 3-D image display using the laser induced breakdown plasma, where a high-power pulse YAG laser beam (0.5mW/pulse) was spatially modulated at fast speed to create a 3-D bitmap image. At that moment, we used a Galvanometric scanner for its durability to the high power laser, and the total optical system was large and bulky. In this paper, we propose a hybrid MEMS scanner platform on which a dielectric coated mirror cube is manually mounted.

2 DESIGN AND OPERATION

Figure 1 schematically illustrates the optical system of the 3-D display. Plasma flushes made by the laser induced breakdown in water are used as 3-D pixels or so-called voxels. The spatial light scanning system for the 3-D display consists of a translating lens on an electromechanical shuttle

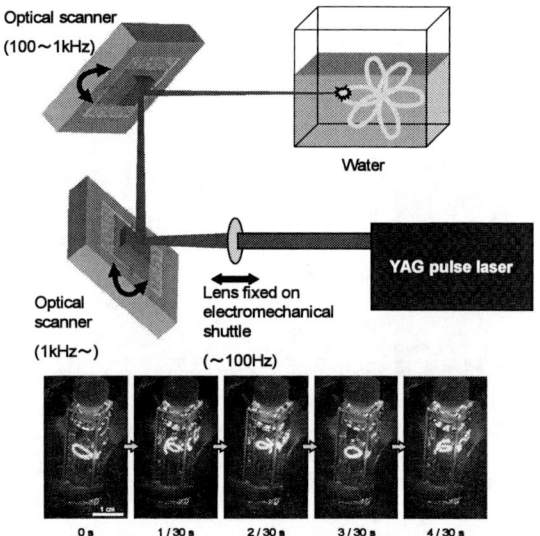

Figure 1 Principle of 3-D display by using laser induced breakdown plasma in water and example of 3-D bitmap image.

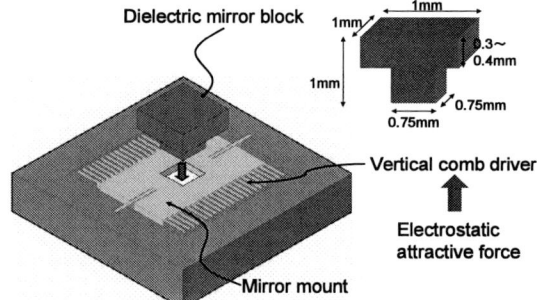

Figure 2 Schematic structure of MEMS optical scanner.

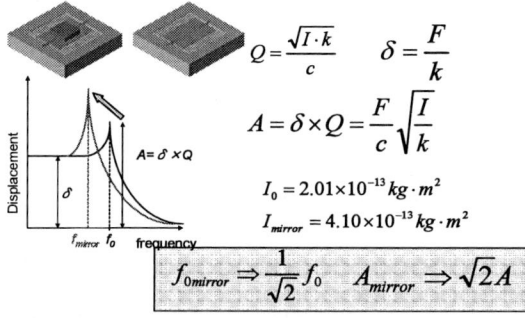

$$Q = \frac{\sqrt{I \cdot k}}{c} \qquad \delta = \frac{F}{k}$$

$$A = \delta \times Q = \frac{F}{c}\sqrt{\frac{I}{k}}$$

$$I_0 = 2.01 \times 10^{-13} kg \cdot m^2$$

$$I_{mirror} = 4.10 \times 10^{-13} kg \cdot m^2$$

$$f_{0mirror} \Rightarrow \frac{1}{\sqrt{2}} f_0 \qquad A_{mirror} \Rightarrow \sqrt{2}A$$

Figure 3 Characteristics change between with and without the mirror block

for the longitudinal Z-direction, and a set of two MEMS optical scanners for the transversal Y and X directions. The inset images are the series of snapshots of a 3-D spiral bitmap images produced by the laser breakdown plasma.

Figure 2 shows the schematic structure of the high-power handling MEMS optical scanner. The proposed device has a MEMS torsion-mirror plate with the electrostatic vertical-comb actuators, and a dielectric coated mirror-cube which is manually mounted on it. The mirror cube is cut out from a dielectric mirror plate in a similar manner to that reported in our previous publication [2].

978-1-4244-1917-3/08/$25.00 ©2008 IEEE

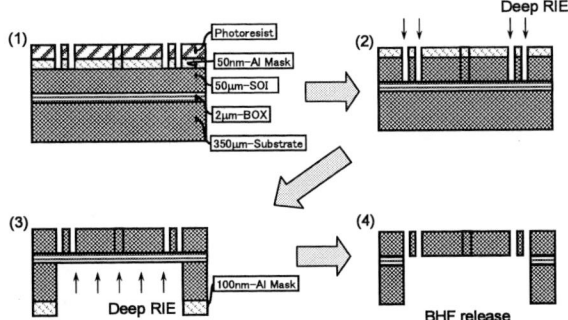

Figure 4 Fabrication of MEMS scanner platform

Figure 5 SEM pictures of high-power handling MEMS optical scanner without mirror block and the photograph comparing a Galvanometric and the MEMS scanner.

Mounting a mirror cube alters the moment of inertia of the MEMS platform, i.e., the resonant behavior, as indicated by the equations in Figure 3. Considering the mass of the mirror cube, the resonant frequency is reduced to nearly 70% of the initial value.

3 FABRICATION

Figure 4 shows the process for the MEMS structures: (1) MEMS structure is made on the SOI layer with an Al mask by photo lithography. (2) Actuator patterns in the SOI layer are defined by silicon DRIE. (3) The backside silicon is etched by DRIE to make actuator. (4) The sacrificial BOX layer is selectively removed by buffered HF.

Figure 5 shows the SEM view of the actuator and photograph of Galvanometric and MEMS scanner. The actuator plate of 2mm by 2mm is made with a square hole 750μm by 750μm; a mirror cube will be later inserted into it.

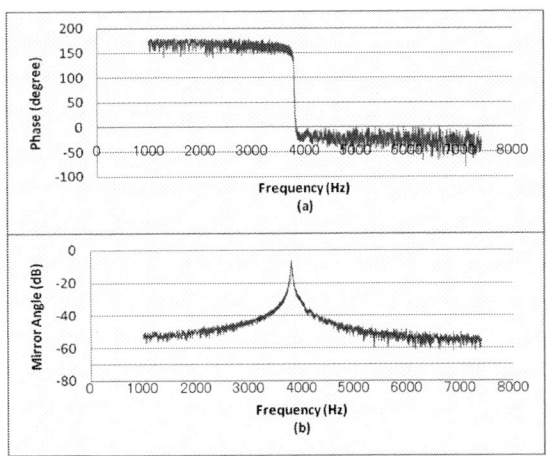

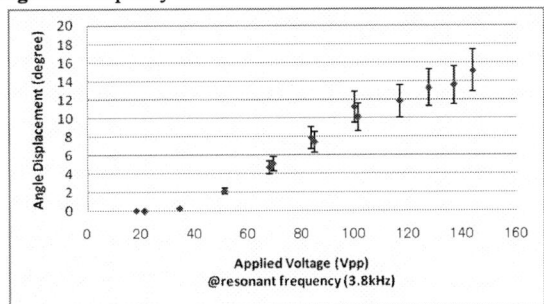

Figure 6 Frequency characteristics of fabricated MEMS scanner.

Figure 7 Measured angle displacement as a function of applied voltage.

Vertical comb driver electrodes are 10μm wide and 200 μm long with a 5-μm air gap.

4 EXPERIMENTAL RESULTS

Figure 6 shows the frequency characteristics of the fabricated MEMS optical scanner without a mirror block. The resonant frequency is seen at 3.8 kHz. Angle displacement measured at resonant frequency is shown in Figure 7. Optical scan angles of 11.2° and 15.1° were obtained at 100Vpp and 150Vpp, respectively. Resonant frequency after mounting a mirror cube is estimated to be 2.7 kHz, and the angle displacement will become 15.8° at 100Vpp. MEMS development results and 3-D image demonstration will be reported.

ACKNOWLEDGEMENTS

The photomasks used in this work were made by using the University of Tokyo VLSI Design and Education Center (VDEC)'s 8-inch EB writer F5112+VD01 donated by ADVANTEST Corporation.

REFERENCES

[1] A.Chekhovskiy et al., IEICE Electronics Express, vol.4, No.14, pp.430-434, Jul., 2007.

[2] Toshio Yamanoi, Takashi Endo, and Hiroshi Toshiyoshi, "A hybrid-assembled MEMS Fabry-Perot wavelength tunable filter," The 14th Int. Conf. on Solid-State Sensors, France, June 10 – 14, 2007 (3EK10.P).

Fast and High-Precision 3D Tracking and Position Measurement with MEMS Micromirrors

Veljko Milanović and Wing Kin Lo

Mirrorcle Technologies, Inc.

828 San Pablo Ave., Ste. 109, Albany, CA 94706

veljko@mirrorcletech.com

Abstract - We demonstrate real-time fast-motion tracking of an object in a 3D volume, while obtaining its precise XYZ co-ordinates. Two separate scanning MEMS micromirror sub-systems track the object in a 20 kHz closed-loop. A demonstration system capable of tracking full-speed human hand motion provides position information at up to 5m distance with 16-bit precision, or $<=20\mu m$ precision on the X and Y axes (up/down, left/right,) and precision on the depth (Z-axis) from $10\mu m$ to 1.5mm, depending on distance.

INTRODUCTION

Obtaining real-time 3D co-ordinates of a moving object has many applications such as gaming [1], robotics and human-computer interaction applications [2-4], industrial applications etc. Various technologies have been investigated for and used in these applications, including sensing via wire-interfaces [2], ultrasound, and laser interferometry. However a simple and low cost solution that can provide enough precision and flexibility has not been available. Recent proliferation of low-cost inertial sensors has not addressed the problem of position tracking. Cassinelli *et al* demonstrated a scanning mirror-based tracking solution [3-4], however their system does not solve the problem of object searching/selecting and does not have adequate depth (Z-axis) measurements.

The objective of this work was to develop and demonstrate an optical-MEMS based, very low cost and versatile platform for tracking and position measurement in a variety of situations. Use of MEMS mirrors [5] with potential for use of wide-angle lenses provides the possibility of tracking in a very large volume, and very far distances. E.g. use of remote-control IR source-detector modules can provide a range of 50m or more.

MULTIPLE TRACKING OPTIONS

We have developed several beam-steering based techniques to track an object inside a conic volume, as depicted in Fig. 1a.

A. Tracking a photo-detector or a retro-reflector

As depicted in Fig. 1b, there are two laser beams scanned by two MEMS mirrors into a common volume. Both systems are pointed in a parallel direction, but are spaced a known distance *d* apart (Fig. 2a.) The devices run a spiral search pattern from origin to maximum angles until they encounter a photo-detector which synchronously relays its readings to the control FPGA. From this point forward the devices renew a search but with an updated origin at the last known position of the photo-detector. The system is therefore in a perpetual search mode, although only in a very small neighborhood of the photo-detector. Full motion tracking (Fig. 2b) was achieved with fast MEMS devices giving at least 2 kHz of motion bandwidth. Since only one device can illuminate the target at a time, we time-multiplex the sub-systems by laser modulation.

In our best-performing setup we use a quadrant photo-detector which provides additional information for tracking, specifically the needed adjustments in X and Y to get centered on the target. Here there are clearly distinct modes: search (spiraling) and tracking. Tracking is a proportional control closed-loop based on the quad-detector X and Y inputs as loop errors. We also implemented a small beam-motion dither on the MEMS scanners that allows us to measure the tilt-orientation of the quad-detector (Fig. 2c,) therefore giving us 4 DoF of the detector and allowing us to use it at any rotational position.

In a nearly identical setup, we placed 2 photo-detectors in close proximity with the MEMS mirrors. The object being searched in the 3D volume is a retro-reflector ("cats eye") or a corner-cube reflector (both were used in our experiments.) In this manner both devices can simultaneously illuminate the target and operate independently.

B. Tracking an LED

As depicted in Fig. 1c, there is a photo-detector near each one of the MEMS scanning units. An optical source such as a near-IR LED is the target object that illuminates the micromirrors. When the mirrors are properly pointed, that illumination is reflected onto each detector. Therefore no time-multiplexing or communication to the target is necessary.

3D POSITION MEASUREMENT

Both devices X and Y axes are driven by separate channels of a 16-bit FPGA system. They achieve angle (negative and positive) maxima ($-\theta_{max}$, $+\theta_{max}$) when the system sends $-K$ to $+K$ to its output DAC, where $K=2^{15}-1$. In most of our experiments we calibrate our devices to provide $\theta_{max}=10°$, giving a total scan angle of 20°. When device 1 successfully tracks the target, the FPGA system records the angle of the device's x-axis and y-axis in terms of the open-loop output values O_{X1} and O_{Y1}. Second device provides knowledge of its open-loop angles O_{X2} and O_{Y2}. The devices are level in y but spaced a known distance d in x. Therefore when both devices are tracking the object they see nearly identical Y readings O_{Y1} and O_{Y2}, but due to motion parallax the X readings are different and depend on the distance of the object. We utilize the X readings to obtain a true distance of the object to the origin (a point directly between the two micromirrors) as:

$$Z = \frac{d \cdot K}{\tan(\theta_{max})} \cdot \frac{1}{(O_{X1} - O_{X2})}.$$

With Z known, X and Y are found from known parameters and by averaging from two devices' readings:

$$X = (O_{X2} + O_{X1}) \cdot Z \cdot \tan(\theta_{max})/(2K) = \frac{d}{2} \frac{(O_{X2} + O_{X1})}{(O_{X1} - O_{X2})}.$$

$$Y = (O_{Y2} + O_{Y1}) \cdot Z \cdot \tan(\theta_{max})/(2K) = \frac{d}{2} \frac{(O_{Y2} + O_{Y1})}{(O_{X1} - O_{X2})}$$

RESULTS

Our MEMS devices provided pointing precision $>=$ the DAC's 16-bit resolution, and therefore our overall system results all demonstrated this 16-bit limitation. When target object was not moving, no single digit of X,Y,Z was changing. Movements of 1mm on an optical-bench micrometer were easily recorded at 5m distance. With the loop-gain and bandwidth capable of tracking full-speed human hand motion, the system provides position information at up to 5m distance with $<=20\mu m$ precision on the X and Y axes (up, down, left, right,) and

978-1-4244-1917-3/08/$25.00 ©2008 IEEE

precision on the depth (Z-axis) from 10μm to 1.5mm, depending on the distance. Precision can be greatly increased with slower tracking settings and lower loop-gain in different applications.

[1] J. Brophy-Warren, "Magic Wand: How Hackers Make Use Of Their Wii-motes," The Wall Street Journal, Apr. 28th, 2007.
[2] P. Arcara, et al, "Perception of Depth Information by Means of a Wire-Actuated Haptic Interface," Proc. of 2000 IEEE Int. Conf. on Robotics and Automation, Apr. 2000.

[3] A. Cassinelli, et al, "Smart Laser-Scanner for 3D Human-Machine Interface," Int. Conf. on Human Factors in Computing Systems, Portland, OR, Apr. 02 - 07, 2005, pp. 1138 - 1139.
[4] S. Perrin, et al, "Laser-Based Finger Tracking System Suitable for MOEMS Integration," Image and Vision Computing, New Zealand, 26-28 Nov. 2003, pp.131-136.
[5] V. Milanović, et al, "Gimbal-less Monolithic Silicon Actuators For Tip-Tilt-Piston Micromirror Applications," IEEE J. of Select Topics in Quantum Electronics, vol. 10(3), Jun 2004.

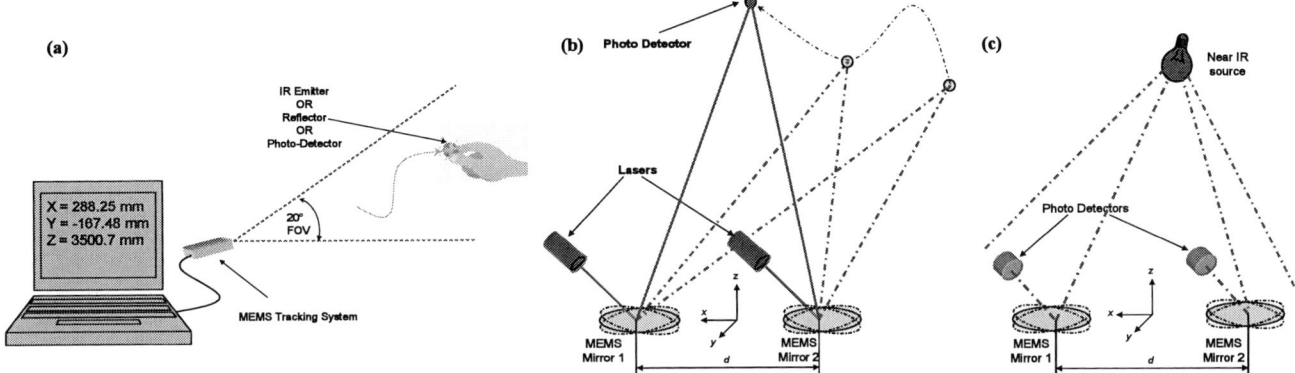

Figure 1. (a) Schematic diagram of 3D tracking of a hand-held object in a 3D volume. (b) Schematic of a 3D Tracking setup with two beam-steering MEMS mirrors aiming their laser sources onto the target. (c) Schematic diagram of 3D tracking and measurement setup with two MEMS devices steering incident light from a (near-IR) source onto their respective photo-detector.

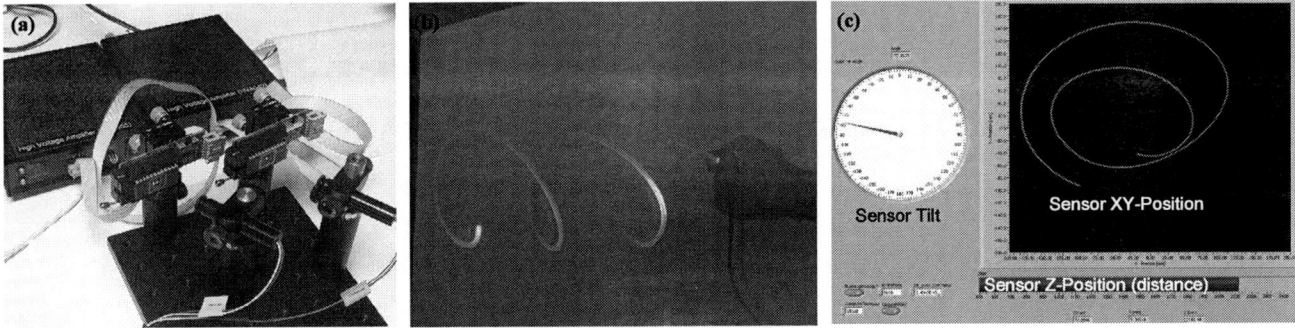

Figure 2. (a)Photograph of the two MEMS scanners and amplifiers. The devices are d=75mm apart and aimed in the same direction. Each amplifier in the background is driven by the FPGA closed-loop controller. (b) A 2s long exposure photograph of quad-detector tracking. Both laser spots are on the detector, and both devices successfully track the target. (c) GUI screen capture showing the measured 4 DoF of the detector: position X [mm], position Y [mm], position Z [mm], and tilt of the quad-detector [deg.]

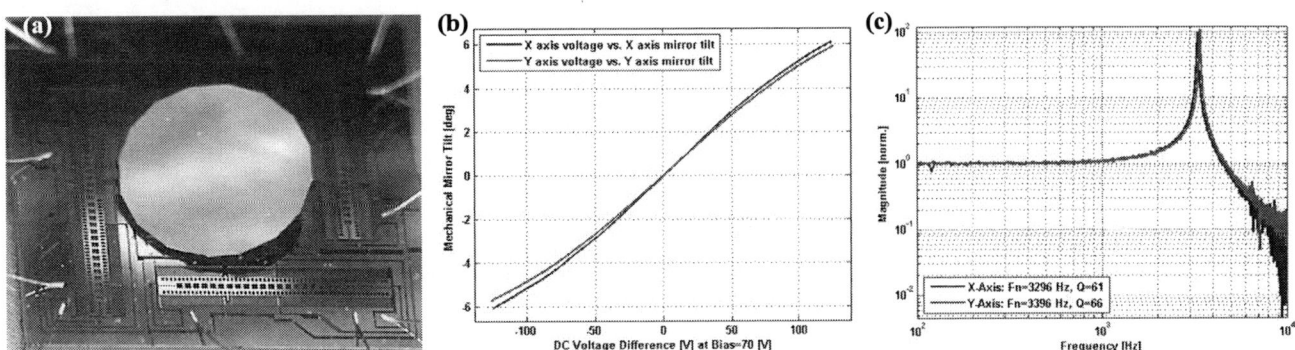

Figure 3. Gimbal-less dual-axis 4-quadrant devices used in this work: (a) typical device which reaches mechanical tilt from -8° to +8° on both axes. Device has a 2mm mirror, this larger aperture being more suitable for the setup of Fig. 1c. (b) Voltage vs. Mechanical tilt angle measurements of a typical 4-quadrant device, linearized by our 4-channel amplifier driving scheme. (c) Small-signal characteristics of fast devices with 0.8mm mirror used in the setup of Fig. 1b, where larger aperture size is not required.

IN-SITU SINGLE CELL ELECTROPORATION USING OPTOELECTRONIC TWEEZERS

Justin K. Valley, Hsan-Yin Hsu, Aaron T. Ohta, Steven Neale, Arash Jamshidi, and Ming C. Wu
Berkeley Sensor & Actuator Center (BSAC) and Department of Electrical Engineering and Computer Sciences,
University of California, Berkeley, CA 94720, USA, E-mail: valleyj@eecs.berkeley.edu

ABSTRACT

Optoelectronic Tweezers are used to achieve light-induced, *in-situ* electroporation of HeLa cells. By controlling electrical bias, patterned light induces either single cell movement or electroporation. Fluorescent dye and dielectrophoretic response are used to monitor electroporation.

INTRODUCTION

Electroporation is a common tool for applications including DNA transfection, drug delivery, and gene therapy. During electroporation, cells are subjected to electric fields which cause the creation of nanometer-scale pores in the cell membrane allowing fluid flow from the medium into the cytoplasm. However, the two most common electroporation techniques employed today are limited either by cell selectivity or small throughput. Much work in recent years has tried to ameliorate these issues by moving towards a microscale electroporation platform [1, 2]. However, these techniques do not provide both the advantage of high throughput and single cell selectivity. Here we report the use of Optoelectronic Tweezers (OET) to achieve *in-situ* electroporation of HeLa cells. Due to the dynamic nature of OET, single cells can be selected, manipulated, and then electroporated in parallel.

DEVICE OVERVIEW

Optoelectronic tweezers uses patterned light to alter the conductivity of a photosensitive film to create localized electric field gradients. These gradients result in a dielectrophoretic (DEP) force on particles in the vicinity. Because of the low light power necessary for actuation, compared to the more traditional optical tweezers, thousands of simultaneous traps can be created and manipulated in parallel [3].

The OET device with optical setup is schematically drawn in Fig. 1. It consists of a piece of ITO-coated glass with a 1 μm layer of photosensitive a-Si:H deposited via plasma enhanced chemical vapor deposition. A top piece of ITO-coated glass is used, in addition to the bottom piece, to sandwich a solution containing the particles of interest. An AC bias is applied between the two ITO electrodes and light patterns are supplied via a computer controlled projector focused through a 20x objective. Observation occurs from the topside with a CCD camera.

Cells are first selected and positioned using the projected light as reported elsewhere [4]. Next, the cells are subjected to an electroporation bias which causes temporary breakdown of the cell membrane to occur. This allows nano-tags (e.g. molecules, nanoparticles) to enter the cell. The porated cells are then moved once again by returning to the normal OET bias and altering the light pattern.

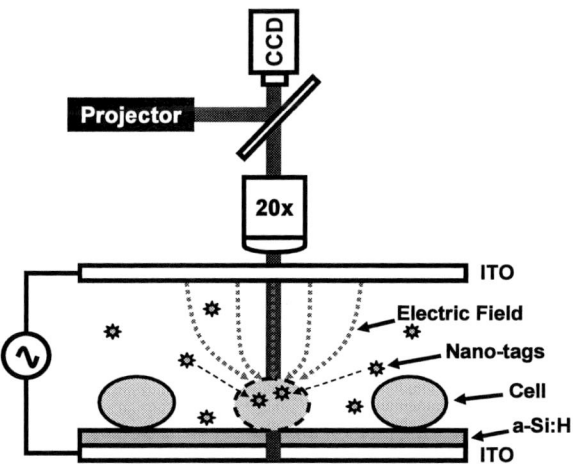

Fig. 1: Overview of OET electroporation platform. Patterned light localizes electric field across a cell of interest resulting in selective electroporation and allowing nano-tags (e.g. molecules, nanoparticles) to enter the intracellular matrix.

MODELING

The DEP force is proportional to the Claussius-Mossotti (CM) factor and the gradient of the square of the electric field. In order to model the expected DEP force on cells, a multi-shell model is used to approximate the CM factor [5]. As shown in the inlay of Fig. 2, this consists of the cell membrane, cytosol, nuclear membrane, and nucleoplasm. This factor is then multiplied by the gradient of the square of the electric field which is extracted by simulating the OET device in a commercially available finite element modeling package (Comsol 3.2a). Using typical cellular parameters, a normalized frequency response of the OET device acting on HeLa cells is shown in Fig. 2. In this figure, we plot the relative DEP force for HeLa cells with varying cytosol and nucleoplasm conductivities. The conductivity varies from that of the cell (0.53 S/m) to that of the experimental medium (10 mS/m). As one can see, the DEP force switches from positive to negative in the kilohertz range as the conductivity is decreased.

978-1-4244-1917-3/08/$25.00 ©2008 IEEE

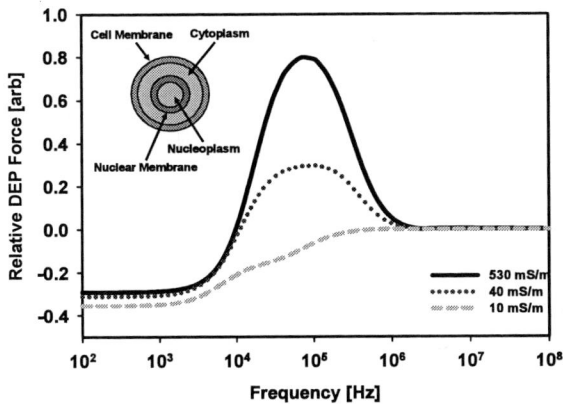

Fig. 2: Relative DEP force for cells with varying cytoplasmic and nucleoplasmic conductivity. As conductivity decreases, DEP force switches from positive to negative over the kHz frequency range. Inlay shows the cell model used for calculating the Claussius-Mossotti factor taking into account both cell and nuclear membrane.

EXPERIMENTAL RESULTS

HeLa cells were suspended in a commercially available electroporation medium (Cytoporation Medium T, Cyto Pulse Sciences, Inc.) with a conductivity of 10 mS/m and density of 1.1 million cells per milliter. Propidium Iodide (PI) was added at a concentration of 1:500 (PI:solution). PI is a membrane impermient dye which fluoresces red only in the presence of nucleic acids. A 1 μL droplet of solution was then placed in the OET device. A 7 Vppk, 100 kHz bias was then applied. When a light pattern (~2 W/cm²) illuminates the cell of interest, it results in a positive DEP force with nominal dye uptake, as evidenced in Fig. 3 and 4. Next, a 5 second 12 Vppk, 100 kHz bias is applied to the cell. This causes substantial uptake of PI by the cell indicating electroporation. However, the cell still exhibits a positive DEP response. This indicates that the electroporation process has not changed the cell's interior conductivity substantially. Lastly, a 5 second 15 Vppk, 100 kHz signal is applied which results in continued uptake of PI by the cell. However, this time the cell exhibits a negative DEP response following the voltage stimulus. This is consistent with the conductivity of the cytosol and nucleoplasm decreasing to that of the surrounding medium as shown in Fig. 2. This further indicates that electroporation of the cell has occurred.

CONCLUSION

The successful light-induced, *in-situ* electroporation of individual HeLa cells has been achieved using Optoelectronic Tweezers. The extent of the electroporation of the cell is controlled by varying the applied voltage. The applied voltage controls how much fluid is exchanged across the cell membrane resulting

either positive or negative DEP response. This approach can easily be parallelized to achieve a high throughput electroporation assay with single cell selectivity.

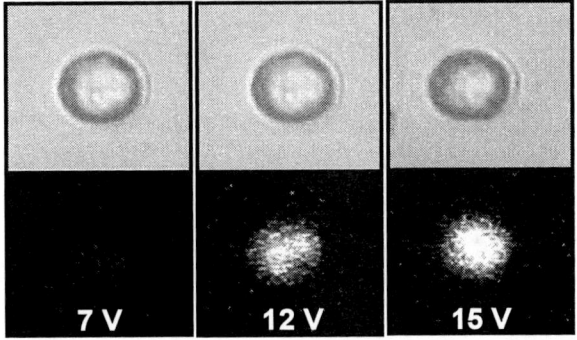

Fig. 3: Bright field (top) and fluorescent (bottom) images of a HeLa cell subjected to 7 Vppk, 12 Vppk, and 15 Vppk at 100 kHz for 5 seconds.

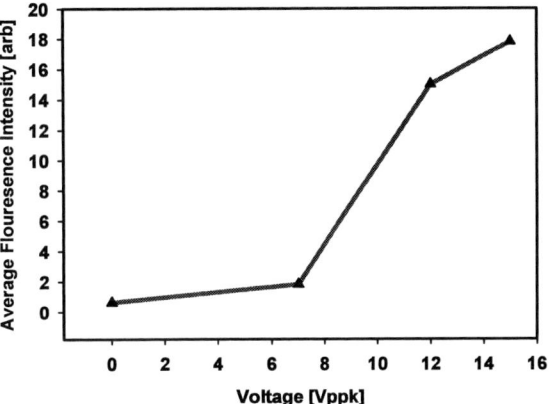

Fig. 4: Average fluorescent intensity of a HeLa cell for varying voltages.

ACKNOWLEDGMENT

The authors would like to thank the UC Berkeley Cell Culture Facility for providing the cells. This work was funded by the NIH through the NIH Roadmap for Medical Research, Grant # PN2 EY018228.

REFERENCES

[1] Y. Huang and B. Rubinsky, "Microfabricated electroporation chip for single cell membrane permeabilization," *Sensors and Actuators a-Physical,* vol. 89, pp. 242-249, Apr 2001.

[2] M. Khine, A. Lau, C. Ionescu-Zanetti, J. Seo, and L. P. Lee, "A single cell electroporation chip," *Lab on a Chip,* vol. 5, pp. 38-43, 2005.

[3] P. Y. Chiou, A. T. Ohta, and M. C. Wu, "Massively parallel manipulation of single cells and microparticles using optical images," *Nature,* vol. 436, pp. 370-372, Jul 21 2005.

[4] A. T. Ohta, P. Y. Chiou, T. H. Han, J. C. Liao, U. Bhardwaj, E. R. B. McCabe, F. Yu, R. Sun, and M. C. Wu, "Dynamic Cell and Microparticle Control via Optoelectronic Tweezers," *Microelectromechanical Systems, Journal of,* vol. 16, pp. 491-499, 2007.

[5] J. S. Crane and H. A. Pohl, "Theoretical models of cellular dielectrophoresis," *Journal of Theoretical Biology,* vol. 37, pp. 15-41, 1972.

Large-Area High-Reflectivity Broadband Monolithic Silicon Photonic Crystal Mirror MEMS Scanner

Il Woong Jung, Shrestha Basu Mallick and Olav Solgaard

E. L. Ginzton Laboratory, Department of Electrical Engineering, Stanford University, Stanford, CA 94305

Tel +1-650-723-9659, Fax +1-650-725-2533, E-mail: iwjung@stanford.edu

Abstract

In this paper we introduce a MEMS scanner with a monolithic silicon 2-D photonic crystal (PC) mirror with broad-band high reflectivity (>90%) in the 1550nm wavelength band. The photonic crystal is generated as an integrated part of the MEMS scanner fabrication process, resulting in a monolithic structure with a mirror that is ultra-flat with ~λ/100 flatness over a 500µm x 500µm area. The reflectance spectrum shows that the PC mirror has a high reflectivity (>90%) band from 1520nm-1620nm. The scanner has a scan range of 22 degrees at an input square wave of 67V with a resonance frequency of 2.13kHz.

Keywords: Photonic crystal, broad-band, MEMS scanner, GOPHER, monolithic

1 INTRODUCTION

We have previously reported on a MEMS scanner with a broad-band high reflectivity photonic crystal mirror [2]. These PC mirrors were fabricated using poly-silicon and low-index oxide on SOI (silicon-on-insulator). Recently S. Hadzialic *et al* have demonstrated a method of fabricating photonic crystal mirrors in monolithic silicon [1]. The monolithic photonic crystals have minimal stress-induced curvature that makes them ideal as large mirrors in scanners or other free-space beam applications. The process is resistant to high temperature and wet chemical etching, making it highly compatible with MEMS processing. In this paper, we demonstrate this by introducing a fabrication process that integrates the monolithic photonic crystal with a MEMS scanner. The fabricated scanner shows broadband, high reflectivity in the communications band of 1550nm and achieves scan angles of 22 degrees.

2 FABRICATION

The fabrication starts with a 460nm thermal oxidation of a SOI wafer with a 25µm silicon device layer and 1.0µm of BOX (buried oxide). The photonic crystals are patterned using a 5x reduction ASM-L i-line stepper. The pattern is transferred to the oxide and silicon using anisotropic RIE (Reactive Ion Etching) processes (Fig. 2a). A second short oxidation is done to protect the sidewalls from Si etchants (Fig. 2b) and then the oxide layer on the bottom of the holes is removed by RIE (Fig. 2c). A second Si RIE is done to make a deeper hole that will later in the process be used for isotropic Si etching to undercut the PC slab (Fig. 2d). Next, the scanner structure is patterned using a contact aligner. The misalignment between the PC patterns (stepper) and scanner patterns (contact aligner) is less than 10µm, resulting in, at most, a 4% reduction in useful mirror area. The oxide is etched using RIE and the SOI device layer is etched using the Bosch process (Fig. 2e-2). A third thin oxidation step is done to protect the combdrives and other device structures (Fig. 2f-2). Again oxide on the bottom of the holes is removed by RIE. The wafer is bonded with resist to a handle wafer and the backside is patterned with 7.0um resist. The substrate trench is etched using the Bosch process and the BOX thinned down by RIE (Fig. 2g-2). After dicing, the PCs are

undercut to create a low-index layer by etching in an isotropic SF6 plasma etch (Fig. 2h-1). Then the PCs are released by vapor HF etching which also removes the oxide on the MEMS structures (Fig. 2i). Finally, the backside of the mirror is patterned by a FIB (focused-ion-beam) to remove reflections at the interface.

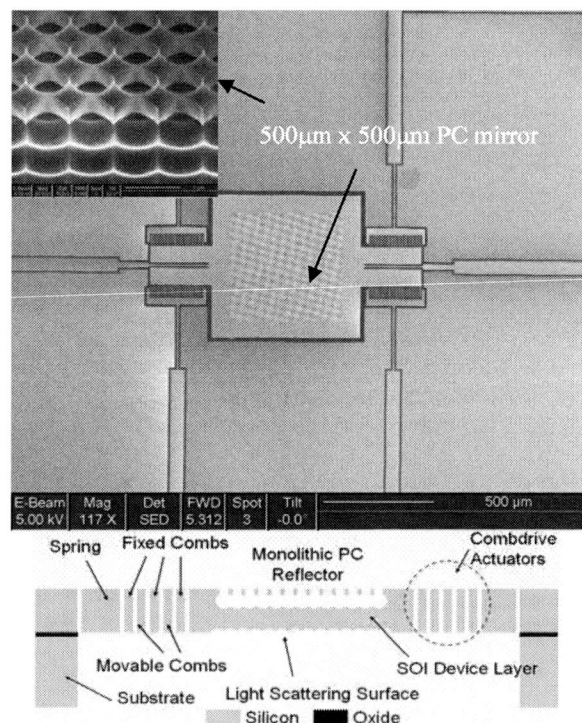

Figure 1. (a) SEM image of the monolithic photonic crystal mirror MEMS scanner (inserts: SEM images of the monolithic PC mirror) (b) Diagram illustrating the cross-section

3 CHARACTERIZATION

The measured surface profile of the PC mirror shows the average peak-to-valley roughness to be ~14nm, a flatness of λ/100 in the 1550nm band. This is a significant improvement in flatness compared to the scanner with a poly-Si/oxide photonic crystal mirror that had a peak-to-valley of ~460nm over the same area.

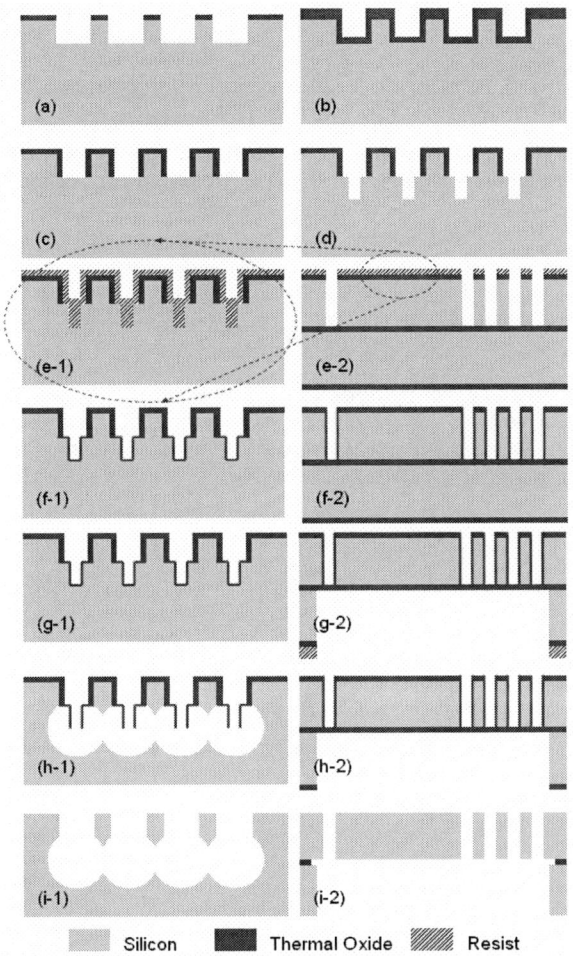

Silicon ▨ Thermal Oxide ■ Resist ▨

Figure 2. Fabrication process: (a)-(d) GOPHER process up to the isotropic Si etch (e-1, e-2) Pattern SOI device layer with resist-oxide mask by DRIE (f-1, f-2) Thin oxidation to protect MEMS parts from isotropic Si etch (g-2) Substrate etch by DRIE (h-1) Undercut by isotropic Si etch (i-1) Oxide release by vapor HF

Fig. 3 shows the reflection spectrum of the photonic crystal mirror. A broad-band reflectance of >90% is shown in the telecommunications band region of 1520nm to 1620nm. The measurement data show close correlation with simulation results for the given parameters. The scanner actuation measurements show a scan range of 22 degrees with an input square wave of 67V and resonance frequency of 2.13kHz. It is expected that these mirrors also have low angular and polarization dependencies of the reflectivity as the poly-oxide mirrors [2].

One difficulty with the demonstrated devices is that the monolithic PC mirror is a free-standing membrane that is likely to dynamically deform if subjected to air damping during actuation. This difficulty can be solved by using a vacuum package, or for applications where viscous damping is necessary, multi-level PCs without complete undercut [3] can be used.

4 CONCLUSION

We have demonstrated fabrication of MEMS scanners with integrated large–area, photonic-crystal mirrors. The monolithic combination of scanners and PC mirrors results in low-stress ultra-flat reflectors that have broadband high reflectivity >90% in the 1550nm region from 1520nm to 1620nm.

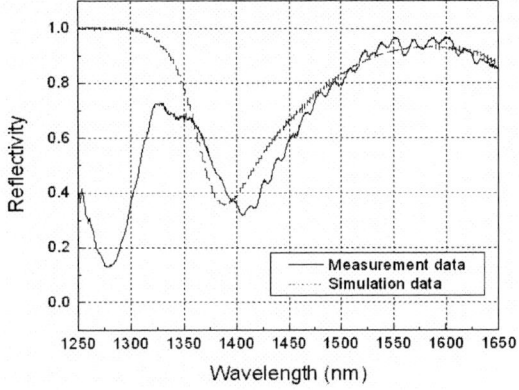

Figure 3. Measured and simulated reflection spectrum of the monolithic photonic crystal mirror MEMS scanner

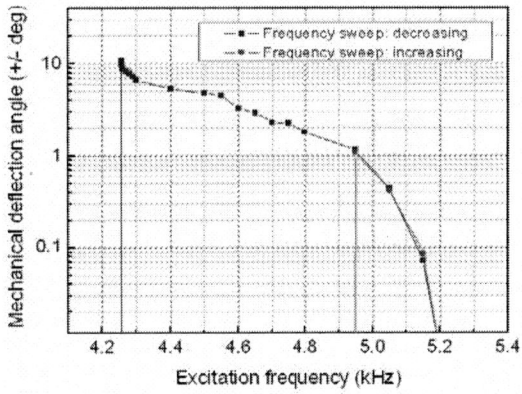

Figure 4. Dynamic deflection measurement at 67V input square wave. Results show a resonance frequency at 2.13kHz and scan range of 22 degrees.

REFERENCES

[1] S. Hadzialic et al, "Monolithic Photonic Crystals," IEEE/ LEOS Conference, Orlando, Florida, Oct 2007, pp. 341-342.

[2] I. W. Jung et al, "High Reflectivity Broadband Photonic Crystal Mirror MEMS Scanner," Transducers '07, Lyon, France, June 2007, pp.1513-1516.

[3] S. B. Mallick et al, "Double-layered Monolithic Photonic Crystals," CLEO, San Jose, CA, May 2008.

3D Imaging Using Resonant Large-Aperture MEMS Mirror Arrays and Laser Distance Measurement

Thilo Sandner, Michael Wildenhain, Thomas Klose,
Harald Schenk
Micro Scanner Devices
Fraunhofer Institute for Photonic Microsystems
Dresden, Germany
thilo.sandner@ipms.fraunhofer.de

Stefan Schwarzer, Vladimir Hinkov, Heinrich Höfler,
Harald Wölfelschneider
Optical Measurement for Production and Processing
Fraunhofer Institute for Physical Measurement Techniques
Freiburg, Germany
stefan.schwarzer@ipm.fraunhofer.de

Abstract—We present a system concept for a scanning laser radar employing a newly developed MEMS mirror array. The array solution permits large reception apertures while preserving outstanding reliability, high scanning speed, compact size and small system weight. We show first results using a single-mirror prototype.

Keywords-MEMS, mirror array, laser radar, scanning, high speed, large aperture

I. INTRODUCTION

Traditional laser scanners for distance measurement involve expensive, heavy and large rotating or vibrating mirrors as a means for light deflection. The mirror size is often dictated by the necessity to collect small amounts of light reflected or scattered by the target. Its replacement by a micromechanical device is not straightforward, since in addition to the large aperture, the device must implement a driver mechanism to achieve sufficiently large optical deflection angles, it has to integrate a position measurement sensor to provide angular position information during the scan process and yet be mechanically stable against large accelerations to be suitable for industrial or outdoor use.

Therefore, we have designed a scalable array MEMS structure composed of comparatively large silicon mirror elements with an dimension of 2.51 x 9.51 mm². We envision that an array of 2 x 7 elements with a resulting usable area of 334 mm^2 provides a satisfactory aperture for many typical security, machine vision and even, to a limited extent, outdoor applications. We further describe the MEMS elements in section III.

The mirror array is used in combination with a laser-phase-delay distance measurement device for which it is of particular importance to provide good separation of the reception from the transmit channel. Thus, the transmitted beam is sent over a separate single mirror of the same mechanical dimensions as the elements of the array (Sec. II).

Since energy is emitted in only one direction, all the receiving mirrors must be synchronized and point in the same direction in order to maximize the optical return of the system. Miniaturized position measurement equipment is integrated into the MEMS mirror devices and permits precise control of the mirror motion so that all receiving mirror elements can be slaved to the motion of the emission mirror (Sec. III).

In Sec. IV we will briefly address situations, in which the system design is advantageously modified in order to illuminate simultaneously separate regions of the imaged area.

II. SYSTEM DESIGN

A sketch of a possible system design is shown in Figure 1.

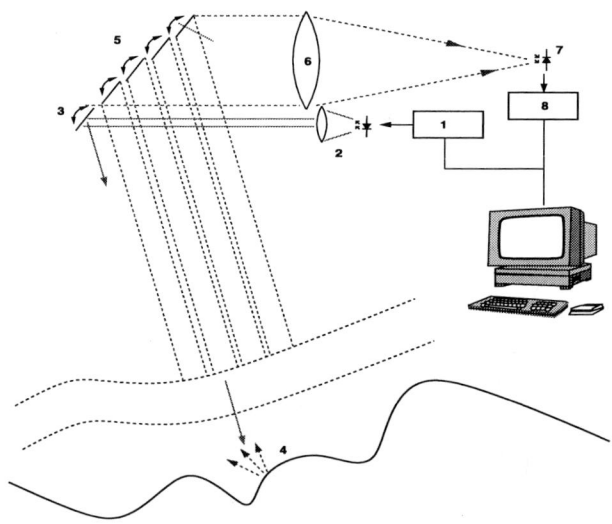

Figure 1. System setup: The modulated emitted laser beam (2) is directed onto the target region via a dedicated mirror (3) and is reflected/scattered on the surface. The light arriving on the receiving mirrors (parallel to the emission mirror, 5) reaches the detector (7) via collecting optical elements (6). The distance to the target point follows from the modulation phase of the received signal with respect to that of the emitted signal (1,8).

A single emitting mirror oscillates in parallel and in phase with all receiving mirrors. Light paths are separated to reduce crosstalk in the system. The areas of the receiving mirrors add up to form a (problem dependent) sufficiently large aperture. For a more general discussion of the distance measurement principle, see [1].

978-1-4244-1917-3/08/$25.00 ©2008 IEEE

III. MEMS MIRROR ARRAY

The MEMS array satisfies at the same time the demand of comparatively large optically active area, 2.51 x 9.51 mm per single mirror element while keeping the resonance frequency of 250 Hz at a value that matches well the properties of current laser distance measurement devices. Typical point rates of measurement modules are 250-1000 kHz. The scan angle is thus split in 500–2000 intervals, the widest intervals being in the center of the field of view about 2π times larger than the average width. The mirrors are driven electrostatically resonantly using in plane comb drives [2] at a frequency close to the mechanical resonance for oscillation around their long symmetry axes. The resulting optical scan angle is ±30 degrees (twice the mechanical deflection angle of ±15 degrees).

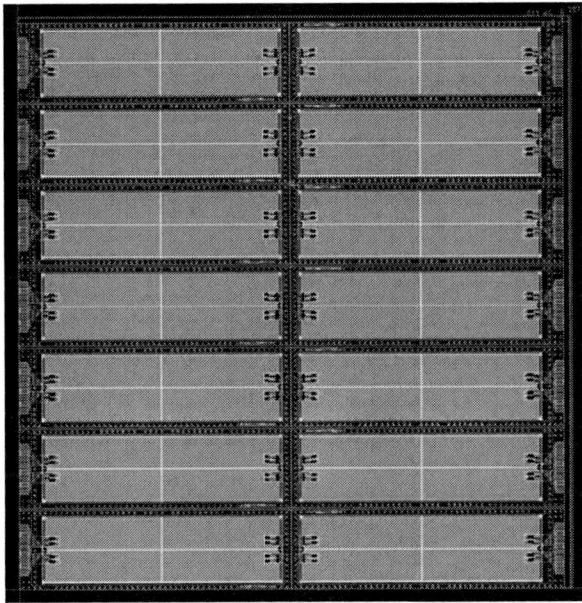

Figure 2. MEMS mirror array (2 x 7 elements) seen from above. The cscillation occurs around the long axis (left to right) and is driven electrostatically resonantly via in-plane comb drive electrodes at the short end of the elements. The depicted chip has an overall size of 22.3 x 21 mm², filling factor about 69 % counting the entire chip and about 80 % if not accounting for the electrical feed structures at the edges.

The mirror units include components to determine path elements (times at given fixed deflection) of the mirror motion and thus permit the precise modeling and prediction of the mirror positions as functions of time. The deflection information enters both the control circuitry for the mirror motion and is necessary to identify uniquely the targeted point.

IV. FIRST RESULTS AND OUTLOOK

We have realized a simplified setup by using a single MEMS scanning mirror in the emission branch, collecting the light with a cylindrical lens from the entire imaged area, setup and measured data are shown in Figure 3.

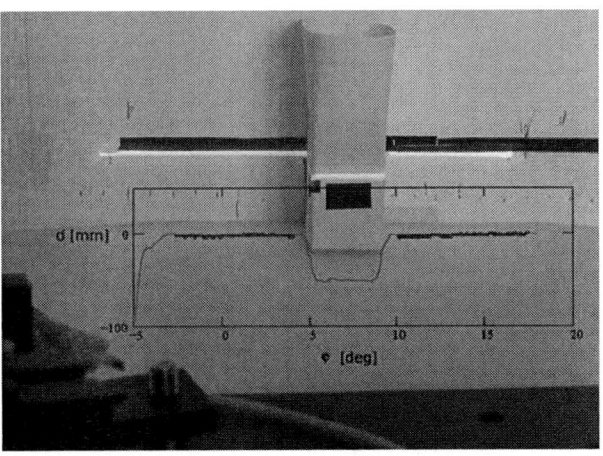

Figure 3. Measured data for a setup with one oscillating mirror for the emitted beam (front left). The reception optics uses a single cylindrical lens to focus the light emanating from a rectangular region of the target onto the detector (not shown, in the projection plane perpendicular to the target).

The spatially segmented structure of the MEMS array offers additional opportunities for system design. Let us imagine splitting the emitted beam over all or a subset of the mirrors in a coaxial design while using the same mirrors again for reception of the scattered radiation. We assume further that the mirror orientations are no longer constrained to be parallel, but that each mirror points in a direction independent of those of the others (relaxing the synchronicity constraint has the potential to reduce the complexity of the array control electronics). Using ideas of compressive imaging [3] we have shown that under these conditions it is still possible to reconstruct the distance information for each target interval from a sufficiently large set of point measurements [4] and performed simulations to prove the feasibility of the concept. By splitting the array, we leverage this property to increase the target scan area or to adapt the system to unusual, possibly disconnected targets.

We have demonstrated a system concept for a scanning laser radar constructed around a novel MEMS mirror array. In comparison to systems with regular optical components for ray deflection, we expect the resulting device to become significantly smaller and more robust.

We acknowledge financial support by the Fraunhofer Gesellschaft in the context of MEF LAMDA.

REFERENCES

[1] H. Wölfelschneider, A. Blug, C. Baulig, and H. Höfler, "Schnelle Entfernungemessung für Laserscanner," Technisches Messen: ATM, TM 72 (2005), no.7-8, pp. 455-467.

[2] H. Schenk, -P. Dürr, T. Haase, D., Kunze, D., Sobe, U., Lakner, H., Kück, H., "Large Deflection Micromechanical Scanning Mirrors for Linear Scans and Pattern Generation", Journal of Selected Topics of Quantum Electronics 6 (2000), no. 5, pp 715- 722.

[3] D. Takhar et al., "A new compressive imaging camera using optical-domain compression," Proc. of Computational Imaging IV, SPIE Electronic Imaging, San Jose, Jan. 2006.

[4] S. Schwarzer, H. Wölfelschneider, V. Hinkov, "Bildgebendes Entfernungsmesssystem mit mehreren gleichzeitig genutzten auch asynchron bewegten Ablenkspiegeln im Lichtweg," pat. pend

Photonic Metamaterials:
Optics Starts Walking on Two Feet

Martin Wegener and Stefan Linden

Institut für Angewandte Physik and DFG-Center for Functional Nanostructures (CFN),
Universität Karlsruhe (TH), D-76128 Karlsruhe, Germany
and
Institut für Nanotechnologie, Forschungszentrum Karlsruhe in der Helmholtz-Gemeinschaft, D-76021 Karlsruhe, Germany

Abstract: We review recent progress in the field of metamaterials for photonics. Examples are artificial magnetism at optical frequencies, negative phase and group velocities, and enhanced nonlinear phenomena.

Electromagnetic light waves have an electric and a magnetic vector component – light is walking on two legs. Thus, complete control of a light wave inside a material requires control of both, its electric and its magnetic component. However, all known natural substances have a negligible magnetic response at optical frequencies, i.e., their magnetic permeability is very nearly unity (μ=1). Hence, optics & photonics have been limited to a direct control of the electric component of light only.

Metamaterials are man-made effective materials composed of metallic sub-wavelength building blocks ("photonic atoms") that are densely packed into an effective material. Split-ring resonators are particularly important "photonic atoms" (see Fig.1). They can be viewed as tiny *LC* circuits that can be excited by the incident light field. Above the *LC* eigenfrequency, the response acquires a 180 degrees phase shift with respect to the excitation, corresponding to a diamagnetic behavior. For pronounced resonances and densely packed split-ring resonators, a negative magnetic response can result (μ<0) [1-4].

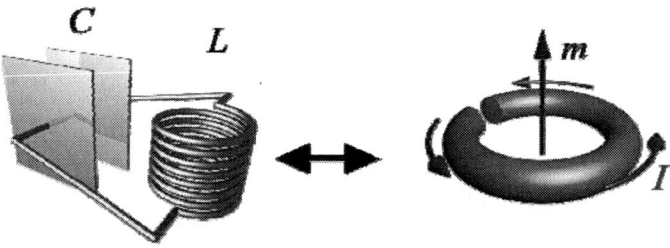

Fig.1: Scheme of a split-ring resonator and its analogy to an electric *LC* circuit.

Combining a negative magnetic response with a negative electric response allows for achieving a negative index of refraction. The corresponding negative phase velocity of light has directly been measured by interferometric time-of-flight experiments [5] at telecom frequencies. Interestingly, phase and group velocity can be negative simultaneously, i.e., be both oriented opposite to the Poynting vector [5]. Using silver rather than gold leads to significantly reduced losses [6] and has also enabled the first visible negative-index photonic metamaterial [7] (precisely 780-nm wavelength).

While most photonic metamaterials have been fabricated via electron-beam lithography, square-centimeter-area metamaterial structures can be made via compact interference lithography [8], using just a single laser beam at 532-nm wavelength impinging onto a suitably designed dielectric object on top of a photoresist.

First steps towards three-dimensional (rather than planar) metamaterials have also been taken. Stacking three functional layers leads to a comparable performance of the negative-index metamaterial [9]. However, truly three-dimensional structures (unpublished) very likely require approaches such as direct laser writing and metal atomic-layer deposition (ALD) or chemical vapor deposition (CVD).

Furthermore, second- and third-harmonic generation from magnetic metamaterials has been reported [10,11]. More recent experiments based on inverse split-ring resonator structures and Babinet's principle give insights into the underlying mechanism (unpublished). In particular, these results are consistent with our original interpretation [10] in terms of the magnetic component of the Lorentz force being a prominent second-harmonic generation source term.

Brief and extensive recent reviews on all of these aspects and several more can be found in Refs.[12] and [13], respectively.

Acknowledgements

We thank Costas M. Soukoulis for stimulating discussions. We acknowledge financial support provided by the Deutsche Forschungsgemeinschaft (DFG) and the State of Baden-Württemberg through the DFG-Center for Functional Nanostructures (CFN) within subprojects A1.4 and A1.5. The research of S.L. is further supported through a "Helmholtz-Hochschul-Nachwuchsgruppe" (VH-NG-232).

References

[1] S. Linden, C. Enkrich, M. Wegener, J. Zhou, T. Koschny, and C.M. Soukoulis, Science **306**, 1351 (2004).
[2] C. Enkrich, M. Wegener, S. Linden, S. Burger, L. Zschiedrich, F. Schmidt, J. Zhou, T. Koschny, and C.M. Soukoulis, Phys. Rev. Lett. **95**, 203901 (2005).
[3] G. Dolling, C. Enkrich, M. Wegener, J. Zhou, C.M. Soukoulis, and S. Linden, Opt. Lett. **30**, 3198 (2005).
[4] M.W. Klein, C. Enkrich, M. Wegener, C.M. Soukoulis, and S. Linden, Opt. Lett. **31**, 1259 (2006).
[5] G. Dolling, C. Enkrich, M. Wegener, C. M. Soukoulis, S. Linden, Science **313**, 502 (2006).
[6] G. Dolling, C. Enkrich, M. Wegener, C. M. Soukoulis, S. Linden, Opt. Lett. **31**, 1800 (2006).
[7] G. Dolling, M. Wegener, C. M. Soukoulis, S. Linden, Opt. Lett **32**, 53 (2007).
[8] N. Feth, C. Enkrich, M. Wegener, and S. Linden, Opt. Express **15**, 501 (2007).
[9] G. Dolling, M. Wegener, S. Linden, Opt. Lett. **32**, 551 (2007).
[10] M.W. Klein, C. Enkrich, M. Wegener, and S. Linden, Science **313**, 502 (2006).
[11] M.W. Klein, M. Wegener, N. Feth, and S. Linden, Opt. Express **15**, 5238 (2007).
[12] C.M. Soukoulis, S. Linden, and M. Wegener, Science **315**, 47 (2007).
[13] K. Busch, G. von Freymann, S. Linden, S. Mingaleev, L. Tkeshelashvili, and M. Wegener, Phys. Rep. **444**, 101 (2007).

Tunable plasmonic nanostructures

Olivier J.F. Martin, Holger Fischer, Gaëtan Lévêque*, and André Christ[†]

Nanophotonics and Metrology Laboratory
Swiss Federal Institute of Technology Lausanne (EPFL)
EPFL–STI–NAM, ELG 240, Station 11, CH–1015 Lausanne, Switzerland
Email: olivier.martin@epfl.ch http://www.nanophotonics.ch

Abstract—We discuss different composite plasmonic nanostructures, which optical properties can be tuned by displacing some of their parts at the nanoscale. The actuation of these structures with MEMs will define new optical functionalities.

*Present address: Tyndall National Institute, Cork, Ireland.
[†]Present address: Carl Zeiss AG, Oberkochen, Germany.

I. INTRODUCTION

A broad variety of plasmonic nanostructures have been introduced over the last few years [1]. These structures can produce strong optical near–fields and exhibit specific spectral responses when they are excited at their plasmon resonance frequency.

When two or more plasmonic nanostructures interact, new plasmonic modes can be created from the coupling of the modes associated with each individual structure [2]. The optical properties of theses new modes strongly depend on the interaction between the individual structures that form the composite. Hence the optical properties of the composite system can be tuned by changing its configuration.

In this communication, we discuss three types of composite plasmonic nanostructures and illustrate how their optical properties can be tuned by modifying their configuration. We hope that this presentation will trigger collaborations with the MEMs community to implement such tunability in realistic devices.

II. TUNABLE PLASMONIC NANOSTRUCTURES

As first example, let us consider a dipolar plasmonic antenna [3]. This structure is made of two gold nanoparticles separated by a 30 nm gap. The typical length of the structure is a few hundreds of nanometers. At the plasmon resonance wavelength, a very strong field is created in the antenna gap. By changing the antenna length l, it is possible to continuously tune the wavelength at which this resonance occurs, as illustrated in Fig. 1.

Another interesting tunable plasmonic system is shown in Fig. 2. It consists of a gold particle placed at a short distance d from a gold thin film. Two types of plasmonic modes are supported by this system: a localized plasmon in the particle and a propagating plasmon on the thin film [4]. The interplay between these two plasmonic modes allows for tuning the response of the system. This can be achieved by changing the spacing distance d, or by modifying the permittivity ε_d of the spacer layer, as illustrated in Fig. 3. In this case, a very broad

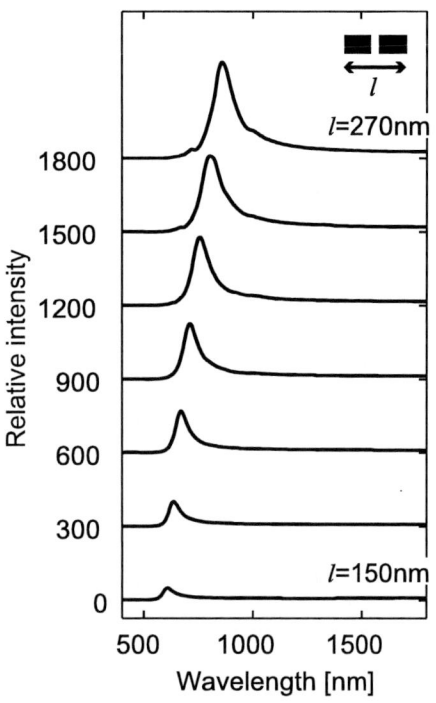

Fig. 1. Field enhancement in the gap of dipolar plasmonic antenna made of gold, as a function of the antenna length l. The resonance wavelength of the system can be tuned continuously.

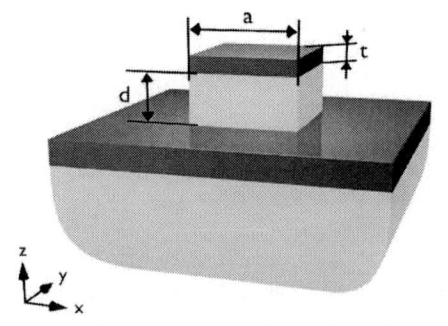

Fig. 2. Two plasmon resonances exist in that system: a localized one on the particle and a propagating one on the film. The interaction between these two modes can be used to continuously tune the optical response of the system.

978-1-4244-1917-3/08/$25.00 ©2008 IEEE

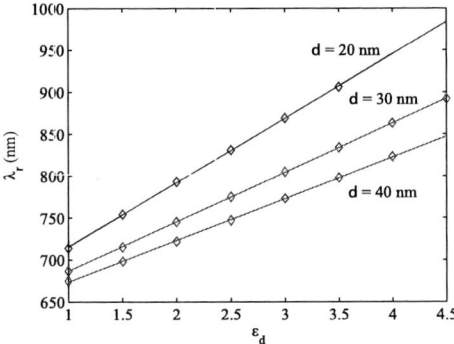

Fig. 3. Resonance wavelength of the system shown in Fig. 2 as a function of the spacer permittivity ε_d for three different spacer thicknesses d. The spectral response of the system can be tuned over a very large range.

range of resonance wavelengths can be achieved. Furthermore, the rate at which the resonance wavelength changes with the spacer permittivity ε_d depends on the spacer thickness d (compare the curves for $d = 20$ nm and $d = 30$ nm in Fig. 3).

Finally, the interaction between two metallic gratings, can produce a highly tunable optical response, as illustrated in Fig. 4. In this case, the near–field interaction between the modes supported by each grating can be controlled by changing the distance between the gratings or by displacing laterally one grating with respect to the other one [5]. In this case, it was possible to fabricate "in phase", Fig. 4(a), and "out–of–phase", Fig. 4(b), gratings in gold with a 200 nm period and 50% duty cycle using X-ray lithography. For practical applications, it would be very interesting to have a device where each grating can move with respect to the other one in a continuous way.

III. CONCLUSION

The optical properties of the different plasmonic nanostructures presented in this communication strongly depend on the position of the different elements that constitute them. By moving one element with respect to the other ones, it is possible to continuously tune the resonance frequency of the system. Furthermore, the field enhancement in the structure is also modified by this process, leading to nanostructures where the field intensity can be adjusted over several orders of magnitude.

The actuation of such plasmonic nanostructures with MEMs will allow for the accurate control of distances and positions. It will open the way for many useful devices with nanometric sensitivity and tunable optical properties.

REFERENCES

[1] W. L. Barnes, A. Dereux, and T. W. Ebbesen, "Surface plasmon sub-wavelength optics," *Nature*, vol. 424, p. 824, (2003).
[2] J. P. Kottmann and O. J. F. Martin, "Plasmon resonant coupling in metallic nanowires," *Opt. Express*, vol. 8, p. 1094 (2001).
[3] H. Fischer and O. J. F. Martin, "Engineering the optical response ofplasmonic nanoantennas," *Opt. Express*, vol. 16, p. 9144 (2008).
[4] G. Lévêque and O. J. F. Martin, "Optical interactions in a plasmonic particle coupled to a metallic film," *Opt. Express*, vol. 14, p. 9971 (2006).
[5] A. Christ, et al., "Controlling the Fano interference in a plasmonic lattice," *Phys. Rev. B*, vol. 76, p. 201405 (2007).

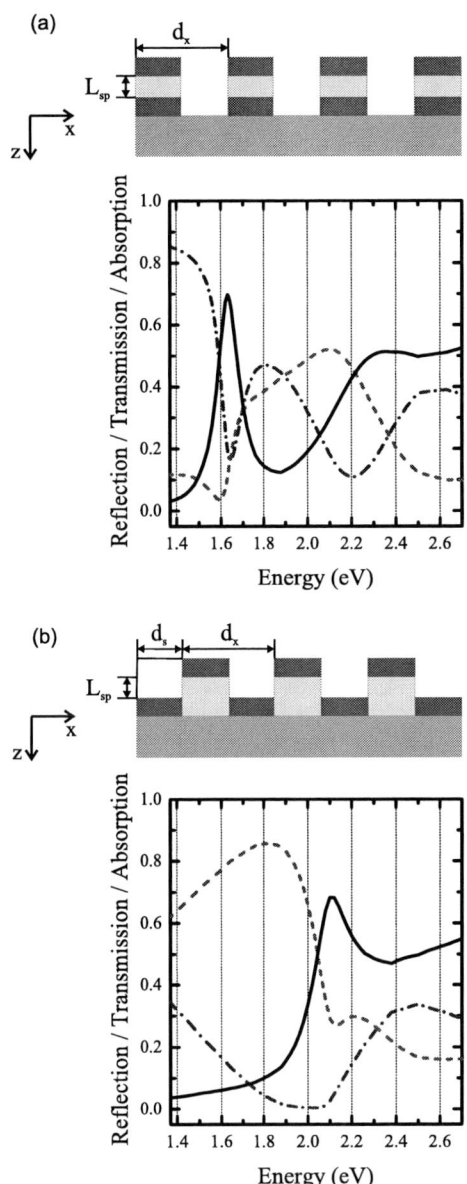

Fig. 4. Optical properties of two coupled metallic gratings. Two different configurations are considered: (a) the gratings are "'in phase'" and (b) the gratings are "'out of phase'", i.e. displaced by half a period.

Clarification of Electromagnetic Responses in Split-Ring Resonators from Electric Excitation

Chia-Yun Chen [1] and Ta-Jen Yen [1,2,*]

1 Department of Materials Science and Engineering, National Tsing Hua University, 101, Section 2, Kuang Fu Road, Hsinchu 30013, Taiwan, R.O.C.
2 National Nano Device Laboratories, Science-Based Industrial Park, Hsinchu 30078, Taiwan, R.O.C.
[*]To whom correspondence should be addressed. E-mail: tjyen@mx.nthu.edu.tw ; Tel: 886-3-5742174; Fax: 886-3-5722366

Abstract—**We demonstrate the electric and magnetic responses from split-ring resonators (SRR). Our quantitative spectroscopic measurements validate in both polarization dependence and multiple resonant characteristics, facilitating to design the desired responses of SRR for practical applications.**

BACKGROUND

Metamaterials are a new class of artificial electromagnetic materials in which the size of building "atoms" and "molecules" is smaller than the wavelength of excitation. Based on the collective resonances in the internal structures, metamaterials enable unprecedented properties to illustrate new physics and potential applications, for example, negative index metamaterials (NIMs) [1], superlensing effect [2] and invisible cloak [3], so that recently this field has increasingly attracted research interests worldwide. Among these striking properties, the foremost challenge is the realization of negative magnetic permeability, a missing property in naturally occurring materials and recently theoretically proposed by J. B. Pendry et al. [4] and experimentally demonstrated by R. A. Shelby et al. [1], respectively.

To date, the split-ring resonator (SRR) remains the most common artificial structure to give rise to the absent negative as well as high-frequency magnetic responses in nature. Composed of the metallic rings with gaps in which the former correspond to the inductance (L) and the latter serve as the capacitance (C), the SRR enables an equivalent LC circuit to couple with the external time-varying electromagnetic excitations as resonating. Such an LC resonator allows negative permeability about its resonant frequency given by $\omega_{LC} = 1/\sqrt{LC}$ while the external magnetic field is applied vertically to the SRR plane. Nevertheless, it is obvious that this H-field coupling to the resonance is quite difficult to perform in particularly beyond microwave frequencies because of the extremely small thickness of the SRR for the grazing-angle incident excitation. Alternatively, the external electric field (E-field) can also excite the resonance of the SRR and further introduce a more feasible method to realize high-frequency responses since such electric coupling can be configured under normal incidence. However, the relevant coupling mechanism wasn't fully understood yet due to the remarkable behavior of electric polarization dependence.

CURRENT RESULTS

In this study, we experimentally demonstrate the arbitrary-polarization resonances in the SRR, as shown in Fig. 1. These sub-wavelength structures fabricated by an electron-beam lithographic process are characterized by performing normal incident reflection measurements, leading the electric excitation to two distinct groups of magnetic and electric resonances. The origin of these resonances is elucidated by both quantitative spectroscopic measurements and the distribution of the simulated surface current density, indicating that the arbitrary-polarization resonances of the SRR stem from the superposition of the horizontally and vertically electric excitations. Moreover, we further experimentally demonstrate the multiple resonances in SRR from direct electric excitations in mid- and near infrared regions, as presented in Fig. 2(a) and (b). The standing-wave plasmonic resonance model is proposed to elucidate this resonant characteristic in SRR. Such expression validates in both cases of electric and magnetic responses in SRR excited by electric field and is further confirmed by examining the SRR with different lengths (Fig. 2(c)). Therefore, our quantitative observations indicate that the multiple resonances can be interpreted by the standing-wave plasmonic resonance and further facilitate to design the desired operation frequencies and responses of SRR for practical applications.

REFERENCES

[1] R. A. Shelby, D. R. Smith, and S. Schultz, Science **292**, 77 (2001).
[2] J. B. Pendry, Phys. Rev. Lett. **85**, 3966 (2000).
[3] D. Schurig, J. J. Mock, B. J. Justice, S. A. Cummer, J. B. Pendry, A. F. Starr, and D. R. Smith, Science **314**, 977 (2006)
[4] J. B. Pendry, A. J. Holden, D. J. Robbins, and W. J. Stewart, IEEE Trans. Microwave Theory Tech. **47**, 2075 (1999).

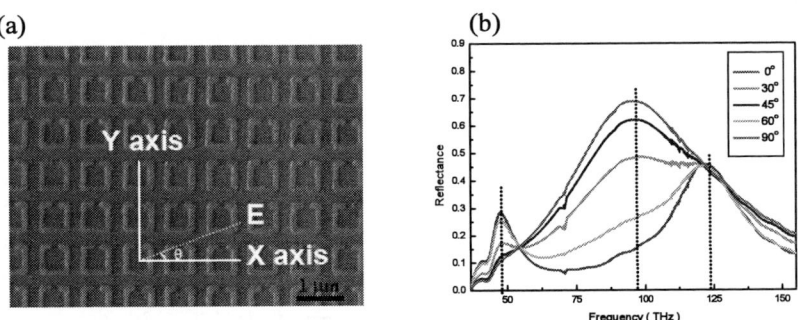

Fig. 1. (a) SEM photograph of the fabricated SRR structure. The sample size is $100 \times 100 \ \mu m^2$ and the detailed dimensions of SRR are 600 nm in length, 100 nm in width, and 150 nm in gap. (b) The measured reflectance under polarized angles: 0°, 30°, 45°, 60° and 90°, respectively. The polarization angles are defined by the angle between the external E-field with respect to the x-axis.

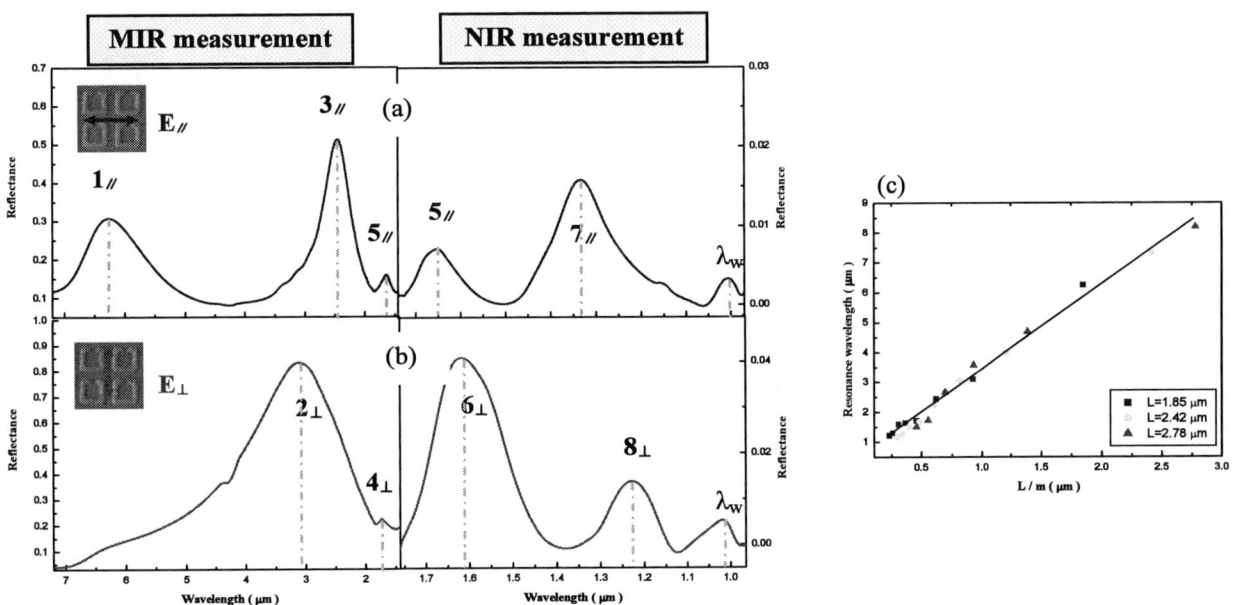

Fig. 2. Two orthogonal polarized excitations ($E_\parallel$ and $E_\perp$) are indicated in blue and red, and their corresponding spectra scanning from mid-infrared to near infrared regions are shown in (a) and (b), respectively. The numbers denoted in the figures represent the order of resonance modes and their subscripts indicate the polarization directions parallel ($E_\parallel$) or perpendicular ($E_\perp$) to the gap side of SRRs. (c) The plot of the measured wavelengths of multiple resonances as a function of L/m from three varied lengths (L) of SRR structures.

Large-Area Monolithic Photonic Crystal Mirrors with High Reflectivity in the 1250-1650nm Band Patterned by Optical Lithography

Il Woong Jung, Shrestha Basu Mallick and Olav Solgaard

E. L. Ginzton Laboratory, Department of Electrical Engineering, Stanford University, Stanford, CA 94305

Tel +1-650-723-9659, Fax +1-650-725-2533, E-mail: iwjung@stanford.edu

Abstract

This paper describes large area (500μm x 500μm) monolithic 2-D photonic crystals (PC) for applications as high-reflectivity, broad-band mirrors in the near-IR (infra-red) spectrum. These large PC mirrors were patterned using an ASM-L i-line stepper to achieve minimum feature sizes of less than 100nm. The reflectivity spectrum of the mirrors show that the high reflectivity (>90%) bands can be shifted in wavelength by varying the hole sizes of the photonic crystal to cover the 400nm near-IR band from 1250nm-1650nm.

Keywords: Photonic crystal, broad-band, GOPHER, monolithic

1 INTRODUCTION

We have previously reported on the fabrication of monolithic photonic crystal mirrors using a SCREAM-like process [1] called the GOPHER (Generation of Photonic Elements by RIE) process. This process uses a combination of isotropic and anisotropic Si etching and oxidation to create 2-D photonic crystals in silicon without the need for deposited films with residual stress. In this paper, we describe the fabrication of an array of large area monolithic 2-D photonic crystal mirrors (Fig. 1) based on optical lithography, *cf.* e-beam lithography, that spans the near-IR spectrum to cover the 1250nm-1650nm wavelength band. The reflection bands are shifted by changing the hole sizes of the photonic crystals. Starting with a pattern with a 2-D array of circles with 800nm pitch and 600nm diameter, we have fabricated PC mirrors with hole diameters from 680nm-780nm. Reflectance spectrum measurements show that individual PCs have anywhere from 20nm to >100nm high reflectivity bands. These large area broadband mirrors are ideal for free-space applications such as reflectors for scanners [2] and may also be used in other applications requiring highly reflective broadband mirrors that are compatible with MEMS processing.

2 FABRICATION

The fabrication follows the procedure for the GOPHER process [1]. Fabrication starts with a 460nm thermal oxidation of the silicon. To fabricate uniform large area PCs, instead of e-beam lithography as previously reported, we use a 5x reduction ASM-L i-line stepper to pattern the photonic crystals. This allows efficient patterning of very large areas (500μm x 500μm) with sub-micron features and allows hole size variation through exposure dose control with a single mask pattern. Using a pattern of 600nm holes at 800nm pitch on the wafer level, we were able to expose hole sizes varying from 620nm to 720nm (Fig. 2). After transferring the pattern to the oxide and silicon by RIE, the sidewalls are protected by a thin oxide. This allows for further increase of the hole diameter, necessary for acquiring broadband characteristics. The oxide on the bottom of the holes is removed by RIE and another Si etch creates deeper holes

into the Si. Next, an isotropic Si etch using SF6 is done to undercut the structure and the oxide is removed by vapor HF to minimize stress during release.

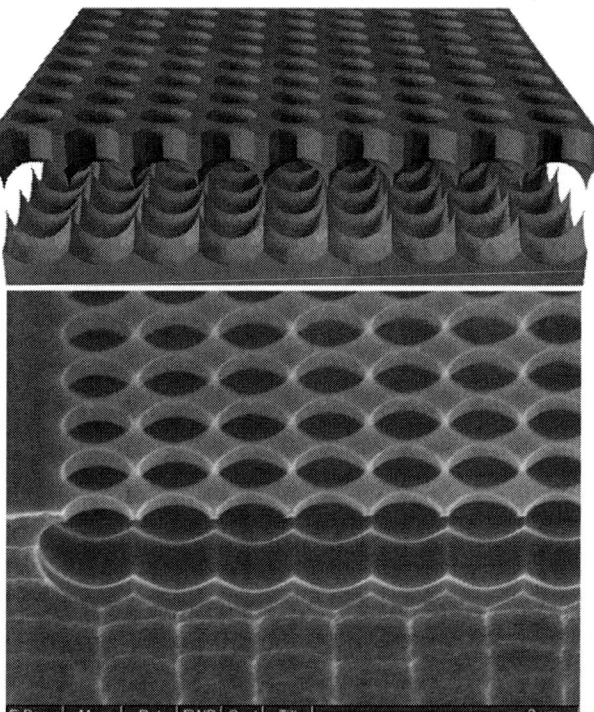

Figure 1. (a) 3-D rendered image of a monolithic photonic crystal (b) SEM image showing an angled cross-sectional view of a fabricated monolithic photonic crystal mirror.

3 CHARACTERIZATION

The final dimensions of the PCs are shown in Fig. 2. The different diameters of the holes in the x and y dimensions show that the holes are somewhat elliptical. This was caused by the fixed addressing grid of the mask writer that shifted some of the points that define the polygon. This is expected to cause some changes in the spectrum compared to simulations of PCs with ideal circles.

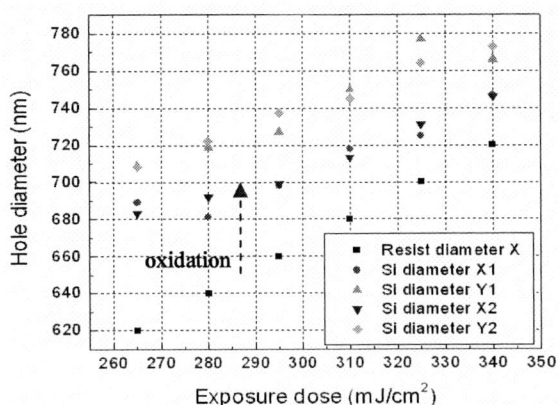

Figure 2. Measured final hole dimensions of the PCs after fabrication with respect to resist exposure dosage. There is an average of 30nm difference between the x and y diameters showing some ellipticity of the holes.

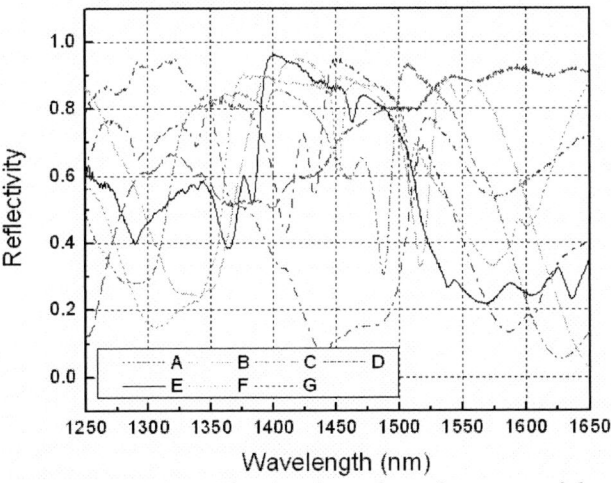

1250nm to 1650nm. The plot shows that a large part of the band is covered with >90% reflectivity. A: 265mJ/cm², B-G: 265, 280, 295, 310, 325, 340mJ/cm²

The reflectivity is measured using an experimental setup with a broadband light source covering the wavelength from 1200nm -1700nm and an optical spectrum analyzer. The reflection spectra of the various PCs from parameters in Fig. 2 are shown in Fig. 3. Curves B through G are for PCs with exposure doses of 265mJ/cm² to 340mJ/cm², respectively. A is also exposed at 265mJ/cm², but with a different amount of undercut than B. The reflection bands for the mirrors are able to cover the wavelengths from 1250nm to 1650nm with >90% reflectivity. Figures 4 and 5 show that experimental data correspond well with FDTD (finite difference

time domain) simulation results for PCs F and G, respectively. The observed discrepancies may be due to a variety of factors such as hole ellipticity, etch profiles and shape of the undercut.

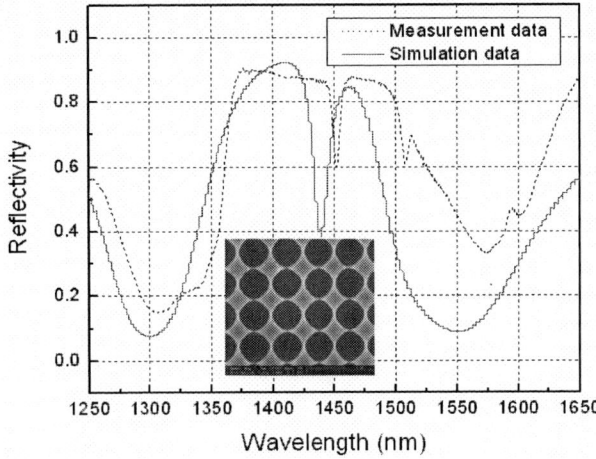

Figure 4. Measured and simulated reflection spectrum of the monolithic photonic crystal mirror F

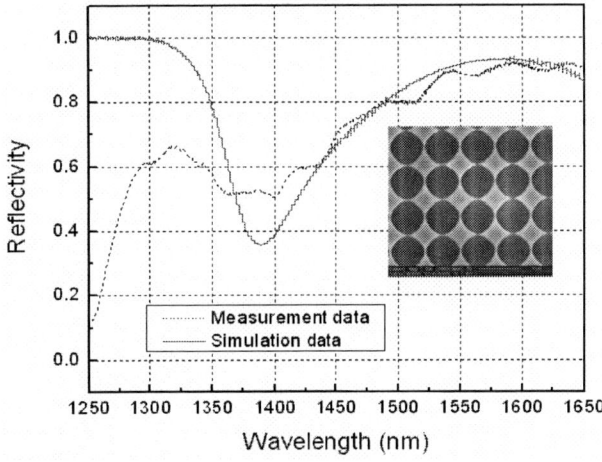

Figure 5. Measured and simulated reflection spectrum of the monolithic photonic crystal mirror G.

4 CONCLUSION

We have shown an array of large area monolithic photonic crystal mirrors with high reflectivity in the near-IR band compatible with MEMS processing. These mirrors have reflectivity of >90% covering the wavelength region from 1250nm to 1650nm and are well suited for applications in telecommunications and MEMS.

REFERENCES

[1] S. Hadzialic et al, "Monolithic Photonic Crystals," IEEE/ LEOS Conference, Orlando, Florida, Oct 2007, pp. 341-342.
[2] I. W. Jung et al, "High Reflectivity Broadband Photonic Crystal Mirror MEMS Scanner," Transducers '07, Lyon, France, June 2007, pp.1513-1516.

Applications of LCoS-based adaptive optical elements in microscopy

Andreas Hermerschmidt
Holoeye Photonics AG
Albert-Einstein-Straße 14, 12489 Berlin
Email: andreas.hermerschmidt@holoeye.com

Jan Haffner, Tobias Haist, and Wolfgang Osten
Institut für Technische Optik, Universität Stuttgart
Pfaffenwaldring 9, Stuttgart, Germany
Email: haist@ito.uni-stuttgart.de

Abstract—Liquid crystal on silicon (LCoS)-based spatial light modulators (SLMs) are versatile adaptive optical elements. In microscopy, among their applications are aberration sensing and correction in wide-field microscopy and also the implementation of holographic optical tweezers. For aberration correction, the required scene-based wavefront sensing can be implemented as a modified correlation-based Shack-Hartmann approach where a high-resolution SLM first senses and then corrects the aberrations. For the implementation of holographic optical tweezers, the SLM serves as a variable optical beam-splitter which is addressed with holograms computed by fast algorithms implemented on the graphics processing unit (GPU) of a common PC almost in real-time.

I. INTRODUCTION

Wavefront sensors are commonly used to determine wavefront aberrations in optical systems and adaptive optical elements are then used to correct them. One option for such correction is a spatial light modulator based on a liquid crystal on silicon microdisplay. This component has the advantage that it can not only be used for correcting, but also for measuring the aberration, which considerably simplifies the setup by sparing the wavefront sensor.

The same component can be used for the implementation of holographic optical tweezers (HOTs), given a careful calibration the device and its capability of producing the full required phase modulation range of 2π. Moreover, a fast implementation of the hologram computation is needed to permit convenient user interaction. The tremendous computational power of recent GPUs makes the programming work for the implementation worthwhile, and is more and more eased by tailored hardware-specific libraries whose development is in continuous progress.

II. SLM CHARACTERIZATION AND CALIBRATION

The spatial light modulator is a device which is capable to create almost arbitrary wavefronts with high spatial resolution, but the device of course requires careful calibration. The phase retardation of the LC pixels can be measured in a rather simple fringe-shift method. For adressing wavefronts with high spatial frequencies it is essential to also investigate the diffraction at addressed gratings and for wavefronts with lower spatial frequencies the aberration introduced by the device itself needs to be measured and compensated for[1].

III. ABERRATION SENSING AND CORRECTION

Fig. 1 shows a video microscope with an additional telescope that images the pupil of the microscope objective onto the LCoS modulator which is used for aberration correction. Instead of employing an additional wavefront sensor making the overall system more complex and expensive, we use the SLM also for sensing the wavefront. Due to the position of the camera we replace the lenslets of the conventional correlating Shack-Hartmann-Sensor[2] (SHS) by local gratings with different grating constants and orientations written into the SLM.

Together with the tube lens of the microscope this leads to multiple (aberrated and bandpass-filtered) copies of the object on the CCD camera of the microscope. Every copy corresponds to a defined pupil zone. The shift of the copy is directly proportional to the local grating period plus the local wavefront tilt. Determination of the shifts by correlation therefore enables us to compute the wavefront given as a Zernike polynomial and finally correct it, also by means of the SLM.

The system simultaneously measures eight pupil zones per frame. That means that we will have eight copies of the image of the object on the CCD. The home position of these copies is given — as already explained — by the local grating constants that we use in the pupil. We chose gratings that lead to a simple tiling of the image on the CCD (Fig. 2). Therefore we measure eight x/y gradients per CCD frame. For sampling the pupil with e.g. 80 zones we use a sequence of 10 grating patterns that we write into the SLM.

The reconstructed wavefront is then inverted, a phase wedge is added in order to implement a carrier frequency and the result is wrapped to 2π. The resulting correction hologram is addressed at SLM. The addition of the carrier frequency leads to a spatially shifted image which makes sure that unwanted diffraction orders due to a non-perfect phase modulation of the LCD won't lead to problems.

For typical objects a Köhler-type illumination is mandatory because otherwise the pupil is not illuminated homogeneously and the limited dynamic range of the CCD would lead to problems.

Fig. 3 shows results of the correction of strong defocus. As can be seen the correction is not perfect but a significant

978-1-4244-1917-3/08/$25.00 ©2008 IEEE

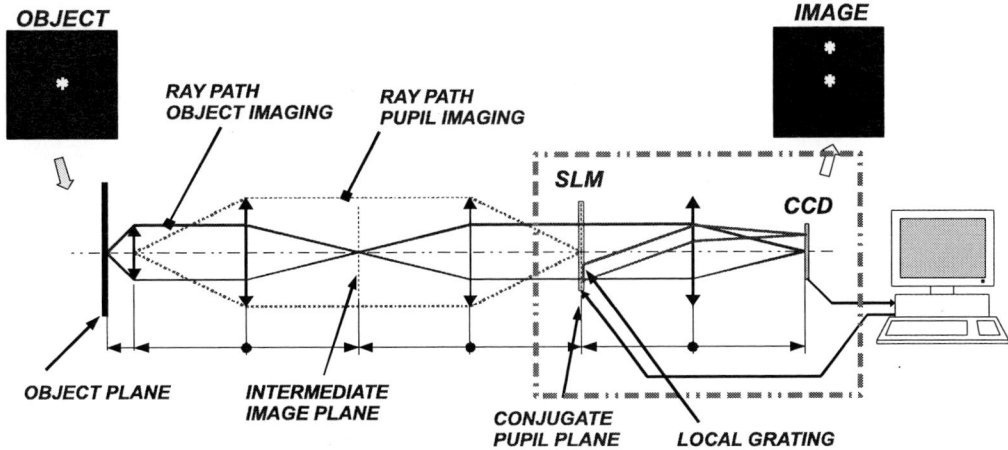

Fig. 1. Principle setup for measuring and correcting aberrations in the wide-field microscope by using an SLM (shown here in transmission). The pupil of the microscope is imaged onto the SLM by an additional telescope (dotted line). For the real implementation we used a reflective LCoS SLM and some additional components. A local grating written into the SLM leads to a shifted and bandpass-filtered copy of the spatially limited object.

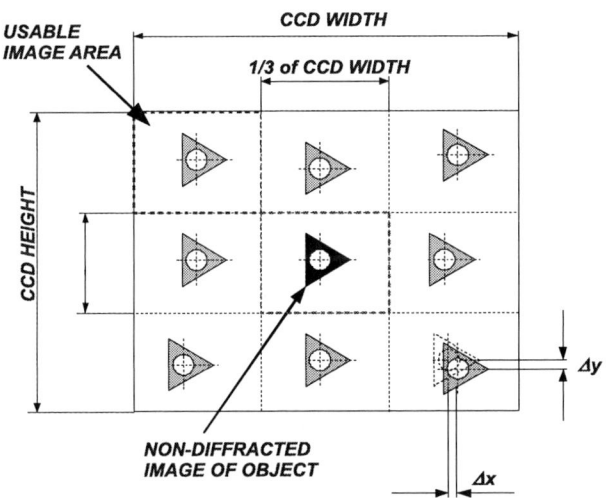

Fig. 2. Arrangement of the copies in the image plane as used within the experiments. Eight pupil zones can be measured simultaneously by limiting the object to 1/9 of the maximum object field of the microscope.

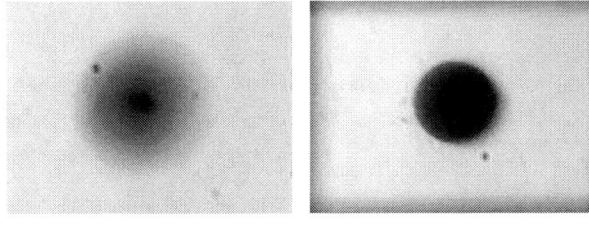

(a) uncorrected 4 μm defocus   (b) 4 μm defocus, corrected

Fig. 3. Experimental results for imaging a hole under strong defocus.

computational speed for the hologram update can now be obtained by implementation of tailored algorithms for HOT generation on the GPU [4], but also the implementation of the IFTA with sufficient speed has become possible. We have achieved hologram update rates of >20 Hz using current NVidia hardware and a tailored implementation of the IFTA.

ACKNOWLEDGMENT

We would like to thank the Bundesministerium für Bildung und Forschung (BMBF) for their financial support under the Project "AZTEK" (FKZ 13N8809).

improvement of the image quality has been achieved. Details of the implementation and more measurement results will be described in a future publication.

IV. HOLOGRAPHIC OPTICAL TWEEZERS

The simultaneous generation of multiple optical traps by an LCoS-based SLM utilizes computational algorithms developed for the design of diffractive optical elements, like e. g. the Iterative Fourier Transform Algorithm (IFTA) [3]. Such computational algorithms are time-consuming and make the desirable real-time update of the holograms addressed on the SLM a challenging task. Today, the computational power of the graphics processing units is by far superior to that of the CPUs of standard PCs. Therefore the desired

REFERENCES

[1] A. Hermerschmidt, S. Osten, S. Krüger and T. Blümel, "Wave front generation using a phase-only modulating liquid-crystal-based micro-display with HDTV resolution," *Proc. SPIE* 6584, p.65840E (2007)

[2] L. A. Poyneer, "Scene-Based Shack-Hartmann Wave-Front Sensing: Analysis and Simulation," *Appl. Opt.* 42, pp. 5807-5815 (2003)

[3] F. Wyrowski and O. Bryngdahl, "Iterative Fourier-transform algorithm applied to computer holography," *J. Opt. Soc. Am. A* 5(7), pp. 1058-1065 (1988)

[4] T. Haist, M. Reicherter, M. Wu, and L. Seifert, "Using graphics boards to compute holograms," *Computing in Science and Engineering* 8(1), pp. 8-13 (2006)

MEMS Deformable Mirrors for Adaptive Optics using Single Crystal PMN-PT

Hyunkyu Park and David A. Horsley
Mechanical and Aeronautical Engineering, University of California
1 Shields Ave., Davis, CA 95616, USA

Abstract — **A MEMS-based deformable mirror constructed using single-crystal PMN-PT for use in ophthalmologic adaptive optics is presented. The fabrication process and the results of characterization of the DM are described. A large stroke and high operating bandwidth assure that the DM can be a promising wavefront corrector.**

I. INTRODUCTION

Deformable mirrors (DMs) have been successfully used in adaptive optics (AO) to correct for optical aberrations in astronomical images by means of actively controlling surface deflection. The advance of AO technology allows expanding its applications to include ophthalmologic instruments and free-space optical communications systems. Since these applications require a more compact and low-cost DM, a variety of different DMs based on micro electromechanical systems (MEMS) technology have been developed [1, 2].

Bimorph DMs are attractive for vision science applications due to their ability to achieve the large mechanical stroke (> 10 μm) required to correct low-order optical aberrations such as defocus and astigmatism. In comparison, electrostatic DMs are limited to ~3 μm stroke [3]. The stroke of a piezoelectric bimorph DM is a strong function of the piezoelectric coefficient (d_{31}). At the same time, the stroke is inversely proportional to the thickness of the DM, suggesting the use of thin-film piezoelectric materials. Unfortunately, the d_{31} of bulk piezoelectric materials is rarely achieved in thin-films. The focus of this paper is on the design and fabrication of a bimorph DM constructed using single-crystal $Pb(Mg_{1/3}Nb_{2/3})O_3$–$PbTiO_3$ (PMN-PT). PMN-PT has a large piezoelectric coefficient ($d_{31} = -1330$ pm/V), more than 3 times larger than bulk $Pb(ZrTi)O_3$ (PZT) [4]. Leveraging the precision lapping technology utilized to create the thick SOI substrates commonly used in MEMS fabrication, a single-crystal PMN-PT layer is bonded to a silicon wafer and lapped to the desired thickness, allowing the best properties of bulk PMN-PT to be achieved in a thin-film MEMS DM.

II. DESIGN

Basically, the DM is composed of two layers; an active PMN-PT layer and a passive single-crystal Si layer. Although a thin membrane is desirable, the precision of the lapping process is approximately ± 1 μm, limiting the minimum PMN-PT thickness to approximately 5 μm. For a given Si layer thickness,

the PMN-PT thickness can be optimized to achieve the maximum stroke. For a 5 μm thick Si layer, a finite element model (FEM) predicted that the optimum PMN-PT thickness is around 5 μm. The preliminary prototype presented here had a 30 μm thick PMN-PT layer; for this thickness, FEM predicts a 20 μm stroke from a single 1.78 mm electrode driven at 10 V.

The electrode pattern is an annular ring type composed of 19 actuators, each having the same surface area as shown in Fig. 1. The DM is intended to be used with a 10 mm aperture; the central disk and the electrodes in the inner ring (channels 1-7) lie within this aperture and will be used to create local curvature, while the electrodes in the outer ring (channels 8-19) are used to provide slope at the edge of the aperture. There are bond pads along the edge of the block for electrical connection to electronics. The Cr/Au and the epoxy layers are used as a common ground.

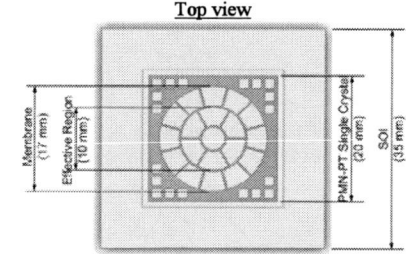

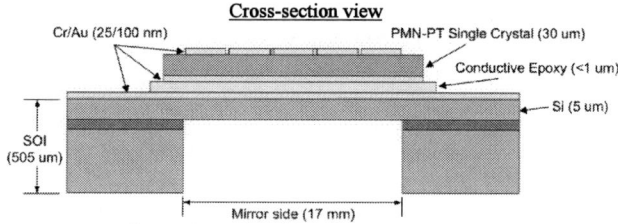

Fig. 1 Schematics of the deformable mirror

III. FABRICATION

Cr/Au (20/100 nm) was deposited on both a 300 μm-thick PMN-PT single crystal block and the device side of an SOI wafer for a common ground and then they were bonded together using a conductive epoxy. The PMN-PT was mechanically lapped to a thickness of 30 μm and then the control Cr/Au (20/100 nm) electrodes were deposited and patterned by lift-off process.

The next step is to remove the handle layer of the SOI wafer by deep reactive ion etch (DRIE) to create the released membrane. Since the buried oxide (BOX) layer has

978-1-4244-1917-3/08/$25.00 ©2008 IEEE

considerable compressive stress, it makes the mirror deformed or even cracked after DRIE. To prevent cracking, we patterned 100 μm width concentric rings on the handle layer to support the stress [5]. The DRIE stops on the BOX layer which was removed with 49 % hydrofluoric acid (HF). To protect the PMN-PT from HF, the actuator side was coated by HF resistant material, ProTEK A2-22 (Brewer Science). When the BOX is removed, the silicon support rings are released from the mirror surface and discarded. The fabricated DM is shown in Fig. 2.

Actuator side   Mirror side

 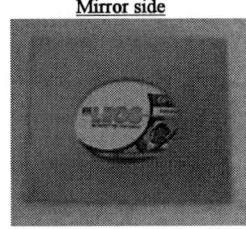

Fig. 2 Photographs of the deformable mirror

IV. CHARACTERIZATION

The DM was characterized using both a commercial laser Doppler vibrometer (LDV) and a custom phase-shifting interferometer. The interferometer is able to measure a surface profile with an RMS accuracy of approximately 6 nm and an absolute accuracy of ± 60 nm. Since PMN-PT is an electrostrictive material which shows unidirectional behavior regardless of the sign of the applied electric field, surface deflection of the DM was measured over a 20 V voltage range with an offset of half of peak-to-peak voltage so that the driving voltage is always in positive region. The displacement was measured by applying voltage to a single electrode at each of 3 points; central disk, inner & outer ring. As shown in Fig. 3, each electrode achieves over 10 um stroke which is comparable to the stroke of a MEMS DM fabricated from sputtered PZT [6]. The stroke of the center electrode at 10 Vpp is smaller than predicted by FEM due to effects of the Cr/Au and the epoxy layers.

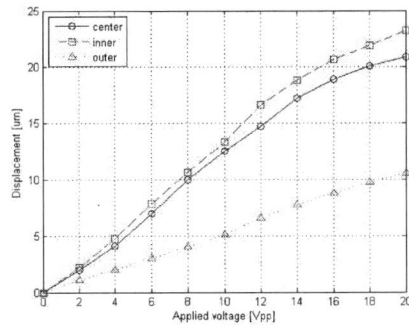

Fig. 3 Voltage-to-displacement measurement using LDV

The frequency response was measured with the LDV and a dynamic signal analyzer (SRS), as shown in Fig. 4. The measured natural frequencies of 874 Hz and 3.07 kHz correspond to the 01 and 02 modes, respectively, and compare favorably to the frequencies of 802 Hz and 3.09 kHz predicted by FEM. This result confirms that the DM can achieve the 100 Hz bandwidth required for ocular AO.

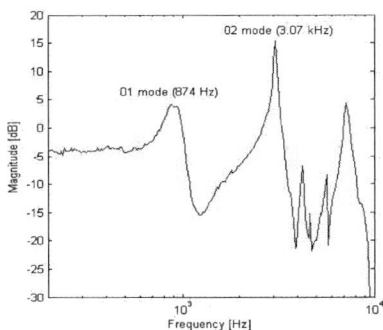

Fig. 4 Frequency response measurement using LDV

The measured interferogram and surface shape of the DM, shown in Fig. 5, exhibit initial deformation with around 17 μm peak-to-valley amplitude over the 10 mm effective region. This deformation is mainly caused by residual stress in the Cr/Au electrode metallization and the epoxy layer. Since the initial deformation is within the DM stroke, it can be compensated by actuating each electrode, resulting in a flat mirror.

Interferogram   Initial deformation

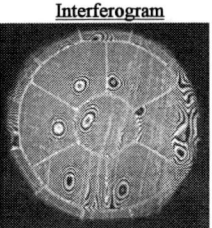 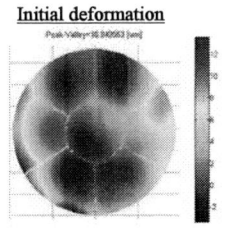

Fig. 5 Surface measurement using phase-shift interferometer

V. CONCLUSION AND DISCUSSION

A MEMS-based DM using single crystal PMN-PT was fabricated and characterized. A large stroke and high operating bandwidth prove that the DM can be a promising candidate of wavefront correctors for ocular AO. We are planning to flatten the mirror surface by controlling the actuators and to evaluate its ability through replicating optical aberrations described by the Zernike polynomials.

REFERENCES

[1] N. Doble and D. Williams, "The Application of MEMS technology for adaptive optics in vision science," Journal of Selected Topics in Quantum Electronics, vol. 10, pp. 629-635, 2004.

[2] E. Dalimier and C. Dainty, "Comparative analysis of deformable mirrors for ocular adaptive optics," Optics Express, vol. 13, pp. 4275-4285, 2005.

[3] T. Bifano, J. Perrault, R. Mali, and M. Hernstein, "Microelectromechanical Deformable Mirrors," IEEE Journal of Selected Topics in Quantum Electronics, vol. 5, pp. 83-89, 1999.

[4] R. Zhang, B. Jiang and W. Cao, "Elastic, piezoelectric, and dielectric properties of multidomain $0.67Pb(Mg_{1/3}Nb_{2/3})O_3$–$0.33PbTiO_3$ single crystals," Journal of Applied Physics, vol. 90, no. 7, pp. 3471-3475, 2001.

[5] M. Sasaki, T. Sasaki, K, Hane and H. Miura, "An optically flat micromirror using a stretched membrane with crystallization-induced stress," Journal of Optics A, vol. 10, no. 4, 2008.

[6] I. Kanno, T. Kunisawa, T. Suzuki and H. Kotera, "Development of deformable mirror composed of piezoelectric thin films for adaptive optics," Journal of Selected Topics in Quantum Electronics, vol. 13, no. 2, pp. 155-161, 2007.

A Varifocal Micromirror with Pure Parabolic Surface using Bending Moment Drive

Ryohei Hokari, Kazuhiro Hane

Department of Nanomechanics, Tohoku University, Sendai, 980-8579, Japan

Tel: +81-22-795-6965, Fax: +81-22-795-6963, E-mail: hokari@hane.mech.tohoku.ac.jp

ABSTRACT

We propose a method to generate a pure parabolic surface of varifocal micromirror by applying a bending moment to the circumference of the micromirror. Pure paraboloid can focus light without aberration as one of the ideal mirror surfaces. In the conventional method, varifocal mirror generates an approximate spherical surface or a paraboloid-like surface by applying a distributed load to the central part of the mirror with parallel plate electrodes. In this study, the micromirror is deformed only by applying a bending moment to the circumference of the mirror in order to generate an ideal paraboloid. The proposed mirror was fabricated from SOI wafer. The deviation of the varifocal mirror from the ideal paraboloid was measured to be smaller than 5nm for the 400μm diameter mirror with the focal lengths from the infinity to 24mm at the voltages from 0 to 215V.

Keywords: micromirror, varifocus, paraboloid, bending moment

INTRODUCTION

Deformable mirror is a key component in optical MEMS. Phase compensation by a deformable mirror has been studied for astrophysical telescopes and precision interferometers [1]. Moreover varifocal mirror also attracts increasing attention to image processing and laser focusing. Since a paraboloid is an ideal shape of mirror for focusing and imaging, several researches have been carried out to generate the mirror deformation close to a paraboloid by adjusting the distribution of electrostatic forces between mirror plate and the bottom electrode [2]. In another technique, the support condition has been modified at the circumference of mirror [3]. However, the deformed profile is an approximate parabola, not purely parabolic. The profile includes the higher order deflection, and thus generates an aberration of mirror.

In this study, we propose a bending moment drive to generate a pure paraboloid surface for varifocal micromirror. The proposed principle is confirmed experimentally. The design and fabrication of the proposed mirror is presented.

DRIVE PRINCIPLE AND DESIGN

The principle of the proposed method is shown in Fig.1. For simplicity, first, we consider the bend of a simple beam as shown in Fig.1 (a). The beam is assumed to be supported at the both ends with the fulcrums. In the case of applying a distributed load to the beam between the fulcrums, which is a conventional method for varifocal mirror, pure parabolic deflection cannot be obtained. It is usually complex to determine the distribution of the load for obtaining an extremely precise parabola. On the other hand, if only bending moments are applied to the beam ends instead of the distributed load, we have found that the deflection curve of the beam is expressed by a pure parabola in the following equation.

$$z = \frac{6(1-v)M_0}{Eh^3}(r^2 - a^2) \qquad (1)$$

where, z is the displacement vertical to the beam, E is the Young's modulus, h is the mirror thickness, v is

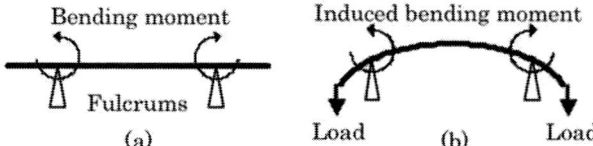

Fig.1. Principle of bending moment drive. (a) Bending moment applied to the beam supported by fulcrums (b) Beam deflection by the loads applied to the outer part of the beam

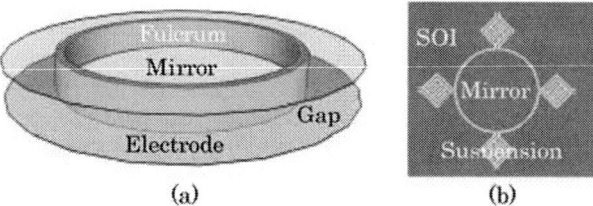

Fig.2. (a) Structure of varifocal mirror (b) mirror pattern

the Poisson ratio, M_0 is the applied bending moment, r is the distance from the center of beam, a is the distance between the fulcrum and the center of beam. Therefore under the condition, a pure parabolic deflection can be obtained theoretically. However, it is practically difficult to apply only a constant bending moment to the beam. From the theoretical analysis, we have found that only bending moment is acting on the beam inside the fulcrums when loads are applied to the beam outside the fulcrums. In other words the beam becomes a parabola theoretically in the area inside the fulcrums. As shown in Fig.1 (b) an electrostatic force is applied to the beam outside the fulcrums to generate the bending moment inside the fulcrums. The one-dimensional analysis of the beam can be expanded into the two-dimensional circular disk.

Figure 2 (a) shows the simplified schematic diagram of the proposed varifocal mirror. A Si circular mirror is placed on a ring fulcrum. The mirror is supported in contact with the fulcrums, and the mirror is rotationally free at the fulcrums. An electrostatic force is applied to the mirror outside the fulcrum.

SOI wafer (10μm/2μm/200μm) is used for the

978-1-4244-1917-3/08/$25.00 ©2008 IEEE

fabrication. Figure 2 (b) shows the schematic diagram of the top view of the designed micromirror which is fabricated from the top layer of SOI wafer. Si circular mirror is suspended at the circumference by the soft spring (suspension) to drive mirrors freely on the fulcrum.

FABRICATION

Process steps for fabricating Si varifocal mirror are shown in Fig.3 (1)-(3). The Si substrate behind the mirror is removed by Deep RIE (1). The top Si layer of SOI wafer is etched to form the mirror (2). The sacrificial layer is etched by vapor HF (3). Process steps for fabricating of glass substrate are also shown in Fig.3 (4)-(8). The fulcrum is formed on the glass substrate (Pyrex glass). A Cr mask is used for the glass etching with HF solution (6). The depth of the glass etching is equal to the gap for the electrostatic mirror drive. An Au film is used as the electrode after the deposition of a Cr film for improving adhesion (7). The metal electrode is patterned on the glass substrate. The glass substrate is bonded to the Si substrate (SOI wafer) by the anodic bonding after sputtering an Au film on the glass fulcrum so as not to bond the Si mirrors to the glass substrate (8). The anodic bonding is carried out at 400 degrees and 100V. The low bonding voltage prevents the sticking of the mirror.

MEASUREMENT RESULTS AND DISCUSSION

In order to measure the deflection of the mirror, the fabricated device was measured with a white-light interferometer (zygo). The optical micrograph of the fabricated device and the measured shape of a circular mirror (diameter of 1000μm, fulcrum diameter of 400μm) are shown in Figs.4 (a)-(c) at the applied voltage of 215V. As shown in Figs.4 (b), (c), a good parabolic deflection is obtained in the inner region of the fulcrum. The displacement of 2.58μm was obtained at the mirror center at 215V. At a low voltage, the mirror is not attracted enough to contact with the fulcrums.

The mirror surface cross-sections measured at the respective applied voltages are shown in Fig.5. Increasing the voltage, deflection increases and the focal length of the parabolic mirror decreases. The root-means-square (RMS) deviation from the parabola of the cross-section and focal length are shown in Table 1. The deviation of the mirror surface profile from the pure parabola is smaller than 5nm in the region inside the fulcrum (400μm diameter) at the voltages from 150V to 215V. Therefore, as predicted from the mechanical strength theory, the mirror surface is very close to a paraboloid. In the outer electrode region of the micromirror, the deviation is about 50nm at 200V. The focal lengths are calculated from the parabolic curves. The focal length can be variable in the range from 24mm to the infinity.

ACKNOWLEGEMENT

The authors thank to Profs. M. Saka and T. Kuriyagawa for the theoretical consideration and the interferometric measurement, respectively.

REFERENCES

[1] G. Vdovin et al., "Technology and applications of micromachined adaptive mirrors", J. Micromech.

Microeng., 1999, R 8-20.

[2] W. Greger et al., "A new approach for focusing deformable mirrors fabricated in polymer technology", Optical MEMS, 2005, pp. 45-46.

[3] Y. Shao et al., "3-D MOEMS mirror for laser beam pointing and focus control", IEEE Journal of selected topics in quantum electronics, Vol. 10, No. 3, 2004, pp.528-535.

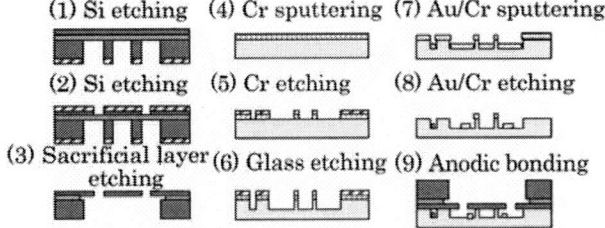

(1) Si etching (4) Cr sputtering (7) Au/Cr sputtering

(2) Si etching (5) Cr etching (8) Au/Cr etching

(3) Sacrificial layer etching (6) Glass etching (9) Anodic bonding

■ Si ■ SiO₂ ▨ Resist ▢ Glass ▢ Cr ▢ Au

Fig.3. Process steps for fabricating Si varifocal mirror.

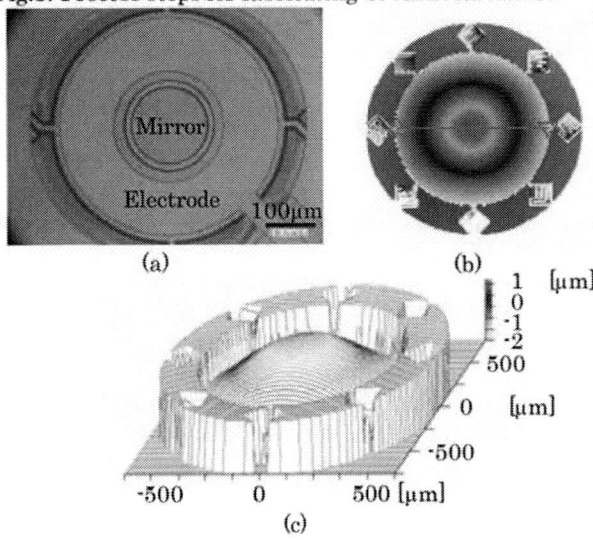

(a) (b)

(c)

Fig.4. (a) Fabricated device (b), (c) Measured mirror surface shapes at 215V

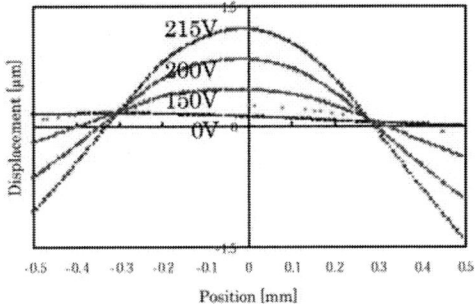

Fig.5. Mirror surface profile along the diameter measured with the applied voltage as a parameter.

Table 1. Deviation from paraboloid and focal length.

Voltage [V]	Deviation (RMS) [nm]	Focal length [mm]
0	—	∞
150	2.73	77
200	3.44	36
215	4.74	24

Simulation and characterization of tunable achromatic micro-lenses

Philipp Waibel, Daniel Mader, David Lämmle, Andreas Seifert, and Hans Zappe

Laboratory for Micro-optics, Department of Microsystems Engineering – IMTEK, University of Freiburg
Georges-Köhler-Allee 102, 79110 Freiburg, Germany
philipp.waibel@imtek.uni-freiburg.de

Abstract—Simulation and measurements of achromatic variable-focus multi-chamber micro-lenses are presented. By using three liquid-filled chambers with well-defined dispersion behavior, separated by highly elastic silicone membranes, this achromatic optical system can be tuned in focal length by adjusting the pressures inside the chambers. Tunable achromatic micro-lenses are thus feasible.

Index Terms—Lens design, achromatic systems, liquid lenses, correction of chromatic aberration.

I. INTRODUCTION

DUE to their wide tuning range, compact size and robustness, liquid micro-lenses may be used in a wide field of applications. Most approaches involve either the electrowetting principle [1] or pneumatically tunable micro-fluidic cavities [2]. The focal length of such membrane-lenses can be tuned by applying a pressure to the fluid-filled chambers causing a change of the curvature of the refractive interfaces.

Single-cavity micro-lenses consist of a liquid-filled chamber confined by a flexible, usually polymer, membrane. As with single macroscopic lenses, these micro-lenses show chromatic aberrations which vary with focus. However, by adding a second chamber filled with another fluid with different dispersive properties, chromatic aberrations can be eliminated by appropriately adjusting the pressure in both chambers [3].

II. DESIGN AND LAYOUT

The outer surfaces of the three-chamber micro-lens system (schematically shown in figure 1) are defined by simple glass plates. The thin membranes, defining the inner interfaces of the lens system, consist of highly flexible silicone. To achieve achromatism, at least two single lenses with different dispersive properties are needed [4]. To achieve this, one chamber is filled with flint-like (high refractive index $n \approx 1.7$ and Abbe-number $V < 50$), one with a crown-like liquid (low refractive index $n \approx 1.5$ and Abbe-number $V > 50$), and one is flooded with air, i.e. the optical system is defined by three different refractive indices n_a, n_b and n_c inside the chambers. The pressure inside the fluid-chambers can be varied by a pressure controller to adjust the curvatures c_1 and c_2 of the membranes.

III. SIMULATIONS

For optical simulation, the distended membranes are approximated as spherical surfaces; thus the aperture of the system

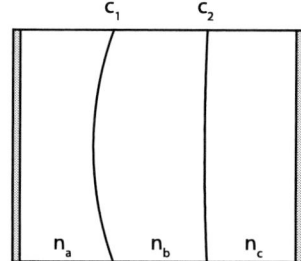

Fig. 1: Three-chamber system; c_1 and c_2 are the curvatures of the membranes, n_a, n_b and n_c the refractive indices inside the chambers. Collimated light coming from the left is focused by the lens system. By varying the pressure on the liquid in the cavities, the resulting focal length of the system can be adjusted.

has to be reduced to avoid aberrations induced by the non-spherical boundary.

The refractive power φ of a cemented system of l thin lenses is simply the sum of the powers of all single elements φ_k:

$$\varphi = \sum_{k=1}^{l} \varphi_k. \tag{1}$$

The condition to achieve achromatism is

$$\sum_{k=1}^{l} \frac{\varphi_k}{V_k} = \sum_{k=1}^{l} (c_k \Delta n_k) = 0 \tag{2}$$

where V_k is the Abbe number, c_k the curvature of lens k and Δn_k the mean dispersion of the material in chamber k [4].

Under this condition the lens system will be free of chromatic aberrations. But other, especially spherical, aberrations remain.

Figure 2 shows the calculated absolute value of the primary longitudinal spherical aberration $|LA'|$ for different chamber configurations (crown-type: $n = 1.48$, $V = 57$; flint-type: $n = 1.52$, $V = 44$), as a function of the focal length of the optical system. The distance between the two membranes is $d = 0.525\,\mathrm{mm}$. Equation 2 is satisfied for all focus values, i.e. for a given value c_1 of the curvature of the first lens, c_2 is suitably adapted. As can be seen, the lowest spherical aberration is achieved when the first chamber is filled with flint-like liquid, the second with crown-like liquid and the third chamber with air; this sequence differs from that of the classic Fraunhofer achromatic doublet, where always crown glass is used in front of flint glass [5].

978-1-4244-1917-3/08/$25.00 ©2008 IEEE

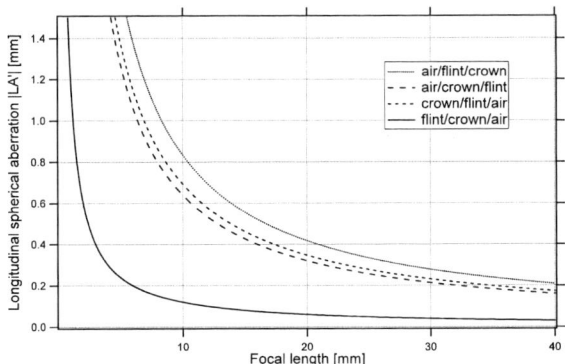

Fig. 2: Simulation results for spherical aberration as a function of focal length of the lens system for different chamber configurations. The lowest spherical aberration is achieved when the first chamber is filled with flint-like, the second with crown-like fluid an the third with air.

IV. MEASUREMENTS

To characterize the lens system, two different measurement setups are used. For qualitative prediction of the chromatic aberration, images of test patterns are taken with the lens systems under Köhler illumination. The pressure in the two chambers is changed in such a way that the chromatic aberration are minimized as the focal length of the lens system is adjusted. The test pattern in figure 3 is sharply imaged, but the color tails indicate that chromatic aberrations remain.

Fig. 3: Image of a test pattern taken with the lens system with small colour tails, which indicate remaining chromatic aberrations.

The measurement of spherical aberration is done by a transmission measurement with a phase-shifting Mach-Zehnder interferometer, schematically shown in figure 4. Figure 5 shows the wavefront error of the micro-lens system measured at a wavelength of $\lambda = 633$ nm. To interpret this data, the Zernike coefficients can be calculated. The pressure inside the chambers is adjusted until the defocus term of the Zernike matrix is minimized. The total amount of spherical aberration can then be extracted from the Zernike matrix.

To determine chromatic aberration quantitatively, the interferometer setup is expanded to employ two wavelengths. The wavefront error of the micro-lens system is measured at $\lambda = 633$ nm and $\lambda = 441$ nm. From the Zernike matrices, the chromatic aberration can be extracted by simply subtracting the two defocus terms. By adding a control unit to the lens system, it should be possible to tune the lens system over the whole focal range, from a few millimeters to infinity, without significant chromatic and spherical aberrations.

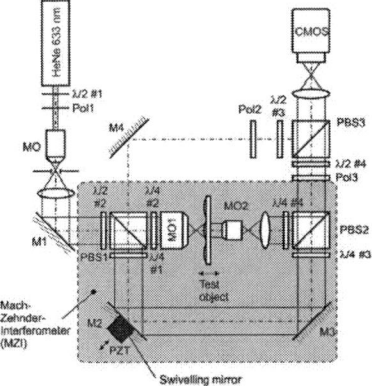

Fig. 4: Schematic view of the phase-shifting Mach-Zehnder-interferometer for measuring the spherical aberration of the micro-lens system.

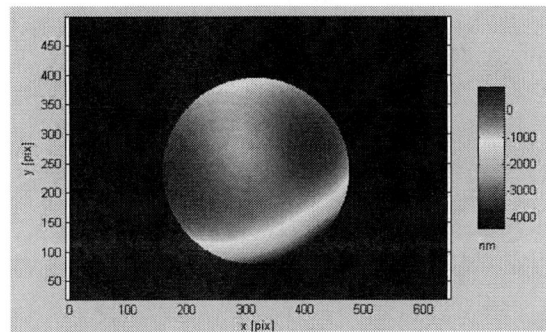

Fig. 5: Wavefront error of a three-chamber micro-lens system measured in transmission with the Mach-Zehnder-interferometer at $\lambda = 633$ nm.

V. SUMMARY AND OUTLOOK

The potential utility of achromatic variable-focus multi-chamber liquid-filled micro-lenses has been shown. The focus can be tuned from several millimeters to infinity without chromatic aberration, but with a significant amount of spherical aberration. These spherical aberrations can be minimized by filling the first chamber with air, the second with a flint-like and the third with crown-like liquid. By simply adding one more chamber to the system, filled with a third liquid with different dispersion properties, the degrees of freedom are further increased and spherical aberrations can also be corrected [3].

REFERENCES

[1] S. Kuiper and B. H. W. Hendriks, "Variable-focus liquid lens for miniature cameras," *Applied Physics Letters*, vol. 85, no. 7, pp. 1128–1130, 2004.
[2] D.-Y. Zhang, V. Lien, Y. Berdichevsky, J. Choi, and Y.-H. Lo, "Fluidic adaptive lens with high focal length tunability," *Applied Physics Letters*, vol. 82, no. 19, pp. 3171–3172, May 2003.
[3] S. Reichelt and H. Zappe, "Design of spherically corrected, achromatic variable-focus liquid lenses," *Optics Express*, vol. 15, no. 21, pp. 14 146–14 154, 2007.
[4] E. Hecht, *Optics*, 4th ed. Addison Wesley, 2001.
[5] R. Kingslake, *Lens Design Fundamentals*. Academic Press, 1978.

Optical scanner with deformable mirror fabricated from SOI wafer

Takashi Sasaki and Kazuhiro Hane

Department of Nanomechanics, Tohoku University, Aramaki 6-6-01 Aoba-ku, Sendai, 980-8579, Japan

Tel: +81-22-795-6965, Fax:81-22-795-6963, E-mail: t_sasaki@hane.mech.tohoku.ac.jp

Abstract

An optical scanner with a deformable mirror is fabricated using SOI wafer. A 1μm thick top silicon layer of SOI wafer is used to fabricate a deformable mirror. Movable comb and fixed comb are fabricated from the silicon substrate. The mirror is rotated by the comb actuator and it is also deformed by an electrostatic force independently. The rotation angle of the mirror is 12 degrees at 80V. The deformation at the mirror center is 3nm at 100V.

Key words: scanner, wave front control, SOI wafer

Introduction

Scannig micromirror is an attractive device, and has many applications such as laser display [1] and optical switch. On the other hand, deformable mirror is a key component in adaptive optics [2]. The combination of the scanning mirror and the deformable mirror is attractive as a functional integrated optical system. Shao et al. fabricated an integrated scanning mirror with focus control [3]. A silicon nitride film was used for the torsion bar and the deformable mirror, which were deposited by LPCVD. A sacrificial layer of phosphosilicate glass was deposited to form a gap for the deformable mirror. The electric wiring was patterned using a gold film.

In this paper, optical scanner with a deformable mirror is fabricated from SOI wafer. A single crystal silicon (Si) layer is used for the torsion bar and the mirror. Therefore, the surface roughness is small. The electronic connections are carried out by separating the Si layers.

Principal and Design

Figure 1 shows the structure of the proposed scanner with a deformable mirror which is divided into three parts. The three parts are connected to the respective electrodes. The electrode 1 is connected to the deformable mirror through a torsion bar which is fabricated from the top Si layer. The electrode 2 on the top Si layer is electrically connected to the moving comb and the counter electrode for the deformable mirror through the contact holes. The contact holes are filled with a metal film. The moving comb and the counter electrode are fabricated from a part of the Si substrate. The electrode 3 is connected to the fixed comb of the Si substrate.

The mirror surface deforms if a voltage is applied between the electrodes 1 and 2. Figure 2 shows the simplified cross sectional schematic diagram across the mirror. The mirror surface can be deflected to change the wave front of reflected light. The mirror frame is rotated by the application of voltage between electrodes 2 and 3. The device design parameters are shown in Table 1.

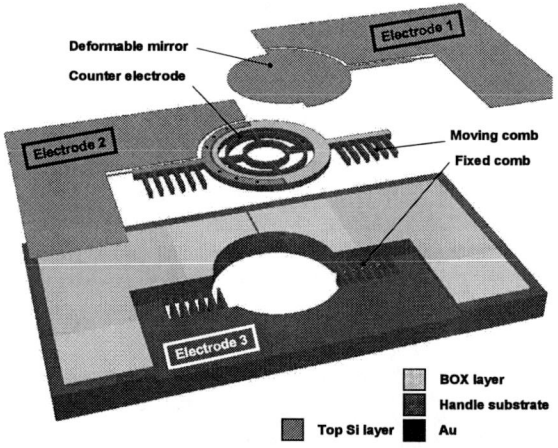

Fig. 1. Structure of scanner with deformable mirror.

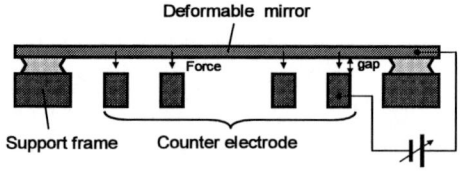

Fig. 2. Cross sectional schematic diagram of the deformable mirror.

Table1. Design parameters

Diameter of the support frame	500μm
Diameter of the deformable mirror	380μm
Thickness of the support flame and moving comb	30μm
Thickness of the mirror and the torsion bar	0.7μm
Length of the torsion bar	100μm
Width of the torsion bar	5μm
Gap between the mirror and the counter electrode	1μm

Fabrication

A SOI wafer with 1µm thick top Si layer, 1µm thick BOX layer and 200µm Si substrate is used. Fabrication process is shown in figure 3. The cross sectional schematic diagrams are drawn in three parts, mirror, torsion bar, moving and fixed comb. (1) Top Si layer is patterned for electric isolation between the electrodes 1 and 2 using Mask 1. The movable comb, fixed comb, mirror and torsion bar are patterned by Mask 2. Next, the wafer is thermally oxidized. The Si dioxide layer (SiO₂) is used to protect the torsion bar from the backside etching. (2) The SiO₂ layer is patterned using Mask 3 for forming the fixed comb. (3) The resist is patterned on the backside to remove Si substrate in the regions of torsion bar and deformable mirror using Mask 4. The Si substrate is etched using ICP RIE. (4) The top Si layer is patterned for the contact holes using Mask 5. (5) The Si substrate is again etched using ICP RIE. The deformable mirror and the torsion bar are formed. Next, the SiO₂ film, the top Si layer and the BOX layer are etched to generate the contact holes. (6) Cr and Au films are coated by the spattering to connect the top Si layer with the Si substrate. (7) The electric wiring is patterned from the Cr and Au films using a lift off process. (8) The SiO₂ layers are removed using HF.

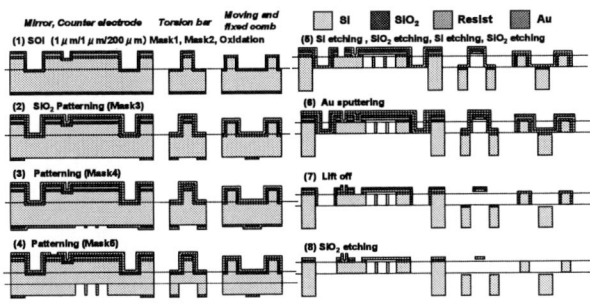

Fig. 3. Schematic diagram of the fabrication sequence

Results and Discussion

Figure 4 shows the fabricated device. The rotation angle of the mirror was measured from the reflection of the laser beam with the increase of the drive voltage. The mirror rotated by 12 degrees at 80V.

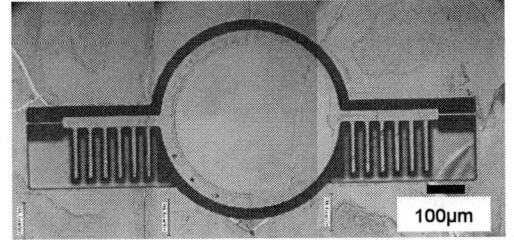

Fig. 4. Optical micrograph of the fabricated scanner with a deformable mirror.

Figure 5 shows the static rotation angle measured as a function of the applied voltage in a different device. The rotation angle increases monotonously with the increase of the voltage in measurement region.

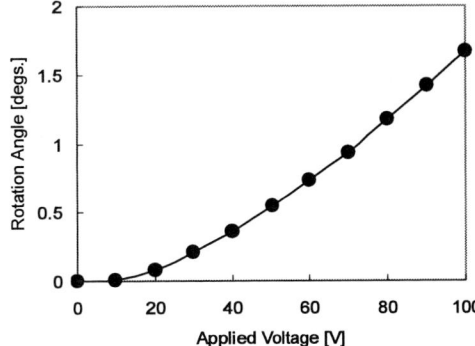

Fig. 5. Experimentally obtained static rotation angle as a function of applied voltage.

The surface profile of the mirror was measured using an optical surface profiler (Zygo NewVeiw). Figure 6 shows the cross sectional surface profiles of the mirror before and after the application of voltage. In the central region of the mirror, a 3nm displacement is obtained with the change of curvature at the voltage of 100V.

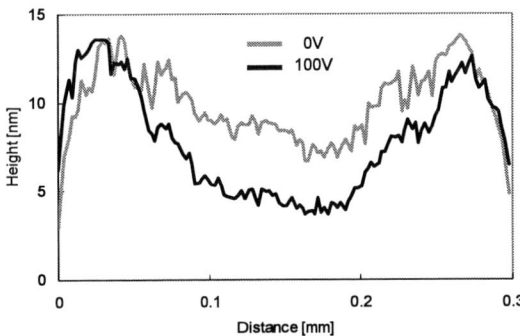

Fig. 6. Surface profiles of the deformable mirror at the voltages of 0V and 100V.

References

[1] P. M. Hagelin and O. Solgaard, "Optical Raster-Scanning Displays Based on Surface-Micromachined Polysilicon Mirrors," *IEEE J. Sel. Top. Quantum Electron.* 5 (1999) 67-74.

[2] I. Kanno, T. Kunisawa, T. Suzuki, and H. Kotera, "Development of Deformable Mirror Composed of Piezoelectric Thin Films for Adaptive Optics," *IEEE J. Sel. Top. Quantum Electron.* 13 (2007) 155-161.

[3] Y. Shao, D. L. Dickensheets and P. Himmer, 3-D MOEMS Mirror for Laser Beam Pointing and Focus Control, *IEEE J. Sel. Top. Quantum Electron.* 10 (2004) 528-535.

A Micromachined Vibratory Sub-Wavelength Diffraction Grating Laser Scanner

[1]Yu Du, [1]Guangya Zhou*, [1]Kelvin K.L. Cheo, [2]Qingxin Zhang, [2]Hanhua Feng and [1]Fook Siong Chau

[1]Dept. of Mechanical Engineering, National University of Singapore [2]Institute of Microelectronics, Singapore

*Corresponding author: Guangya Zhou, Tel: 65-65161235; Fax: 65-67791459; E-mail: mpezgy@nus.edu.sg

Abstract

A novel MEMS based in-plane vibratory sub-wavelength diffraction grating scanner is reported. Diffraction efficiency of more than 75%, optical scan angle of 13.7° and scanning frequency of 20.35 kHz are experimentally achieved

Keywords: Diffraction grating, Micro-scanners, Micro-opto-electromechanical systems (MOEMS)

1. INTRODUCTION

MEMS technology based vibratory diffraction grating scanners [1-2] have the potential to scan at very high frequency without optical performance degradation due to dynamic aberration. This is because the *rotational in-plane motion* rather than *rotational out-of-plane motion* (as in a micro scanning mirror [3-4]) is adopted to cause the laser beam to scan. However, the optical efficiency of grating scanner is lower than micro-mirror scanners, because the incident light is diffracted to multiple orders and the zeroth-order diffraction beam can not be used for scanning. Although a grating can theoretically achieve 100% diffraction efficiency for a selected diffraction order at a given wavelength, the required continuous surface profile is difficult to achieve using silicon-based micro-fabrication process. This paper demonstrates that high optical efficiency is achievable by using a properly designed sub-wavelength binary diffraction grating with a proper polarization.

2. GRATING PROFILE OPTIMIZATION OF HIGH DIFFRACTION EFFICIENCY

In this work, we adopt a binary grating profile with a fixed duty cycle of 50%. The grating period is set at 500 nm and for He-Ne laser beam at wavelength 632.8 nm, the incident angle for bow-free scanning [2] is estimated to be around 28.4°. According to the well-know grating equation, only two propagating diffraction order exist, the zeroth- and the first-order. Since sub-wavelength grating is adopted, the diffraction efficiency can not be predicted using the scalar diffraction theory. Numerical simulations using rigorous coupled-wave analysis (RCWA) are then used to investigate the relationship between diffraction efficiency and the details of the binary grating profile. Simulation results show that the grating efficiency is polarization dependent and is a function of grating groove depth. Figure1 shows the simulated and measured grating efficiency (the ratio of optical power of the first diffraction order to the total incident power) as a function of grating groove (a) for TM-polarization and (b) for TE-polarization.

It is observed that a diffraction efficiency of more than 75% was experimentally achieved for TM polarization with shallow grating grooves depth of 125 nm.

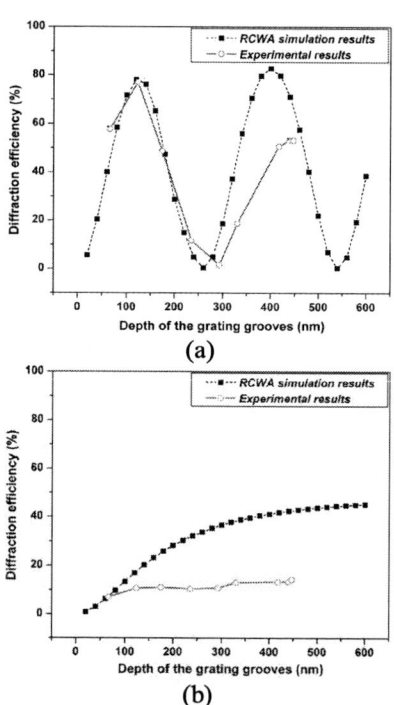

Figure 1: Diffraction efficiency as function of grating depth for (a) TM- (b) TE-polarized incident light.

3. MEMS VIBRATORY GRATING SCANNER DESIGN AND FABRICATION

The prototype grating scanner was fabricated using Silicon-on-insulator (SOI) micromachining technology. Figure 2 and 3 show the schematic illustration and SEM image of the proposed grating scanner. Grating platform with sub-wavelength diffraction grating is supported by two flexure beams and each of them consists of a platform suspension beam and several stress alleviating beams. The platform is driven into a pure in-plane rotational vibration about its geometric center by two

978-1-4244-1917-3/08/$25.00 ©2008 IEEE

sets of electrostatic comb drive resonators vibrating at the same frequency and amplitude but 180 degrees out-of-phase relative to each other.

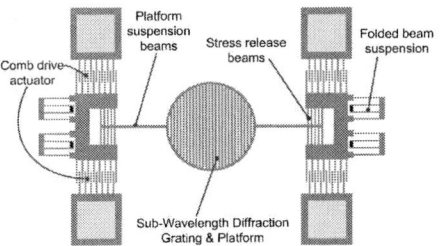

Figure 2: Schematic view of the MEMS based in-plane vibratory diffraction grating scanner

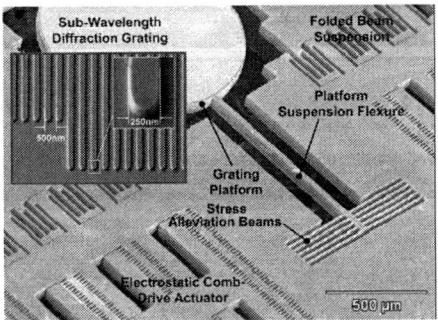

Figure 3: SEM image of the MEMS based in-plane vibratory diffraction grating scanner

4. EXPERIMENTAL RESULTS

The optical performance of the MEMS vibratory grating scanner was tested using a linearly polarized 632.8 nm wavelength He-Ne laser beam at the optimal bow-free incident angle of 28.4 degrees. The experimental results show that our prototype MEMS vibratory grating scanner can achieve high-speed, high-optical-efficiency (more than 75%) laser scanning with the TM-polarized laser beam. Figure 4 shows a photo of scanning trajectory on a projection screen at a distance of 200 mm from the scanner.

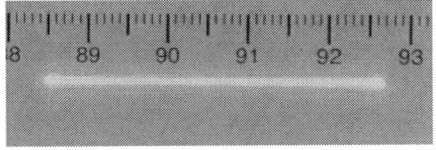

Figure 4: Scanning trajectory on a projection screen at a distance of 200 mm from the scanner

The MEMS grating scanner was operated in air and each comb-drive resonator was driven by a push-pull mechanism [5] with 45 V DC bias voltage and 84 V AC peak-to-peak voltage at a driving frequency of 20.35 kHz, which is close to the resonant frequency. The optical scanning angle is estimated to be around 13.7 degrees. Figure 5 shows the frequency response of the MEMS grating scanner at the regions near the resonant frequencies.

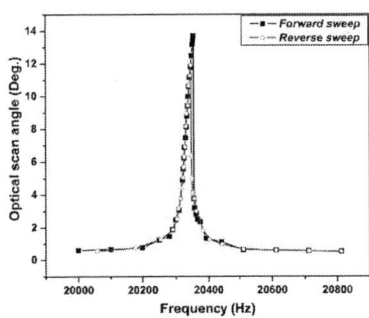

Figure 5: Frequency response of the MEMS grating scanner at the regions near the resonant frequencies.

The scanned-beam quality is tested using stroboscopic method. The experimental setup is shown in Figure 6.

Figure 6: Experimental setup to investigate the scanned beam quality and optical resolution using stroboscopic method

The intensity profiles of the laser focal spots are almost the same as the theoretical diffraction pattern calculated assuming a uniform illumination of the diffraction grating with no dynamic deformation. The full–width–half-maximum (FWHM) diameter of the focal spots is measured to be about 155 μm along the scanning direction so that the overall scanned optical resolution of roughly 310 pixels per unidirectional scan.

REFERENCES

1. G. Zhou, L. Vj, F.S. Chau, and F.E.H Tay, "Micromachined in-plane vibrating diffraction grating laser scanner" *IEEE photonic technology letters*, 16 (2004), pp. 2293-2295.
2. G. Zhou and F. S. Chau, "Micromachined vibratory diffraction grating scanner for multi-wavelength collinear laser scanning", *Journal of Microelectro mechanical Systems*, 15 (2006), pp. 1777-1788.
3. P. M. Hagelin and O. Solgaard, "Optical raster-scanning displays based on surface micro- machined poly-silicon mirrors" *IEEE Journal of Selected Topics in Quantum Electronics*, 5 (1999), pp. 67-74.
4. R. S. Muller and K. Y. Lau, "Surface-micromachined microoptical elements and systems" *Proceedings of the IEEE*, 86 (1998), pp. 1705-1720.
5. W. C. Tang, T. C. H. Nguyen, M. W. Judy and R. T. Howe, "Electrostatic combdrive of lateral polysilicon resonators," *Sensors and Actuators. A* 21/23 328 – 331,1990.

ZnO nanorod-based polymer solar cells with optimized electrodes

Chen-Yu Chou[1], Jing-Shun Huang[1], Chun-Yu Lee[1], and Ching-Fuh Lin[1,2], *Senior Member, IEEE*

[1]Institute of Photonics and Optoelectronics, National Taiwan University; [2]Graduate Institute of Electronics Engineering, and
Department of Electrical Engineering, National Taiwan University
Taipei, 10617 Taiwan, Republic of China
cflin@cc.ee.ntu.edu.tw

Abstract—The selection of electrodes in ZnO nanorod-based polymer solar cells was investigated. Increases in the work function of metal electrode result in the increases in open-circuit voltage up to 120 mV, leading to improved performance.

Keywords- nanostructures, zinc oxide, polymer solar cell, open-circuit voltage

I. INTRODUCTION

Inexpensive solar cells with a low energy budget are interesting due to the need for alternative energy sources in the future. Organic solar cells may become low-cost promising candidates for solar energy conversion because they have low cost, light weight, and flexibility [1]. However, it is still limited by several factors such as low-charge mobility and short excition diffusion length, leading to the "thickness dilemma" of active layer. It could be solved by using a nanostructured oxide such as ZnO nanorods due to their excellent electron mobility and the long diffusion length [2]. Recently, ZnO nanorod arrays can be grown vertically on substrates at low temperature ($<$100 ℃) using a hydrothermal method.[3] Therefore, it is suggested that ZnO nanorods can be used to improve charge carrier collection and transport in the polymer solar cells because the nanostructures offer the direct and ordered path for carriers to electrodes. Moreover, ZnO nanorods allow for the use of thicker active layers, leading to increases in power conversion efficiency (PCE). As a result, ZnO is well suited to the application for polymer solar cells.

On the other hand, the PCE is correlated to a short-circuit current (J_{sc}), an open-circuit voltage (V_{oc}), and a fill factor (FF). It is hard to increase V_{oc} and J_{sc} simultaneously in polymer solar cells, because a larger V_{oc} needs a larger band gap absorber, resulting in a smaller photocurrent due to fewer absorption regions. It is believed that the V_{oc} is proportional to the energy gap between the highest occupied molecular orbital (HOMO) of the donor and the lowest unoccupied molecular orbital (LUMO) of the acceptor. Moreover, V_{oc} also scales with the work function difference between the cathode and the anode. In order to enhance the performance of polymer solar cells, selection of the both electrodes is important. In conventional BHJ solar cells, a lower work function metal, such as Al, is used as the cathode, while a transparent conducting oxide (TCO), such indium tin oxide (ITO), is used as the anode due to its higher work function. However, in ZnO nanorod-based polymer solar cells, the direction of the current flow across the solar cells is inverted, compared to the conventional BHJ polymer solar cells. Hence, ZnO nanorod-based polymer solar cells need a higher work function metal as the anode and the TCO as the cathode. Although an optimized cathode in conventional polymer solar cell have been studied [4, 5], the research on the selection of electrodes in the ZnO nanorod-based polymer solar cells has not been reported. Here, we have investigated the role of various electrodes on the performance of ZnO nanorod-based polymer solar cells. The increase in V_{oc} is up to 120 mV by replacing the lower work function (φ) anode of gold (Au, φ =5.1 eV) with platinum (Pt, φ =5.36 eV). Moreover, the increased V_{oc} of 20 mV has been observed when varying the work function of TCO from 4.7 eV (ITO) to 4.4 eV (fluorine-doped tin oxide, FTO). As a consequence, ZnO nanorod-based polymer solar cells with large work function difference between electrodes have enhanced performance.

II. CURRENT RESULT

In this study, we have successfully fabricated an inverted bulk-heterojuction solar cells using ZnO nanorod arrays between TCO, poly(3-hexylthiophene) (P3HT): methanofullerene [6,6]-phenyl-C_{61}-butyric acid methyl ester (PCBM), and high work function metal as a hole-collecting electrode. The cross sectional morphology of ZnO nanorod arrays was obtained by a field-emission scanning electron microscope (FESEM), as shown in Fig. 1. A schematic view of the device geometry (TCO / ZnO nanorod array / P3HT:PCBM blend / poly(3,4-ethylenedioxylenethiophene): polystyrene sulfonic acid (PEDOT:PSS)/metal) is shown in Fig. 2.

First, we studied ZnO nanorod-based polymer solar cells using Au and Pt as anodes while keeping the ITO cathode constant. The positive electrodes were deposited under high vacuum (~ 1×10^{-6} torr) using an e-beam evaporation. Current density–voltage (J–V) curves were measured with a Keithley 2400 source meter, under illumination at 100 mW/cm^2 from a 150 W Oriel solar simulator with AM 1.5G filter. For two devices utilizing Au (φ = 5.1 eV) and Pt (φ = 5.36 eV) as the positive electrodes, the measured V_{oc}, J_{sc}, and PCE values are summarized in Table 1, with values for the respective work function difference between electrodes. The enhanced V_{oc} of the device with Pt is correlated to its larger work function

978-1-4244-1917-3/08/$25.00 ©2008 IEEE

difference between electrodes (0.66 eV). A total variation of 120 mV of the Voc was observed for a variation of the positive electrode work function by 0.26 eV. The high work function of Pt ($\varphi > 5.2$ eV) will favor ohmic contacts to P3HT (highest occupied molecular orbital, HOMO ~ 5.2 eV), and the work function of Au is slightly smaller than the HOMO of P3HT. Therefore, the higher interface barrier for holes is formed in the devices with Au. Moreover, the difference in J_{sc} between the different metals originates from the change in the electric field in each device. As a result, the device with Pt electrode has larger V_{oc} and J_{sc} than that with Au electrode due to the larger work function difference between electrodes.

Second, Au was chosen as anode for investigating the influence of different work function TCO as cathodes on the built-in potential of ZnO nanorod-based polymer solar cells. A device was produced on an ITO-coated glass substrate, and another device was on a FTO-coated substrate for comparison. A very weak variation of the V_{oc} of only 20 meV has been observed when varying the negative electrode from ITO ($\varphi = 4.7$ eV) to FTO ($\varphi = 4.4$ eV). Fig. 3 shows the J-V curves of two devices using ITO and FTO as the negative electrodes. The device with FTO has larger work function difference between electrodes resulting in slightly higher V_{oc}. The change in V_{oc} with different cathodes is smaller than that with different anodes. The reason is suggested that the ZnO seed layer and nanorod arrays are sandwiched between the negative electrode and the fullerene. Therefore, the built-in potential is less sensitive to the mismatch between the work function of TCO and the LUMO level of fullerene. Nevertheless, the V_{oc} still scales with work function difference between electrodes, and the built-in potential correlates to the extraction of photocurrent. The devices with FTO perform well due to the large work function difference.

III. CONCLUSIONS

In conclusion we have shown that by variation of both negative electrodes and positive electrodes the open-circuit voltage of the ZnO nanorod-based polymer solar cells varies by more than 100 mV. Using low work function FTO as cathodes and high work function metal (Pt) as anodes produces large work function difference between electrodes and then leads to large V_{oc}. Therefore, it is important to optimize electrodes to

enhance the performance of the devices for the design of ZnO

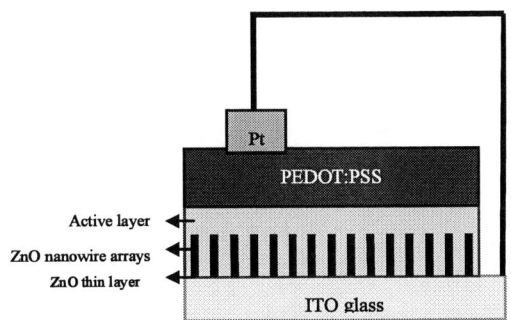

Figure 2 Schematic of a device structure of ZnO nanorod-based polymer solar cells fabricated in this study.

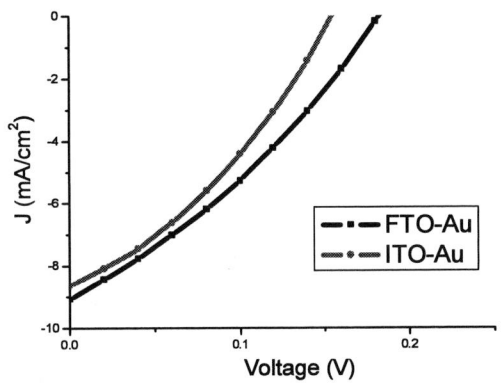

Figure 3 J –V curves of ZnO nanorod-based polymer solar cells under illumination, with different TCO electrodes.

nanorod-based polymer solar cells.

TABLE I. ZnO NANOROD-BASED POLYMER SOLAR CELLS PARAMETERS WITH DIFFERENT TOP ELECTRODES

	Jsc (mAcm⁻²)	Voc (V)	PCE (%)	workfunction difference between electrodes (eV)
ITO-Pt	10.28	0.30	1.00	0.66
ITO-Au	9.08	0.18	0.53	0.4

REFERENCES

[1] J. Y. Kim, K. Lee, N. E. Coates, D. Moses, T. Q. Nguyen, M. Dante, and A. J. Heeger, "Efficient Tandem Polymer Solar Cells Fabricated by All-Solution Processing," Science, vol. 317, pp. 222-225, 2007.

[2] D. C. Olson, Y. J. Lee, M. S. White, N. Kopidakis, S. E. Shaheen, D. S. Ginley, J. A. Voigt, and J. W. P. Hsu, "Effect of polymer processing on the performance of poly(3-hexylthiophene)/ZnO nanorod photovoltaic devices," Journal of Physical Chemistry C, vol. 111, pp. 16640-16645, Nov 2007.

[3] J.-S. Huang and C.-F. Lin, "Influences of ZnO sol-gel thin film characteristics on ZnO nanowire arrays prepared at low temperature using all solution-based processing," Journal of Applied Physics, vol. 103, p. 014304, 2008.

[4] V. D. Mihailetchi, L. J. A. Koster, and P. W. M. Blom, "Effect of metal electrodes on the performance of polymer : fullerene bulk heterojunction solar cells," Applied Physics Letters, vol. 85, pp. 970-972, Aug 2004.

[5] M. O. Reese, M. S. White, G. Rumbles, D. S. Ginley, and S. E. Shaheen, "Optimal negative electrodes for poly(3-hexylthiophene): [6,6]-phenyl C61-butyric acid methyl ester bulk heterojunction photovoltaic devices," Applied Physics Letters, vol. 92, p. 3, Feb 2008.

Figure 1 Cross-sectional SEM image of ZnO nanorods

Wafer Level Batch Fabrication and Assembly of Small Form Factor Optical Pickup Head

Sheng-Yi Hsiao[1], Chih-Chun Lee[1], Yi Chiu[2], Hsi-Fu Shih[3], Jin-Chern Chiou[2],
Han-Ping D. Shieh[4] and Weileun Fang[1,5]

[1]Dept. Power Mechanical Engineering, [5]NEMS Institute, National Tsing Hua University, HsinChu, Taiwan
[2]Dept. Electrical & Control Engineering, National Chiao Tung University, HsinChu, Taiwan
[3]Dept. Mechanical Engineering, National Chung Hsing University, Taichung, Taiwan
[4]Department of Photonics and Display Institute, National Chiao Tung University, Hsin Chu, Taiwan

Abstract-A MEMS batch assembly process for fabricating small form factor optical pickup head is proposed to minimize the complexity of assembly. Electrical and optical components are sealed in a chamber. A silicon optical bench with a packaged laser diode and crystalline-plane mirrors is demonstrated.

Keywords-Small form factor, optical pickup head, crystalline plane mirror, wafer level packaging.

I. INTRODUCTION

The growing market for digital information technology (IT) and handheld consumer products, such as MP3 music players, digital still cameras and personal digital assistants, have stimulated the development of high density mobile storage media and their I/O drives. Optical storage systems are portable and exchangeable. High numerical aperture (NA) and short wavelength is the main solution to achieve high density optical data storage. For instance, the Blu-ray Disk (BD) technology has a storage capacitance up to 50GB. DPHI has proposed a small form factor (SFF) optical drive with a disk capacity up to 1GB using a lens of NA 0.85 [1]. However, SFF optical pickup heads (OPH) [1, 2] require complex assembly of the micro components such as microprisms, micromirrors, microlenses, laser diodes (LD), and photo diodes (PD). The assembly remains the major challenges in fabricating SFF OPH.

In this study, a batch assembly process using MEMS technology for fabricating SFF OPH is proposed. A silicon optical bench (SiOB) substrate, a wet etched silicon mirror wafer, and a cover glass are wafer-level bonded together to form sealed chambers. The components inside the chambers can then be protected from the dicing process. Wet etched <111> crystalline mirrors act as the mirror components. After dicing, a die-level assembly of the holographic optical element (HOE) is used to compensate for the dimensional errors and misalignment in all the previous assembly processes. A laser diode packaged in the chamber with wet etched micromirrors is demonstrated using the proposed batch assembly process.

II. DESIGN CONCEPT

SiOB can be used to achieve miniature optical system by planar optical design and implementation on a single chip. Various configurations of the SFF OPH design have been published [3]. A compact SFF OPH design is shown in Fig. 1.

The dimension of the OPH is 6.5mm×3mm×3.2mm (L×W×H). It consists of a SiOB substrate, microprisms, micromirrors, a LD, a PD, a HOE, and an objective lens. The alignment and assembly of the micro components with μm accuracy are needed to ensure the proper function of the OPU.

In this study, a simplified SFF OPH fabrication and assembly process is proposed. The integrated OPH is shown in Fig. 2. Instead of discrete glass micro prisms, wet etched <111> crystalline plane mirrors serve as the parallel mirrors. As a result, wafer level silicon bonding can be used to greatly simplify the processes of assembly of micro optical components. Moreover, instead of a single thick HOE component, two thinner glass substrates are used for wafer level packaging and die level HOE component, respectively. The optical and electrical components are placed and sealed in the chambers formed by the SiOB substrate, the mirror wafer, and the cover glass. Therefore they can be protected in the dicing process. After dicing, the HOE glass and the objective lens are attached to the OPH for compensation and calibration.

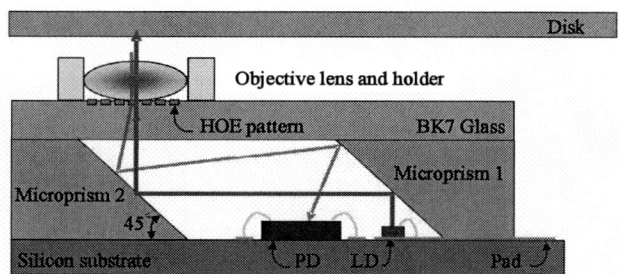

Fig.1 Configuration of a SFF OPH design.

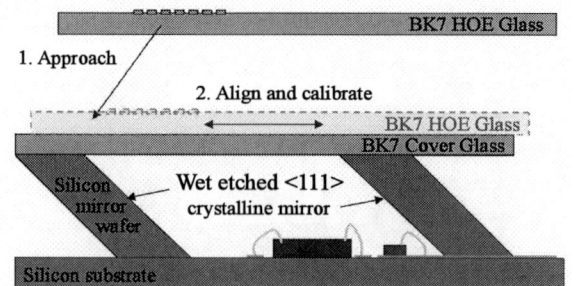

Fig.2 Configuration of batch fabricated OPH design.

978-1-4244-1917-3/08/$25.00 ©2008 IEEE

III. FABRICATION PROCESS

The integrated OPH consists of a SiOB substrate wafer, a wet etched mirror wafer with parallel crystalline mirrors, and a cover glass wafer. Fig. 3 shows the fabrication processes. In Fig. 3a, LPCVD insulation nitride layer and aluminum electrical routing are prepared on the SiOB substrate wafer. Meanwhile, a 9.7° off-cut silicon wafer is also deposited with LPCVD nitride film. The etching windows are patterned on both sides. As shown in Fig. 3b, a laser diode and other electronics are placed and bonded to the substrate wafer. The mirror wafer is wet etched to form the parallel mirrors. Finally, the cover glass wafer, the mirror wafer, and the substrate wafer are bonded and then diced, as shown in Fig. 3c. The electrical components and their bonding wires are sealed by the wafers in chambers. Thus, the dicing process will not damage or contaminate the components and the mirrors. Moreover, two dicing steps are proposed to make the contact pads on an extended part in the SiOB substrate, as indicated in Fig. 3c.

IV. DEMONSTRATION RESULT

A fabricated module with a laser diode in a chamber is show in Fig. 4. The packaged LD remains functional and lases at an 18mA threshold current. Fig. 5 shows the cross-sectional photograph of the wet etched parallel mirrors. The surface

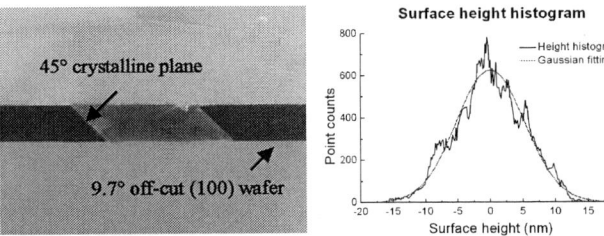

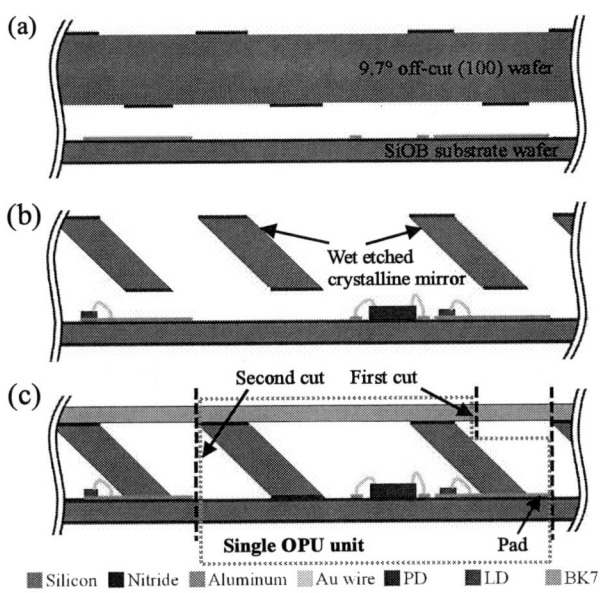

Fig.3 Batch fabrication processes for the SFF OPU.

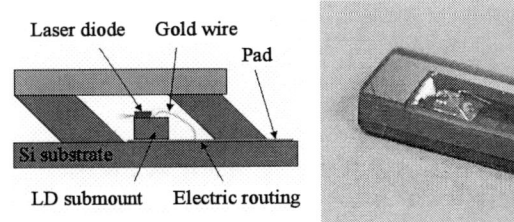

Fig.4 Fabricated module with a sealed LD package.

Fig.5 Cross section view of the wet etched 45° parallel mirrors and the surface roughness measurement result.

Fig.6 (a) Collimation of the packaged laser beam, and (b) beam steering achieved by moving the collimating lens.

roughness is lower than 10nm within a 100μm×100μm area. The 45° mirrors are used to reflect the in-plane laser beam to an out-of-plane direction. An objective lens is mounted on the packaged module to collimate the out-of-plane laser beam, as shown in Fig. 6. Fig. 6b shows the steering of the laser beam in different directions by moving the collimating lens.

V. CONCLUSIONS

This study presents a solution for batch fabricating SFF OPH using MEMS technique. Wet etched crystalline mirrors act as the mirror components. The electronics and bonding wires are sealed by the wafers in a chamber. As a result, the electronics and microoptical components are protected from the dicing process. Moreover, two dicing steps are proposed to keep the pads on the extended part of the SiOB for further PCB connection. A demonstration of a packaged LD device and crystalline mirror is presented. The preliminary result has proved the feasibility of the proposed integration and packaging concept.

ACKNOWLEDGMENTS

This work was supported in part by the Ministry of Economic Affairs, Taiwan, under contract no. 96-EC-17-A-07-S1-011 and by the Nation Science Council, Taiwan, under contract NSC-96-2628-E-007-008-MY3.

REFERENCES

[1] D. L. Blankenbeckler, B. W. Bell, Jr., K. Ramadurai, and R. L. Mahajan, "Recent advancements in DataPlay's small form-factor optical disc and drive," Jpn. J. Appl. Phys., **45(2B)**, 1181-1186, 2006.

[2] J.-S. Sohn, E.-H. Cho, M.B. Lee, H.-S. Kim, S.-D. Suh,S.-M. Kang, N.-C. Park, Y.-P. Park, "Development of integrated optical pickup for small form factor optical disc drive," Proc. SPIE, **6282**, 61-67, 2006.

[3] H.-F. Shih, Y.-C. Lee, Y. Chiu, and G.-D. Lin, "Optical head design using prism-type holographic optical element for small form factor applications," Proc. SPIE, **6827**, 161-166, 2007.

Evaluation of X-ray reflectivity of a MEMS X-ray optic

I. Mitsuishi[1], Y. Ezoe[2], M. Koshiishi[1], M. Mita[1], Y. Maeda[1], N. Y. Yamasaki[1], K. Mitsuda[1],
T. Shirata[2], T.Hayashi[2], T. Takano[3], and R. Maeda[3]

[1] Institute of Space and Astronautical Science (ISAS), Japan Aerospace Exploration Agency (JAXA)
[2] Tokyo Metropolitan University
[3] National Institute of Adbanced Industrial Science and Technology (AIST)

ABSTRACT

X-ray reflectivity of an ultra light-weight X-ray optic using MEMS technologies was measured in two different energies (0.28 keV and 1.49 keV). The obtained reflectivities can be understood by considering the mirror surface structures.

1. INTRODUCTION

Because X-rays are difficult to focus refractively, grazing-incidence optics are often utilized. In such system, X-ray reflectivity depends on energy, reflection angle and mirror surface roughness. Because the allowable reflection angle is typically less than several degrees, at most thousands of mirrors are prepared to collect X-rays from astronomical objects in X-ray astronomical satellites. Mirror surfaces must be smooth on the order of the incident X-ray wavelength, i.e., nm or less. Thus, next generation satellites need novel light-weight and low-cost X-ray optics.

We have been developing ultra light-weight and low-cost X-ray optics using MEMS technologies [1-4]. We use anisotropically etched Si (111) side walls as X-ray mirrors. We have confirmed X-ray reflection [1] and imaging [4]. Here we report on X-ray reflectivity measurements of our first MEMS X-ray optic in two different energies.

2. X-RAY EXPERIMENTS

We have fabricated a MEMS X-ray optic using anisotropic wet etching and deep dry ion etching. Fabrication process flows of the optic is described in [4]. Figure 1 shows the optic which consists of 14 mirror chips and an optic mount. Incident X-rays are reflected upon the Si (111) side walls within the mirror chips.

Previously we have measured X-ray reflectivity of these mirror chips at Al K_α 1.49keV at ISAS 30 m beam line in 15-19 Jan 2007. The experimental setup is shown in figure 2. Collimated quasi-parallel X-rays by the Mo slit hit the MEMS X-ray optic in the detector chamber. We radiated a pencil X-ray beam upon each mirror chip and compared the reflection image with the direct beam.

Figure 3 shows the obtained X-ray reflectivity from 7 innermost mirror chips. We also plot the theoretical reflectivity assuming the mirror surface roughness of 1 nm (black) which roughly represents the data [4]. However, we could not check reflectivity from outer chips because their expected reflectivities were below the detection limit. Furthermore, below 1 degree, the measured X-ray reflectivity significantly separates from the theoretical one.

One hypothesis to explain the decrease is the Si (111) step structures upon the side walls as shown in figure 4.

Preliminary discussion is done in [3]. Such structures are caused by misalighnment of a photomask to the Si (111) crystal plane. Since the step structure works as an obstacle against the reflected X-rays (see the concept in figure 3 bottom), this can decrease the X-ray reflectivity. We estimated the occultation angle θ_0 from the surface measurements using the Dektak stylus-based profiler. Red and blue curves in figure 3 includes this occultation effect. Clearly, the curve with the rms roughness of 1 nm and the occultation angle of 15' explains the data very well.

In order to fully evaluate X-ray reflectivity from the mirror chips, we newly measured the reflectivity at C K_α 0.28 keV in 11-21 March 2008. We changed the X-ray detector from the X-ray CCD to the gas-flow type proportional counter, to avoid the absorption in the optical blocking filter on the CCD and detect X-rays in this extremely low energy band. We used a thin polypropylene film with a thickness of about 10 μm as an X-ray entrance window of the counter.

Figure 5 is the obtained X-ray reflectivity at C K_α 0.28 keV. In contrast to figure 3, the expected X-ray reflectivity is still higher above 1 degree thanks to the lower X-ray energy. As seen from the two theoretical curves, the X-ray reflectivity depends on the surface roughness rather weakly than Al K_α. We found that the data are not well represented by the same parameters (red, rms 1 nm and θ_0=15') as those in Al K $_\alpha$ but needs larger roughness (green, 5 nm). The reason for this increase may be due to change of the X-ray coherent length. The coherent lengths at Al K_α and C K_α are $\sim$ 5 and 10 μm, respectively. In our mirror chips, the roughness increases in proportional to the measured scale as shown in figure 6. The measured roughness at 5 μm and 10 μm is consistent with the estimated one from the X-ray reflectivity.

REFERENCES

[1] Y. Ezoe et al., "Micro pore X-ray optics using anisotropic wet etching of (110) silicon wafers", Appl. Opt., 45, pp. 8932-8938, 2006

[2] M. Koshiishi et al., "Design and Fabrication of a MEMS X-ray Optic using Anisotropic Wet Etching of Si wafers, IEEE/LEOS Int. Conf. on Optical MEMS and Their Applications, pp. 84-85, 2006

[3] M. Koshiishi et al., "The first light of a single-stage MEMS X-ray optic", Proc. of SPIE, 6688, 668814, 2007
[4] Y. Ezoe et al., "A micromachined X-ray collector for space astronomy", Sens. & Act. A, in press.

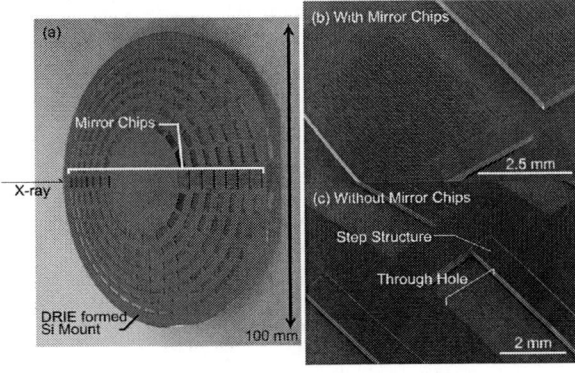

Figure 1: Photographs of a single-stage MEMS X-ray optic. (a) Overview of the optic with 14 mirror chips. (b) Closeup view of the mount. (c) The same as (b) but without the mirror chips. Taken from [4].

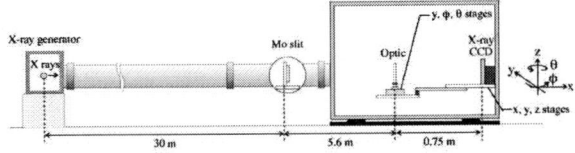

Figure 2: The experimental setup in the ISAS 30 m beamline.

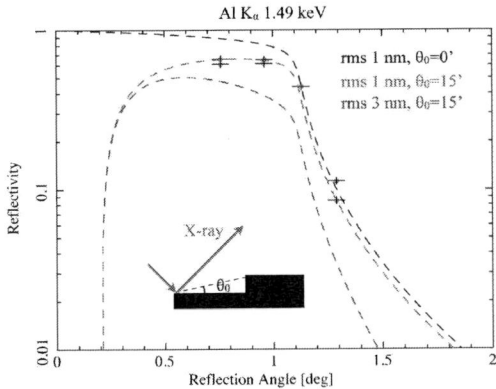

Figure 3: X-ray reflectivity of mirror chips at Al K_α 1.49 keV. A black line indicates a theoretical reflectivity curve with rms mirror roughness of 1 nm. Red and blue ones include the geometrical occultation effect.

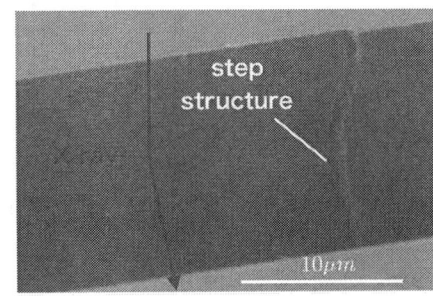

Figure 4: SEM view of the Si (111) side walls in the mirror chip.

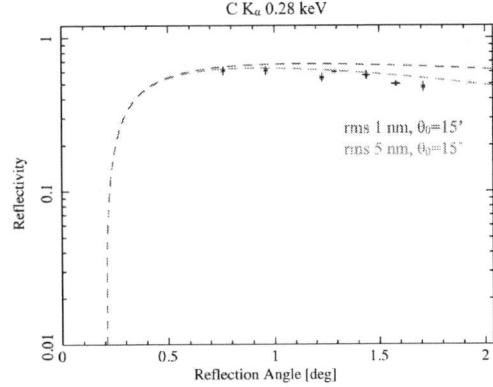

Figure 5: X-ray reflectivity of mirror chips at C K_α 0.28 keV. Red and blue green lines are theoretical X-ray reflectivity with the rms roughness of 1 and 5 nm, including the occultation effect, respectively.

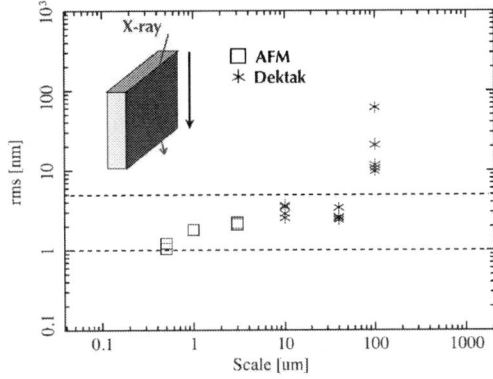

Figure 6: Examples of the surface roughness of the mirror chip measured with AFM and the Dektak profiler. Two dashed lines indicate the rms roughness of 1 and 5 nm.

Design and Fabrication of CMOS-Integrated Thermoelectric IR Microsensors

Keng-Shuen Lin, Rongshun Chen
Department of Power Mechanical Engineering, National Tsing Hua University
Hsinchu City, 30013, TAIWAN

Abstract- **This work presents a thermoelectric infrared microsensor which is designed and fabricated with TSMC CMOS-MEMS processes. The proposed device can achieve the responsivity of 432.3 V/W and time constant of 2.49 ms at 1 atm.**

I. INTRODUCTION

Infrared sensors have developed for several hundred years due to their ability of detecting the invisible light. Most high performance infrared sensors have been presented so far and used for thermal imaging in the fields of military detection systems, industrial automation, and home security monitoring [1-3]. This paper proposes a novel device which has been fabricated using a standard CMOS IC processes with subsequent bulk-micromachining techniques. The advantage offered by the CMOS compatibility is the cost-effective, on-chip integration of the sensor elements with signal processing circuitry.

II. A HIGH FILL-FACTOR INFRARED SENSOR DESIGN

The proposed thermoelectric infrared microsensor features high fill-factor and small pixel size due to a novel two-level structure design through removing sacrificed metal layers between infrared absorber and thermocouple cantilever beam. Fig. 1 is the schematic view of the sensor which consists of infrared absorber, high heat conductive materials, thermocouple cantilever beam and silicon substrate which acts as heat sink.

III. THERMAL MODELING OF INFRARED SENSOR

Numerical simulation is usually inefficient and time-consumption, especially for time-dependent transient response. Therefore, a complex thermal model has been simplified and converted to an equivalent circuit model through the equivalence between thermal and electrical parameters. Fig. 2 shows the simplified RC circuit model of the sensor.

Fig. 3 is the simulation results of equivalent RC circuit model using HSPICE software in vacuum and atmosphere environment. In a given 50x50 μm^2 sensing area, the design with thermocouples length = 50 μm, wide = 0.5 μm and series of numbers = 14 pairs of N/P type polysilicon will achieve outstanding system performance through optimal geometric consideration. The results show a responsivity of 1001.7 V/W and time constant of 5.8 ms in vacuum as well as a responsivity of 432.3 V/W and time constant of 2.49 ms at 1 atm.

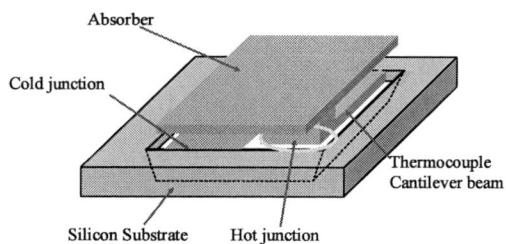

Fig. 1. Schematic view of the proposed two-level thermoelectric infrared sensor.

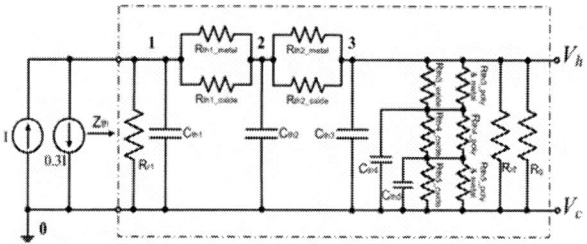

Fig. 2. Equivalent RC circuit model of the proposed two-level thermoelectric infrared sensor.

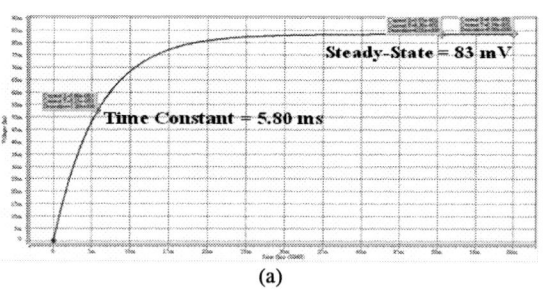

(a)

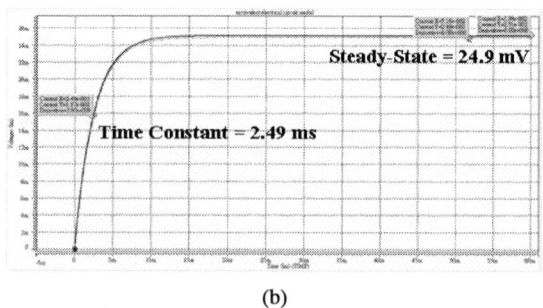

(b)

Fig. 3. Simulation results of equivalent RC circuit model with (a) ideal vacuum environment and (b) 1 atm environment.

IV. NUMERICAL SIMULATION

Numerical simulation software ConventorWare is used and executed thermal transient simulation. Temperature distribution and thermal transient response of the sensor are shown in Fig. 4. Almost most temperature gradient($\triangle T = 82.8$ mK) occurs in thermocouple cantilever beam, which proves that good thermal isolation only appears between hot and cold junctions and thermoelectric materials inside cantilever beam always generate efficiently large Seebeck voltage.

V. OPTIMIZATION OF GEOMETRIC PARAMETERS

This study implemented explicit analysis for several geometric parameters, including length, width and number of the thermocouple pairs in a given sensing area (50×50 μm^2). Fig. 5 shows that the proposed sensor possesses optimal responsivity in 50×50 μm^2 active area when width = 0.5 μm, length = 50 μm and N = 14(maximum series pairs with L = 50 μm) are chosen.

VI. EXPERIMENTAL RESULTS

In this paper, some preliminary experimental results have been measured so far. Fig. 6 shows the two-level thermoelectric infrared sensor with integrated heater and its sensing area is about $120\times78.5\mu m^2$. Moreover, Measured Seebeck voltages of this sensor are depicted in Fig. 7(a) and Fig. 7(b) for N/P type polysilicon and N-ploy/Al thermoelectric materials respectively when various heater powers are applied to on-chip heater. The measured results show a responsivity of 36.23 V/W for P/N type materials, a responsivity of 163.9 V/W and a thermal time constant of 7 ms for N-ploy and aluminum materials.

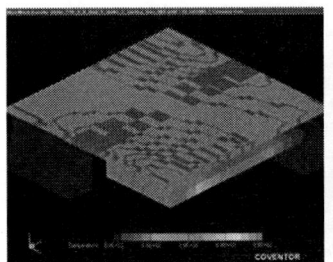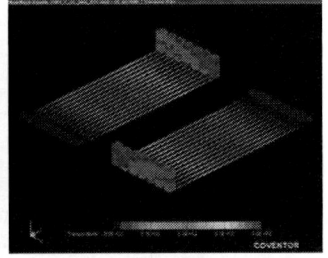

Fig. 4. Temperature distribution of the proposed sensor when constant heat flux(100 pω/μm^2) is applied to the top of IR absorber.

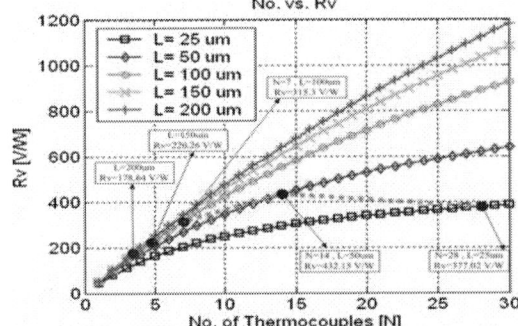

Fig. 5. Optimal responsivity analysis of the sensor for 50x50 μm^2 sensing area.

Fig. 6. The two-level thermoelectric infrared microsensor with integrated heater.

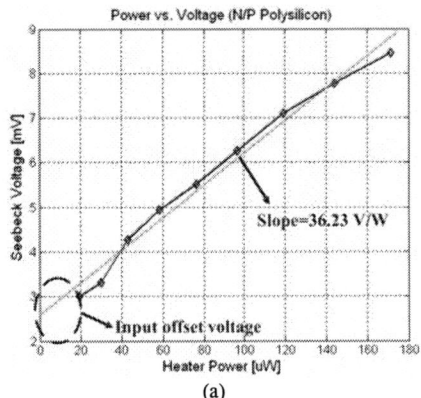

(a)

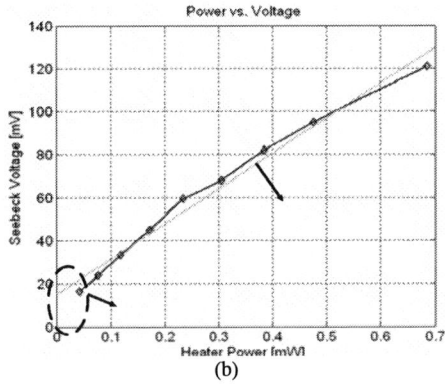

(b)

Fig. 7. Measured responsivity results of thermoelectric infrared microsensors with 120 x 78.5 μm^2 sensing area using (a) N/P type polysilicon and (b) N type polysilicon and aluminum as thermopile respectively.

REFERENCES

[1] Radford, W., Murphy, D., Ray, M., Propst, S., Kennedy, A., Kojiro, J., Woolaway, J., Soch, K., Coda, R., Lung, G., Moody, E., Gleichman, D., and Baur, S., "320×240 silicon micro- bolometer uncooled IRFPA's with on-chip offset correction," *Proc. SPIE Infrared Detectors and Focal Plane Arrays IV*, 2746, pp. 82–92, 1996.

[2] Jahanzeb, A., Travers, C. M., Çelik-Butler, Z., Butler, D. P., and Tan, S., "A semiconductor YBaCuO micro- bolometer for room temperature IR-imaging," *IEEE Trans. Electron Devices*, 44, pp. 1795–1801, 1997.

[3] Tezcan, D. S., Eminoglu, S., and Akin, T., "A low-cost uncooled infrared microbolometer detector in standard CMOS technology," *IEEE Electron Devices*, 50, pp. 494–502, 2003.

Performance Improvement of a Two-Axis Radial-Vertical-Combdrive Scanner by Using a Symmetric Spring Design

Tien-liang Hsieh[1], Yao-tien Chang[1], Sheng-jie Chiou[1], Jui-che Tsai[1], Dooyoung Hah[2], and Ming C. Wu[3]

[1] Graduate Institute of Photonics and Optoelectronics and Department of Electrical Engineering
National Taiwan University
No. 1, Sec.4, Roosevelt Rd., Taipei 10617, Taiwan
Tel: +886-2-3366-3700 Ext. 247, Fax: +886-2-3366-3686, E-mail: jctsai@cc.ee.ntu.edu.tw

[2] Department of Electrical and Computer Engineering, Louisiana State University, Baton Rouge, LA 70803, USA
[3] Department of Electrical Engineering and Computer Sciences and Berkeley Sensor and Actuator Center
(BSAC), University of California, Berkeley, CA 94720-1774, USA

Abstract

A symmetric cross-bar spring structure is employed to improve the performance of a two-axis gimbal-less micromirror driven by radial vertical combdrive actuators. The actuators are hidden underneath the mirror to minimize the device form factor. The device is fabricated by the SUMMiT-V surface micromachining process. The entire improved cross-bar spring structure is made of the same layer with a 1-μm thickness and the torsion springs are shortened to suppress the lateral instability. The mechanical rotation angles are ±5.33° (50.7V) and ±6.04° (52.8V) for rotation about the x- and y-axes, respectively.

Keywords: Symmetric spring, two-axis scanner, radial comb-drive actuators, small form factor, surface micromachining

1 INTRODUCTION

Two-axis MEMS scanners have been intensively studied thanks to their compactness and capability of two-dimensional (2-D) optical beam steering, which are suitable for various applications such as optical fiber communication [1], biological imaging [2], and display technologies [3]. Electrostatic actuation is an attractive mechanism for driving the MEMS device due to its structural simplicity and easy fabrication process. Particularly, vertical combdrive actuators are of great interest as they are free from the pull-in effect and also offer larger force densities; as a result larger displacements or rotational ranges can be obtained. Typically, a gimbal structure exists in a dual-axis vertical combdrive micromirror to achieve two rotational degrees of freedom (DOFs). It is desirable to eliminate the gimbal so that the device form factor can be reduced.

In our previous study [4,5], we utilized the radial vertical combdrive and the cross-bar spring structure to obtain a two-axis gimbal-less micromirror. The combdrive actuators and cross-bar torsion springs were hidden underneath the device to obtain a small form factor. Several designs were evaluated and devices with different parameter values have been characterized. Mechanical scan angles of ±5.4° and ±2.3° for x- and y-axis rotation respectively were achieved. The imbalance of the scan pattern was mainly due to the dissimilar spring constants, which was caused by the thickness difference between the lower and upper torsion springs [5]. They are 1-μm (mmpoly1) and 2.25-μm

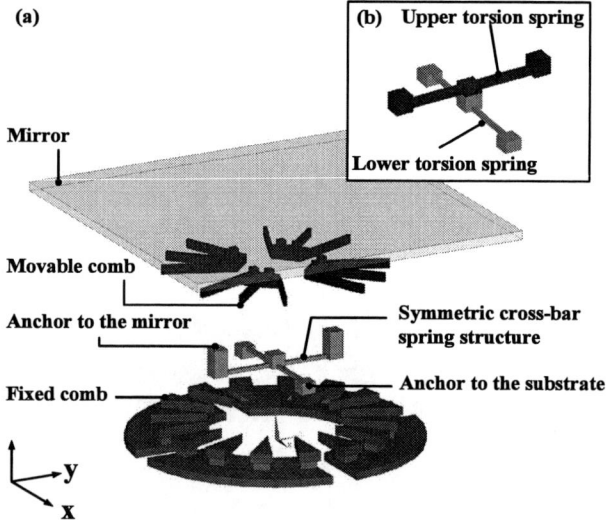

Figure 1. (a) Schematic drawing of the improved two-axis MEMS scanner with a symmetric cross-bar spring structure. (b) The cross-bar spring structure of our previous device [4,5].

(mmpoly3) thick respectively, fixed by the SUMMiT-V surface micromachining process. To achieve the same scan angle, a higher voltage was required for y-axis rotation due to its larger torsion spring constant. Additionally, the rotation about the y axis was more susceptible to lateral instability [5] and therefore had a smaller scan range. In this paper, the cross-bar spring is modified into a symmetric structure so

978-1-4244-1917-3/08/$25.00 ©2008 IEEE   108

that orthogonal rotational modes become degenerate. This results in identical x- and y-axis scan ranges. The lateral instability is also suppressed by reducing the length of the torsion springs. Figure 1 is the schematic drawing of the device.

2 DESIGN AND FABRICATION

The device is fabricated using SUMMiT-V surface micromachining process offered by Sandia National Laboratory. It has five polysilicon layers, including one nonreleasable ground/shield layer (mmpoly0) and four structural layers (mmpoly1 to mmpoly4). Figure 1 shows the main difference of the cross-bar spring structure between the previous device and this improved design. In our previous study, the lower and upper torsion springs which provide restoring torques for x- and y-axis rotations have different thicknesses and a height offset exists between them. In this paper, these torsion springs are made of the same layer (1-μm thick mmpoly1); consequently the spring constants of both rotational modes are the same. The gap between the torsion spring and the ground layer is fixed at 2 μm. As a result the maximum room for rotation about the x axis depends on the spring length. Therefore, a device with shorter springs can reach a larger tilt angle. It also provides better lateral stability. Figure 2(a) is the optical microscope image of the device with the mirror and movable combs removed intentionally and figure 2(b) is the device 3D profile obtained with WYKO, a white light interferometric profiler, during actuation.

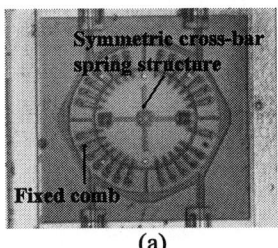 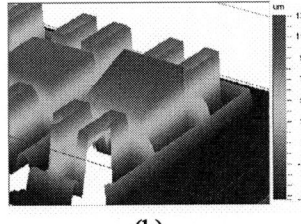

(a)   (b)

Figure 2. (a) Optical microscope image of the device with the mirror and movable combs removed intentionally, and (b) 3D profile of the micromirror during actuation.

3 DEVICE CHARACTERIZATION

Two devices with different parameter values are demonstrated in this paper (Table 1), where g, l_s, l_o and l_f are the finger gap, spring length, overlap length between the movable and fixed fingers, and finger length, respectively. With the modal analysis using ANSYS, the resonant frequency of the degenerate modes is calculated to be 21 kHz and 26 kHz for device I and II, respectively. The resonant frequency of the in-plane twist mode, which is related to the lateral stability, is 36 kHz and 49 kHz for device I and II, respectively. The ratios k_{twist}/k_x and k_{twist}/k_y for both devices are higher than those in our previous study, where k_x, k_y, and k_{twist} are the spring constants of the x- and y-axis rotations, and the spring constant of the in-plane twist

motion, respectively. This indicates that the lateral stability can be improved. Figure 3 is the DC characteristic for the degenerate rotational modes of the improved devices. For device I, the mechanical rotation angles are ±5.93° (95.1V) and ±5.55° (102.7V) for rotation about the x- and y-axis, respectively; for device II, the measured angles are ±5.33° (50.7V) and ±6.04° (52.8V). A higher voltage is required for device I because of the shorter spring length and the larger finger gap, which also means fewer fingers. The maximum scan angles are limited by the rotational pull-in effect, except that the x-axis tilt of device II is limited by the maximum rotation room as discussed above.

Table 1. Device parameters

Device #	g (μm)	l_s (μm)	l_o (μm)	l_f (μm)
I	3	8	12.80	15
II	1	12	12.07	15

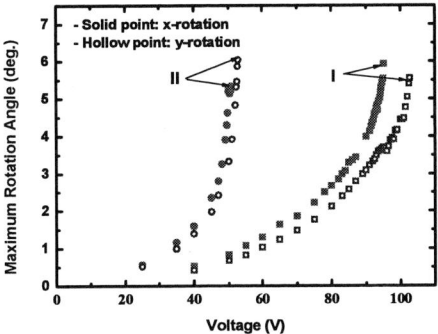

Figure 3. DC characteristic of the devices.

4 CONCLUSION

We propose to modify the cross-bar spring of a two-axis gimbal-less scanner into a symmetric structure. Almost identical scan ranges about the x- and y-axes are obtained. The mechanical rotation angles are ±5.33° (50.7V) and ±6.04° (52.8V) for rotation about the x- and y-axis, respectively.

ACKNOWLEDGEMENTS

This work was supported by National Science Council of Taiwan under grants NSC 95-2221-E-002-053 and NSC 96-2221-E-002-198-MY2, and Excellent Research Projects of National Taiwan University, 95R0062-AE00-06.

REFERENCES

[1] M. C. Wu et al., *J. Lightw. Technol.*, vol. 24, pp. 4433-4454.

[2] A. D. Aguirre et al., *Opt. Express* vol. 15, Issue 5, pp. 2445-2453.

[3] M. O. Freeman *Proc. of SPIE* vol. 4985, pp. 56-62

[4] S. J. Chiou et al., *Proc. of IEEE/LEOS Optical MEMS and Nanophotonics, 2007*, pp. 83 – 84.

[5] J. C. Tsai et al., *J. Opt. A: Pure Appl. Opt.*, 10 (2008) 044006.

Assembly of Micro Mirrors on SOI Wafers Using SU-8 Mechanisms and One-Push Operation

Yi Chiu, Wei-Zhi Huang, Jhong-Wei Wu, Jin-Chern Chiou, Han-Ping D. Shieh*

Department of Electrical and Control Engineering, National Chiao Tung University, Hsin Chu, Taiwan, R.O.C.
* Department of Photonics and Display Institute, National Chiao Tung University, Hsin Chu, Taiwan, R.O.C.

Abstract- **Micro mirrors are assembled on silicon-on-insulator wafers using one-push operation. The proposed technique can reduce the overall complexity of micro system assembly by using automated equipments. Novel SU-8 mechanisms are also demonstrated.**

I. INTRODUCTION

Micro-assembled three-dimensional (3D) structures are used in many MEMS applications such as mirrors and planar optics in micro optical systems [1] and coil inductors for high-frequency applications [2]. Most of these components are fabricated in thin films by surface micromaching. They are then flipped up to form the 3D micro systems. Micro hinges are a common mechanism to anchor the flip-up components [3]. Many techniques can be used to assemble the 3D components. Aside from the manual assembly using probes, powered assembly processes have been demonstrated by using magnetic force, electrostatic force, centrifugal force, ultrasonic agitation, or on-chip actuators. Pre-stressed bimorph beams and surface tension are also used for self assembly. Recently, automated assembly was demonstrated using standard or specially designed equipments [4, 5]. The automated assembly is particularly attractive in system-in-package (SIP) design where multiple chips need to be bonded and wired. If the pick-and-place and wire bonding equipments in standard electronic packaging can be used to assist the assembly of MEMS components, a more reliable, flexible and systematic assembly and packaging process can be designed and achieved.

In this paper we present a novel design and assembly of micro mirrors on SOI wafers using simple one-push operations. SU-8 is used to fabricate locking and positioning mechanisms. A new hinge design is also introduced that can improve the positioning accuracy.

II. PRINCIPLE AND DESIGN

One of the major difficulties in manual or automated assembly of 3D MEMS structures is the control and positioning of probes or pick-up tips. Since the gap spacing between the released components and the substrates is very small, inserting the probes into the gap involves motion control in multiple degrees of freedom with high precision. A novel automated assembly with simple one-push operation in the vertical direction and large probe positioning tolerance in both vertical and lateral directions are proposed and discussed.

A. One-push assembly process

The schematic and layout design of the proposed micro mirrors are shown in Fig. 1. The blue substrate is the handle wafer of the SOI substrate. The red micro mirror is fabricated in the device layer. The yellow locking mechanisms are fabricated using SU-8. A through-wafer hole is etched under the push pad. When a micro probe pushes the pad down, the mirror is flipped up by 90° and locked by the side latch and the V-shaped hinge. The through-wafer hole provides plenty room for positioning tolerance in the vertical direction. The probe can push deeper than needed without affecting the final angular position. The lateral positioning is also non-critical as long as the probe is within the area of the push pad. Therefore the assembly process is reduced to a simple push operation. The demand of equipment accuracy is also greatly reduced.

B. Locking mechanism

When the push pad is pushed down, the mirror plate rotates about the pin axis. As the mirror plate moves out of the plane, the wing of the mirror plate contacts the bottom of the side latches and rotate the spring loaded latches. As the mirror plate reaches the upright position, it slides into the V-shaped notch in the side latches and be firmly locked in place.

Flip-up mirrors are usually fixed to the substrate by hinges. Conventional hinges need play space in the staple for the pins to rotate. The play greatly affects the accuracy of the pin position and the flip-up mirror angle. Fig. 2 shows the schematic of a novel play-less V-shaped hinge. As shown in Fig. 2(b), the downward force of the bent beams can eliminate the vertical play and fix the hinge pin on the substrate. Further-more, the hinge pin is locked between the two sides of the V-

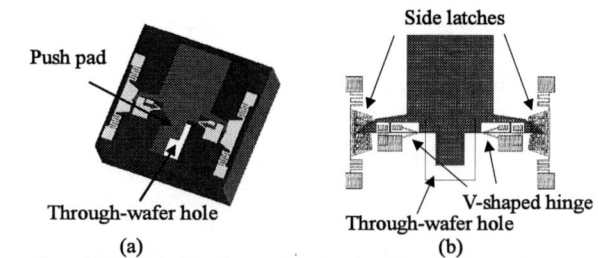

Fig. 1 (a) Schematic view and (b) layout design of the micro mirror.

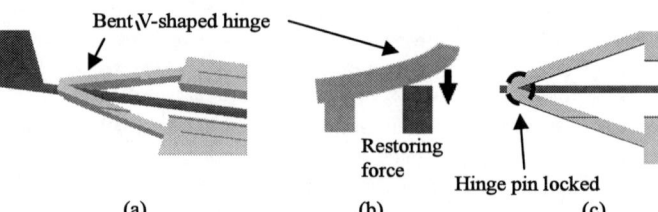

Fig. 2 (a) 3-D solid model of the V-shaped hinge after assembly; (b) cross-section view of the playless hinge pin; (c) top view of the mechanism.

shaped structure when the mirror is in the upright position, as shown in Fig. 2(c). Hence the transverse play can also be eliminated.

III. FABRICATION AND MEASUREMENT

The fabrication process for the proposed micro mirror in SOI substrates is illustrated in Fig. 3. The micro mirror was first patterned and dry-etched in the device layer (Fig. 3(a)) PECVD oxide was then deposit as the sacrificial layer in which anchors were defined (Fig. 3(b)). SU-8 photoresist was coated and patterned for side latches and hinges (Fig. 3(c)). Subsequently the silicon substrate was dry-etched from the backside to form the through-wafer holes, which can be extended to the mirror region if no release holes are desired in the mirror (Fig. 3(c)). Finally the sacrificial and buried oxide was etched by HF vapor to release the structure; the mirror was then pushed up by a micro probe to the up-right position (Fig. 3(d)). The advantage to use SOI wafers and SU-8 structural layers include simple processes, thick and stress free mirror plates, and low processing temperature that enables potential integration of photo detectors and circuits [6].

Fig. 4 shows optical and SEM micrographs of the fabricated and assembled mirrors. Fig. 4(a) is the top view of a mirror before assembly; the image of the push probe can be clearly seen. There are no etching holes in the mirror, as evidenced by the through hole in the substrate under the mirror plate. Fig. 4(b) is the mirror after the one-push assembly. Fig. 4(c) is an array of assembled mirrors with etching holes. For manual assembly by probes in our laboratory, it took less than 30 seconds to finish the entire one-push assembly of one micro mirror. Fig. 4(d) is the side view of the mirror, showing the angular position. Fig. 4(e) shows the V-shaped hinge mechanism. In our preliminary tests, the average angle of the flip-up mirrors was 89.2±0.3°, which was less than satisfactory. The angular accuracy is determined in part by the locking mechanisms of the side latches (Fig. 4(d)) and the V-shaped hinges. We are currently conducting more experiments to investigate the causes of the angular errors.

VI. CONCLUSION

A novel fabrication and assembly process that utilizes SU-8 and SOI wafers was demonstrated for 90° micromirrors. The concept of assembly with one-push operation was verified. This novel technique greatly reduced the assembly complexity and time. V-shaped hinges were used to address the issues of play in conventional hinge design. More experiments are being conducted to improve the angular accuracy.

ACKNOWLEDGMENT

This project was supported in part by the Ministry of Economic Affairs, Taiwan, R.O.C., under contract no. 96-EC-17-A-07-S1-011. The authors were grateful to the use of facilities at the National Center for High-performance Computing and the National Nano Device Laboratory, Taiwan, R.O.C..

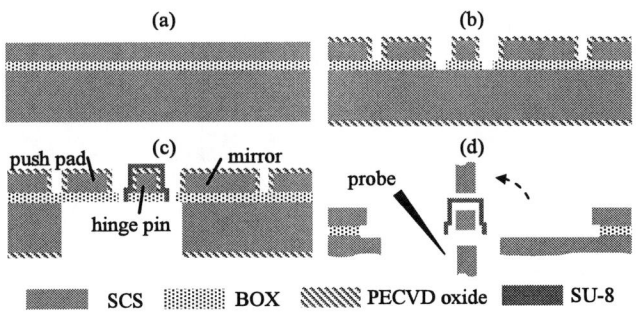

Fig. 3 Fabrication process of the micro mirror

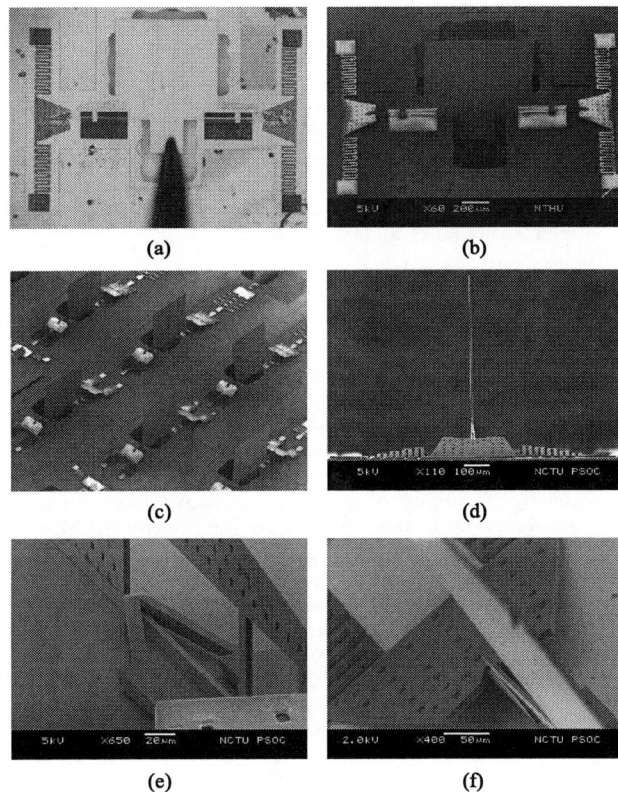

Fig. 4 Optical and SEM micrographs of fabricated and assembled micro mirrors.

REFERENCES

[1] M. C. Wu, "Micromachining for optical and optoelectronic systems," *Proceedings of the IEEE*, vol. 85, no. 11, pp. 1833-1856, 1997.

[2] G. W. Dahlmann and E. M. Yeatman, "High Q microwave inductors on silicon by surface tension selfassembly," *Electronics Letters*, vol. 36, no. 20, pp. 1707-1708, 2000.

[3] K. S. J. Pister, M. W. Judy, S. R. Burgett and R. S. Fearing, "Microfabricated hinges," *Sensors and Actuators*, vol. A, no. 33, pp. 249-256, 1992.

[4] S. H. Tsang, D. Sameoto, I. G. Foulds, R. WJohnstone and M Parameswaran, "Automated assembly of hingeless 90° out-of-plane microstructures," *J. Micromech. Microeng.* vol. 17, 1314–1325, 2007

[5] N. Dechev, W. L. Cleghorn and J. K. Mills, "Microassembly of 3-D microstructures using a compliant, passive microgripper," *J. Microelectromechanical Syst.*, vol. 13, no. 2, pp. 176-189, 2004.

[6] Y. Chiu, J. C. Chiou, W. Fang, Y. J. Lin, M. Wu, "Design, fabrication, and control of components in MEMS-based optical pickups," *IEEE Trans. Magnetics*, vol. 43, no. 2, pp. 780-784, 2007.

Spatially resolved optical characterization of photonic crystal slabs using direct evaluation of photonic modes

Yousef Nazirizadeh, Ulf Geyer, Uli Lemmer
Light Technology Institute
Universität Karlsruhe (TH)
76131 Karlsruhe, Germany
Yousef.Nazirizadeh@lti.uni-karlsruhe.de

Martina Gerken
Institute of Electrical and Information Engineering
Christian-Albrechts-Universität Kiel
24143 Kiel, Germany

Abstract— **Transmission measurements with crossed polarization filters are performed in a confocal microscope setup for spatially resolved evaluation of photonic crystal modes. The homogeneity of samples fabricated with electron-beam lithography and laser interference lithography is investigated.**

Keywords: photonic crystal slabs, transmission, guided-mode resonances, polarization

I. INTRODUCTION

Photonic crystal slabs are promising structures for modern optical devices. The fabrication of such photonic crystals is realized employing different methods, for example, electron beam lithography or laser interference lithography. For proper quality monitoring, imaging methods such as scanning electron microscopy (SEM) or atomic force microscopy (AFM) are not sufficient, since they only provide geometric parameters such as the periodicity or the hole radius. Performing spectrally and spatially resolved transmission measurements employing orthogonally oriented polarization filters allows for additional investigation of the optical properties of fabricated photonic crystal slabs [1]. Here we characterize and compare the optical homogeneity of photonic crystal slabs fabricated by electron beam lithography and by laser interference lithography.

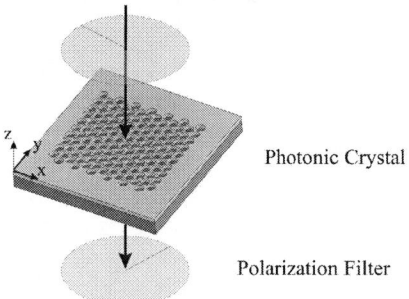

Figure 1. Setup for transmission measurements with orthogonally oriented polarization filters.

II. EXPERIMENTAL SETUP

We perform transmission measurements with two orthogonally oriented polarization filters – one placed before and one behind the photonic crystal (Fig. 1) [1]. With this configuration direct light from the broadband light source is suppressed. Only light interacting with the photonic crystal modes passes the second polarization filter [2]. Using a confocal microscope in combination with a spectrometer, we are able to measure the spectral distribution at any position on the photonic crystal slab with a spatial resolution of 2.5 μm [3]. This configuration allows for spatially and spectrally resolved access to the photonic crystal modes and characterization of fabricated structures to an accuracy of better than 1% [1].

III. COMPARISON OF FABRICATION METHODS FOR PHOTONIC CRYSTAL SLABS

We present the comparison of two fabrication methods for photonic crystal slabs using the introduced characterization method. As a first example Fig. 2(a) shows the transmission image observed with orthogonally oriented polarization filters for a hexagonal photonic crystal slab fabricated by electron beam lithography [4]. A sample was chosen where two main disadvantages of this fabrication method are clearly visible: stitching errors, which appear as squares in the structure, and drift in the exposure dose, which shows up as a color shift. Figure 2(b) illustrates transmission measurements at three positions of the photonic crystal slab shown in Fig. 2(a). Due to drift in the exposure dose the hole radii differ in size, which effects the photonic mode distribution in the photonic crystal (point A and B). However, also small changes can be visualized analyzing the transmission spectra (point B and C). Inset of Fig. 2(b) shows a spectral zoom of point B and C, a spectral drift of 1 nm is observed for photonic modes.

We acknowledge financial support by the Volkswagen Foundation and the German Federal Ministry for Education and Research BMBF (Project No. 03X5514).

978-1-4244-1917-3/08/$25.00 ©2008 IEEE

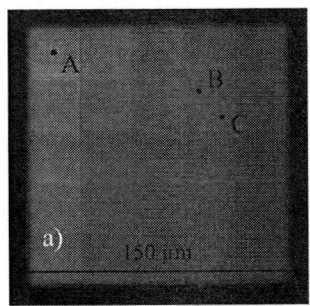

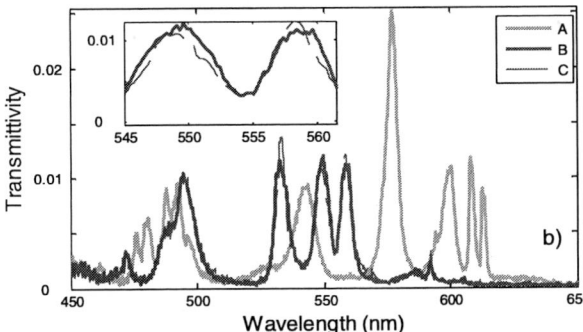

Figure 2. a) Transmission image of a hexagonal photonic crystal slab with orthogonally oriented polarization filters. b) Transmission spectra in A, B and C at normal incidence. Inset: Spectral zoom for B and C.

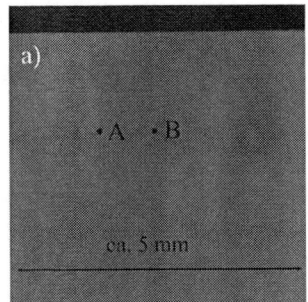

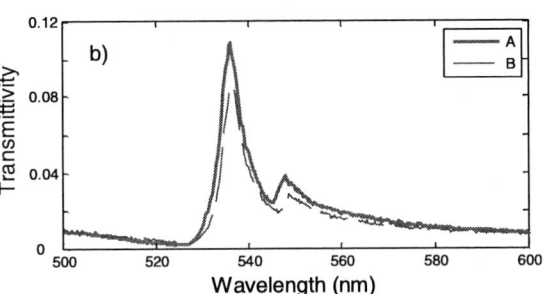

Figure 3. a) Transmission image of linear photonic crystal slabs with orthogonally oriented polarization filters. b) Transmission spectra in A and B at normal incidence.

In Fig. 3(a) a transmission image of a linear photonic crystal slab with crossed polarization filters is shown for a structure fabricated using laser interference lithography. A good homogeneity over a large area photonic crystal slab is observed, which is approved in Fig. 3(b) showing the transmission spectra. However, our characterization method visualizes overmodulations in these photonic crystal slabs, which are due to undesirable interferences in the photo resist.

IV. CONCLUSION

The experimental method presented allows for direct observation of photonic crystal modes and for improvement of photonic crystal slab fabrication methods. The investigated photonic crystal slabs fabricated by laser interference lithography exhibited better large-area homogeneity than structures fabricated by electron beam lithography.

REFERENCES

[1] Y. Nazirizadeh, J. G. Müller, U. Geyer, D. Schelle, E.-B. Kley, A. Tünnermann, U. Lemmer, M. Gerken, "Optical characterization of photonic crystal slabs using orthogonally oriented polarization filters," Opt. Express **16**, 7153-7160 (2008).

[2] A. D. Bristow, V. N. Astratov, R. Shimada, I. S. Culshaw, M. S. Skolnick, D. M. Whittaker, A. Tahraoui, T. F. Krauss, "Polarization conversion in the reflectivity properties of photonic crystal waveguides," IEEE J. Quantum Electron. **38**, 880-884 (2002).

[3] Y. A. Vlasov, M. Deutsch, D. J. Norris, "Single-domain spectroscopy of self-assembled photonic crystals," Appl. Phys. Lett. **76**, 1627-1629 (2000).

[4] M. Augustin, H.-J. Fuchs, D. Schelle, E.-B. Kley, S. Nolte, A. Tünnermann, R. Iliew, C. Etrich, U. Peschel, F. Lederer, „Highly efficient waveguide bends in photonic crystal with a low in-plane index contrast," Opt. Express **11**, 3284-3289 (2003).

Micro-mirror array for multi-object spectroscopy in cryogenic environment

Severin Waldis[1], Frederic Zamkotsian[2], Patrick Lanzoni[2], Wilfried Noell[1], Michael Canonica[1], Nico de Rooij[1]

[1] IMT, University of Neuchatel, Rue Jaquet-Droz 1, CH-2007 Neuchatel, Switzerland
[2] Laboratoire d'Astrophysique de Marseille (LAM), 38 rue Frederic Joliot Curie, 13388 Marseille Cedex 13, France

Corresponding author: F. Zamkotsian, Tel +33 4 95 04 4151, e-mail: frederic.zamkotsian@oamp.fr

Abstract

Next-generation infra-red astronomical instrumentation for space and ground-based telescopes requires MOEMS-based programmable slit masks for multi-object spectroscopy (MOS) usable in a cryogenic environment. A $100 \times 200 \mu m^2$ micro-mirror array was successfully designed, fabricated and tested, with tilting angle of 20° at 100V. In parallel we have developed and implemented a high-resolution Twyman-Green interferometer and a cryo-chamber for full surface and operation characterization. The micromirrors could be successfully actuated before, during and after cryogenic cooling. Surface deformation of gold coated mirrors at 300K and below 100K shows a slight increase from 35nm to 50nm PtV, still well suited for MOS application.

Keywords: Multi-Object Spectroscopy, programmable slit masks, micro-mirror array, cryogenic MOEMS.

1 INTRODUCTION

Multi-Object Spectrographs (MOS) are the major astronomical instruments for studying primary galaxies and remote and faint objects, whose light spectra are strongly shifted to infrared wavelengths. Current object selection systems are limited and/or difficult to implement in next generation MOS for space and ground-based telescopes. Programmable multi-slit masks are a very suitable solution for this task. A promising solution is the use of MOEMS devices such as micromirror arrays (MMA) or micro-shutter arrays (MSA), which both allow the remote control of the multi-slit configuration in real time.

Within the framework of the JRA Smart Focal Planes (European FP6 Opticon), the LAM and the IMT have engaged a collaboration in order to demonstrate a European MOEMS-based slit mask. We develop and microfabricate a novel MMA suited for MOS. The requirements are: high contrast, optically flat mirrors in operation, high fill factor, uniform tilt angle over the whole array and low actuation voltage. In order to fulfil these requirements we use a combination of bulk and surface micromachining in silicon, compatible for cryogenic operation. The mirrors are actuated electrostatically and a system of multiple landing beams has been developed, which passively locks the mirror at a well defined tilt angle when actuated. The mechanical tilt angle obtained on 100 x 200 μm^2 micromirrors (prototypes of 5x5 mirrors array) is 20° at a pull-in voltage of 90V. The tilt angle of the actuated and locked mirror is stable with a precision of one arc minute over the whole array. The surface quality of the mirrors in actuated state is better than 10 nm peak-to-valley (PtV) and the local roughness is

around 1 nm rms [1]. For future MOS, MMA will be mostly designed for infra-red applications, leading to instruments operating in vacuum at cryogenic temperatures. Large arrays of 20'000 micromirrors per array are under development.

2 HIGH SURFACE-QUALITY MICRO-MIRROR ARRAYS

Due to its location at the focal plane of the spectrograph, the surface quality of each micromirror must be very high, i. e. better than λ/20. In the ON position, any surface aberration will result in an image quality degradation on the detector of the spectrograph. The surface quality or flatness is characterized by the local roughness and the deformation of the surface. Intrinsically a mono-crystalline and polished silicon surface is supposed to be flat. There are a variety of factors that degrade the flatness of the micromirror surface and contribute to the deformation budget: 1. Initial non-uniformities of the silicon substrate due to polishing errors; 2. Partial plastic deformation during fabrication; 3. Stress at the interface between single-crystalline micromirror and poly-cristalline suspension; 4. External stress coupled into the micromirror via the suspension; 5. Intrinsic stress of the reflective layer on top of the mirror; 6. Thermal stress due to the mismatch of the coefficient of thermal expansion (CTE) between silicon and reflective layer in cryogenic environment; 7. Stress induced in the ON state of the micromirror by the stopper beams. The factor #1 depends on the choice of the substrate, factors #2 through #4 are process dependent and factor #7 depends on the design of the actuator and the operation condition.

978-1-4244-1917-3/08/$25.00 ©2008 IEEE

Using Stoney's formula, we have calculated the surface deformation of a Si substrate coated with a thin film layer. Calculation parameters are the thicknesses of the substrate (mirror) and the coating layer, nature of the materials, intrinsic stress in the thin film layer and finally thermal effect, when the mirror is cooled down to cryogenic temperatures. For microfabrication reasons, a 10 μm thick mono-crystalline Si mirror was chosen.

The surface quality of 100x200 μm² uncoated mirrors is 7 nm PtV in the OFF state as well as in the ON state - the micromirrors remain flat when operated (Fig. 1). The reflective coating on the optical side consists of a 10 nm chrome adhesion layer and a 50nm gold layer. The PtV deformation increases to about 35 nm; adding the same coating on the backside of the mirror decreases the PtV deformation to below 20 nm. Note that this backside coating also changed the sign of the curvature of the mirror. These values are in close agreement with the simulated deformations.

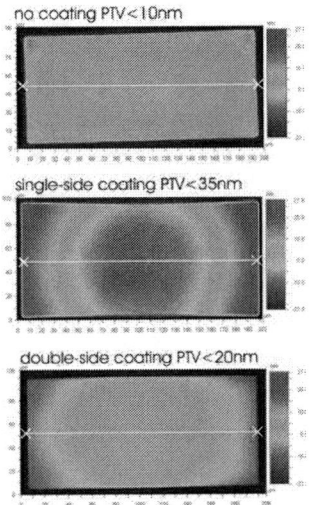

Fig. 1: Deformation maps of uncoated and coated micro-mirrors.

3 CRYOGENIC OPERATION AND CHARACTERIZATION

The cryogenic compatibility is crucial for the application in an infrared (IR) MOS. The operating temperature must be below 100 K for near and mid IR and below 40 K for far IR. Our MMA is designed such that all structural elements have a matched coefficient of thermal expansion (CTE) in order to avoid deformation or even flaking within the device when cooling it down to the operating temperature. The mirrors are covered with a gold layer for IR operation, whereas gold has a different CTE than silicon. As the silicon mirror is 10 μm thick and the coating is 60 nm thin, we estimate that the induced deformation will be small.

For characterising the surface quality and the performance of our MMA's at low temperature, we have developed a cryo chamber optically coupled to a high-resolution Twyman-Green interferometer [2]. The interferometer provides a sub-nanometer accuracy by using phase-shifting technique with a low-coherence source. The cryo-chamber allows pressure down to 10^{-6} mbar and temperatures down to 60 K. A dedicated PCB interface and mechanical support were developed, which allows working in vacuum and which avoids additional stress on the MMA device, which itself is packaged in PGA chip carrier. The PGA is inserted in a ZIF-holder integrated on the PCB board (Fig. 2).

The micromirrors could be successfully actuated before, during and after cryogenic cooling. We could measure the surface quality of the gold coated micromirrors at room temperature, below 100K and being actuated: There is a slight increase of the deformation from 35 nm to 50nm PtV, due to CTE mismatch between silicon and gold layer (Fig. 2). This small deformation is still well below the requirement for MOS application at IR. This value could be decreased if needed by using double-side coated mirrors, easily feasible in our process flow.

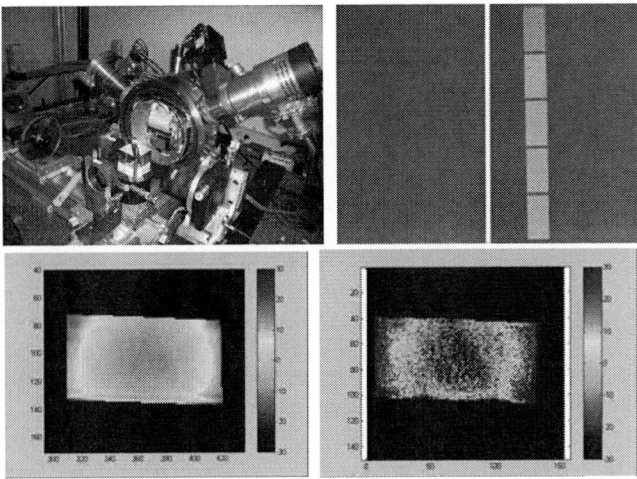

Fig. 2: Cryogenic set-up; functional testing of a micro-mirror array at 92K (0V and 90V applied); 100x200 μm² mirrors surface deformation at room temperature and at 92K

ACKNOWLEDGMENT

The authors would like to thank the SAMLAB at IMT and the Service Essais at LAM.

REFERENCES

[1] S. Waldis, F. Zamkotsian, P.-A. Clerc, W. Noell, M. Zickar, and N. de Rooij, "Arrays of high tilt-angle micromirrors for multiobject spectroscopy, " *IEEE Journal of Selected Topics in Quantum Electronics* **13**, pp. 168–176, 2007

[2] F. Zamkotsian, E. Grassi, S. Waldis, R. Barette, P. Lanzoni, C. Fabron, W. Noell and N. deRooij, "Interferometric characterization of MOEMS devices in cryogenic environment for astronomical instrumentation," in Proc. SPIE **6884**, San Jose, USA (2008).

Low Operation Voltage Non Self-Emissive MEMS Color Filter Pixels

Cheng-Yao Lo, Jukka Hast[*], Olli-Heikki Huttunen[*], Jarno Petäjä[*], Johanna Hiitola-Keinänen[*], Arto Maaninen[*], Harri Kopola[*], Hiroyuki Fujita and Hiroshi Toshiyoshi

Institute of Industrial Science, University of Tokyo, 4-6-1 Komaba, Meguro-ku, Tokyo 153-8505, JAPAN
E-mail. chengyao@iis.u-tokyo.ac.jp, Tel. +81-3-54526276, Fax. +81-3-54526250
[*]VTT Technical Research Centre of Finland, Kaitoväylä 1, Oulu, FI-90571, Finland

Abstract

A 50% reduction of operation voltage improvement was achieved on a non self-emissive color filter pixels based on MEMS (micro electro mechanical system) Fabry-Perot interference device by minimizing its Newton's rings' size. Newly designed air grooves in spacer layer were proved to be efficient to evacuate air trapped inside pixels which in turn effectively lowered its operation voltage. A seesaw effect was also found if air groove occupied too large area which degrades the operation voltage lowering benefit.

Keyword: Fabry-Perot, color filter, MEMS (micro electro mechanical system), air groove, seesaw effect

1. Introduction

Color filters have been widely used in display industry and decoration fields nowadays. Compared to traditional color-unchangeable color filters prepared by multilayer deposition [1], recent researches show possibilities of color-tunable ideas by either reflective [2] or transmissive [3] ways. The idea of making traditional multilayer color filter is to prepare a high-low-high-low stack of layers' index of refractions (n). Precise control of multilayer's thickness and material characteristics will provide a narrower transmissive bandwidth for high quality communication; however, by doing so, the loss of transmittance and process difficulties are also drawbacks. In this work, we modified previous work [3] with least multilayer design for broader visual band and by changing its spacer layer's structure. Partial coverage of spacer layer with air groove design provides air evacuation routes under operation and hence provided a successful way to suppress operation voltage.

2. Device Concept

Fabry-Perot interference cavity was designed in a multilayer structure as shown in Fig. 1. Under its OFF state, this multilayer structure does not meet the requirement for Fabry-Perot interference condition and the output light will be only influenced by two metal layers as grayish light. When electronic charges are supplied to both electrodes, Coulomb electrostatic force will pull the upper layer in contact with lower layer and the 6-layer (including air gap) structure becomes a 5-layer (without air gap) structure which satisfies Fabry-Perot interference condition. In such case, the output color will be controllable by adjusting either each layer's thickness or their optical characteristics of index of refraction.

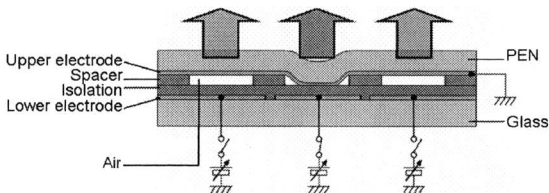

Fig. 1. Schematic plots of pixels in ON (center) and OFF states.

3. Experimental

Operation voltage is one of the most important factors which limit MEMS Fabry-Perot display pixels' commercial realization. A straightforward solution is to lower spacer's height for larger Coulomb electrostatic force with a parallel-plate model. Simulation results with lower spacer heights done with CoventorWare® in Fig. 2 typically followed the prediction of modified model [4] but the uncontrollability of electrostatic movement of lower spacer degrades the advantages gained from lower applied voltage.

In addition to initial gap, air inside a pixel is also believed to be a bottleneck due to the Newton's ring's width change along the operation voltage change. A new improvement solution was proposed with air groove in spacer layer as shown in Fig. 3. The purpose is to evacuate air in the pixel-under-operation, which reduce the air pressure when upper layer comes down to lower layer, and thus reduce the operation voltage. The spacer coverage were designed with 60% (wider air groove), 80%, 90% (narrower air groove), and 100% (no air groove, control group sample). These experimental group samples helped not only on the evaluation of structure shape but also on the evaluation of air evacuation efficiency.

978-1-4244-1917-3/08/$25.00 ©2008 IEEE   116

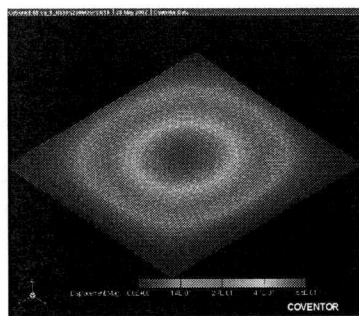

Fig. 2. Simulation results of Newton's ring under operation.

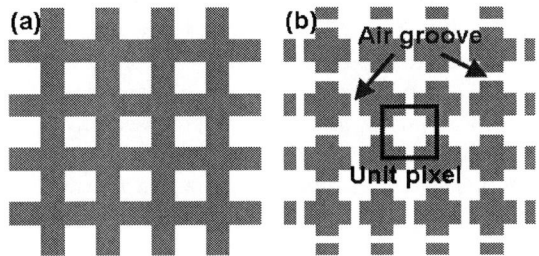

Fig. 3. Top-view plots of (a) 100% and (b) 60% spacer coverage.

4. Analysis

The demonstration of air groove is shown in Fig. 4. The display boundary (Newton's ring) was still located at the edges of a pixel but its width was successfully reduced. When we continued to increase the applied voltage, we found Newton's ring's width decreased dramatically at first and stopped decreasing finally. Newton's ring's width was measured at each point and was used for comparison as shown in Fig. 5.

The decrease stopped at higher applied voltage is believed to be the influence of competition force. When the applied force is strong enough, the upper layer above air groove will also be pulled down to lower layer. Since the total area of upper layer is fixed and somehow still stiff, a seesaw effect takes place: when one wants to push one end of seesaw (upper layer on air groove area) down with a fulcrum (spacer) will raise another end (upper layer on pixel area) unintentionally. We also found the area of display intruded into air groove area deeper and deeper when the applied voltage was raised. This proved that the opening width of air groove was wide enough for upper layer to be pulled down by electrostatic force. In other words, a seesaw effect takes place. We believe this is the root cause of the reverse trend in Fig. 5. Thus, we found that a wider air groove is not necessary to be able to provide a smaller Newton's ring under the same operation voltage. Instead, Newton's ring's width increased when air groove's width was not optimized.

5. Summary

We proposed a new spacer structure with air groove design for air evacuation during operation to lower the operation voltage. This design reduced the contact area of spacer and upper/lower layers without changing pixel's size and structure. Newton's ring's width at pixel edge was used to judge the improvement and air groove structure was proved to be effective to reduce Newton's ring's width. When air pressure trapped inside the pixel-in-operation was alleviated, operation voltage was smoothly lowered. With the same setting of Newton's ring's width, we cut down 50% operation voltage with optimized air groove design. A seesaw effect was used to explain the reverse trend of relationship between air groove size and Newton's ring's width with spacer coverage between 60% and 80%.

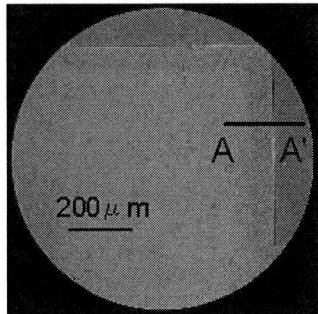

Fig. 4. 2000µm pixel demonstration with 60% spacer coverage..

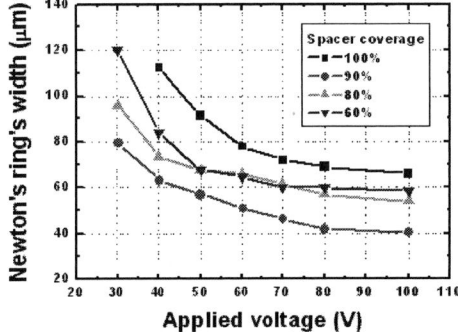

Fig. 5. Seesaw effect takes place when spacer coverage is not optimized..

Acknowledgment

This project is supported by Industrial Technology Research Grant Program in 2006 (ID: 06D48522d) from New Energy and Industrial Technology Development Organization (NEDO) of Japan and has been supported in part by the International Training Program (ITP) of the Japan Society for the Promotion of Science (JSPS).

Reference

1. P. Yeh, "Optical waves in layered media", John Wiley & Sons, pp. 144-149, 1986.
2. M. Miles, "Toward an iMOD ecosystem", Proc. 13th International Display Workshop, Otsu, Japan, MEMS4-2, pp. 1583-1586, Dec., 2006.
3. C. Lo, "Toward realization of transmissive display by MEMS etalon", IEICE Electronics Express vol. 5, no. 9, 2008 (in press).
4. Y. Taii, "A study on transparent flexible color display device using plastic MEMS technology", Master's thesis, Dept. Elec. Eng., University of Tokyo, pp. 19-21, 2006.

White-light electroluminescence from ZnO nanowires/polyfluorene heterojunction diodes

Chun-Yu Lee[a], Jing-Shun Huang[a], Sheng-Hao Hui[b], Wei-Fang Su[c], and Ching-Fuh Lin*[d]

Abstract—The characteristics of a nanocomposite consisting of the blue-emitting polymer polyfluorene and ZnO nanowires are reported. The electroluminescence spectrum of the white light emission is from about 400 nm to 750 nm.

Index Terms—zinc oxide, electroluminescence, polyfluorene, nanowires, light emitting device, white light

I. BACKGROUND

Zinc oxide (ZnO) semiconductor is considered a promising material for advanced optoelectronic applications since it is an environmentally friendly material grown at low temperatures on cheap transparent substrates [1-2] and has a direct wide band gap of 3.37 eV, a very large exciton binding energy of 60 meV, and a strong cohesive energy of 1.89 eV, important for robust light emission. Recently, one-dimensional oxide semiconductor materials have received great attention due to their excellent properties [3]. In this regard, ZnO nanostructures such as nanoparticles, nanobelts, nanowires, nanotips, and nanosheets have been synthesized for optoelectronic devices. However, only a few ZnO-based electroluminescence (EL) devices have been constructed using ZnO nanowires, ZnO nanorods, and ZnO epitaxial film [4-5]. Here, we report the use of ZnO nanowires to fabricate the ZnO EL devices of white-light emission by hydrothermal method. The method has the prominent advantage of making the cost of devices less expensive.

II. CURRENT RESULTS

In this study, the fabrication of ZnO nanowires/polyfluorene (PF) heterojunction diode LEDs and their EL characteristics are introduced. Two types of EL devices have been fabricated. One has the ZnO nanowire layer. The device structure is ITO/ZnO nanowires/PF/Al (200nm) (device 1). The other has no ZnO nanowire layer, and the device structure is ITO/PF/Al (200nm) (device 2). We have employed hydrothermal method to grow ZnO nanowires on an indium tin oxide (ITO) coated glass substrate. Hydrothermal growth of ZnO nanowires was achieved by suspending the ZnO seed-coated substrates upside down in an aqueous solution of zinc nitrate and hexamethylenetetramine at low temperature of 90 ℃. Then the blue-emitting polymer, PF, is fabricated on ZnO nanowire layer for device 1 and ITO substrate for device 2 by the spin-coating method respectively. Afterwards, the composite film is sandwiched between the indium tin oxide and the negative electrode of aluminum to form a light emitting device (LED).

We investigate the ZnO nanowires/polyfluorene composite films by field emission scanning electron microscope (FESEM). Figure 1 shows the FESEM images of the surface of ZnO nanowires with (a) no PF, (b) PF, and the depth profile of ZnO nanowires/PF composite (c). The average diameter and length of ZnO nanowires formed on the surface of the glass coated with ITO was about 42±15 nm and 500 nm, respectively. Figure 1(b) and Figure 1(c) show a very thick coating on top of the ZnO nanowires, while the space between wires is solidly filled with the PF. According to the spin-coating parameter and electron micrograph, we estimate the PF thickness on the ZnO nanowire tips to be in the range of 50-100 nm.

The I-V characteristics from ZnO nanowires EL devices are shown in Figure 2 (a). The device with ZnO nanowires (curve a) shows much higher current density at the same voltage than the device without ZnO nanowires (curve b). We attribute this phenomenon to two reasons. The first reason is higher carrier mobility of ZnO NWs/PF composite than that of PF film only [6]. The second reason is more injection area of ZnO NWs/PF composite than that of PF film only. Based on the FESEM micrograph after PF spin-on process, the device has about 175 nanowires per μm^2 on average to successfully form the heterojunction. Therefore, it is estimated that a total of about 2-3 billions nanowires exist in an area of 0.15 cm2. Hence there are more contact interface in small area, which could therefore produce more current injection.

Figure 2(b) shows the EL spectra of the two kinds of the EL devices under forward bias of 10 V. The device without ZnO nanowires (device 2) shows three emission peaks at 425 nm, 449 nm, and 491 nm. In comparison, the emission of the device with ZnO nanowires (device 1) appears at 480 nm and 568 nm, which

---

This work was supported by the National Science Council, Taiwan, Republic of China, with Grant No.: NSC96-2221-E-002-277-MY3, and NSC96-2218-E-002-025.

The authors are with [a]Graduate Institute of Photonics and Optoelectronics, National Taiwan University, Taipei, Taiwan, 10617 R.O.C.;

[b]Institute of Polymer Science and Engineering, National Taiwan University, Taipei, Taiwan, 10617 R.O.C.;

[c]Department of Materials Science Engineering, Center for Condensed Matter Sciences, National Taiwan University, Taipei, Taiwan, 10617 R.O.C.

[d]*Corresponding author: C. -F. Lin is also with *Graduate Institute of Photonics and Optoelectronics, Graduate Institute of Electronics Engineering and Department of Electrical Engineering*, National Taiwan University. (phone: 886 2 3366 3540; fax: (886) 2 2364 2603; e-mail: cflin@cc.ee.ntu.edu.tw)

is red-shift from that of the device without ZnO nanowires (device 2), with a broader shape. Both spectra suggest that the emissions should take place in the PF layer because the emission from the ZnO nanowires with band-gap emission of 380 nm and defect emission of 560 nm are not observed. In addition, the broad luminescence of the device with ZnO nanowires results from inhomogeneous broadening transition mechanism in the PF [7]. This phenomenon is probably due to the interaction between the surface hydroxyl groups of the ZnO nanowires and the PF [8].

In summary, we demonstrate that ZnO nanowires/polyfluorene composite films produce white light electroluminescence. The electroluminescent spectrum of the white light emission spans from 400 nm to 700 nm, which is attributed to the interaction between the surface of ZnO nanowires and polyfluorene components of the hybrid film.

REFERENCES

[1] C. -Y. Lee, Y. -T. Haung, W. -F. Su, and C. -F. Lin, "Electroluminescence from ZnO nanoparticles/organic nanocomposites," Appl. Phys. Lett., vol. 89, pp. 231116-1-231116-3, Dec. 2006.

[2] Z. -Z. Ye, J. -G. Lu, Y. -Z. Zhang, Y. -J. Zeng, L. -L. Chen, F. Zhuge, G. -D.Yuan, H. -P. He, L. -P. Zhu, J. -Y. Huang, and B. -H. Zhao, "ZnO light-emitting diodes fabricated on Si substrates with homobuffer layers" Appl. Phys. Lett., vol. 91, pp. 113503-1-113503-3, September. 2007.

[3] C. -P. Chen, M. -Y. Ke, C. -C. Liu, Y. -J. Chang, F. -H. Yang, and J. -J. Huang, "Observation of 394 nm electroluminescence from low-temperature sputtered n-ZnO/SiO2 thin films on top of the p-GaN heterostructure" Appl. Phys. Lett., vol. 91, pp. 091107-1-091107-3, August. 2007.

[4] Y. Ryu, T. -S. Lee, J. -A. Lubguban, H. -W. White, B. -J. Kim, Y. -S. Park, and C. -J. Youn, "Next generation of oxide photonic devices: ZnO-based ultraviolet light emitting diodes," Appl. Phys. Lett., vol. 88, pp. 241108-1-241108-3, June. 2006.

[5] W. -Z. Xu, Z. -Z. Ye, Y. -J. Zeng, L. -P. Zhu, B. -H. Zhao, L. Jiang, J. -G. Lu, H. -P. He, and S. -B. Zhang, "ZnO light-emitting diode grown by plasma-assisted metal organic chemical vapor deposition," Appl. Phys. Lett., vol. 88, pp. 173506-1-173506-3, April. 2006.

[6] D. J. D. Moet, L. Jan Anton Koster, B. –D. Boer, and P. W. M. Blom, "Hybrid Polymer Solar Cells from Highly Reactive Diethylzinc: MDMO–PPV versus P3HT," Chem. Mater., vol. 19, pp. 5856-5861, July. 2007.

[7] D. Vak, B. Lim, S. –H. Lee, and D. Y. Kim, "Synthesis of a Double Spiro-Polyindenofluorene with a Stable Blue Emission," Org. Lett., vol. 7, pp. 4229-4232, July. 2005.

[8] J. Zheng, R. Ozisik, and R. –W. Siegel, "Disruption of self-assembly and altered mechanical behavior in polyurethane/zinc oxide nanocomposites," polymer, vol. 46, pp. 10873-10882, September. 2005.

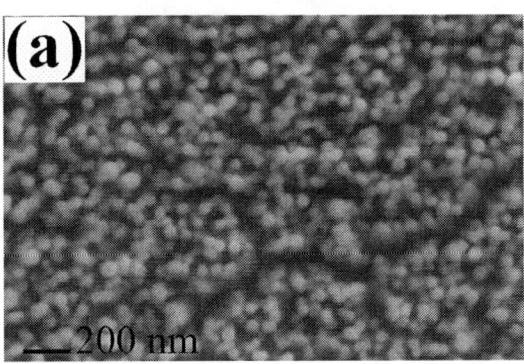

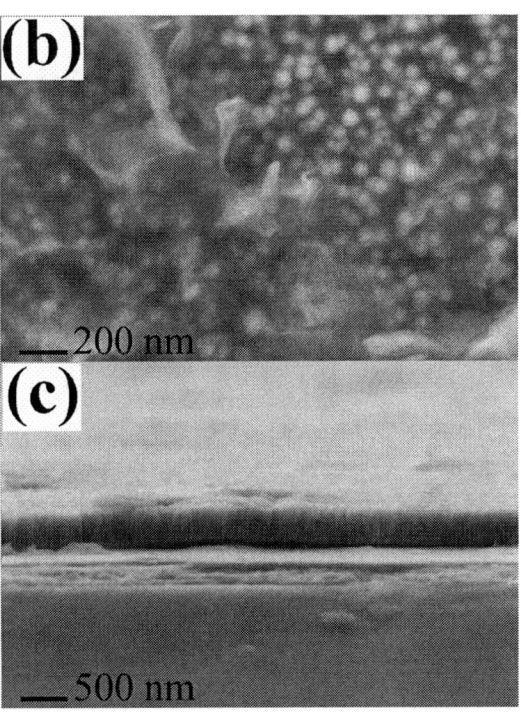

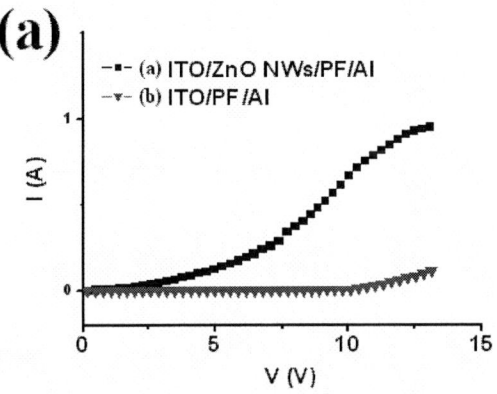

Fig. 1. FESEM images (top view) of the (a) ZnO nanowire layer; (b) ZnO nanowire layer after deposition of polyfluorene; and (c) FESEM image (cross-sectional view) of the ZnO nanowire layer after deposition of polyfluorene.

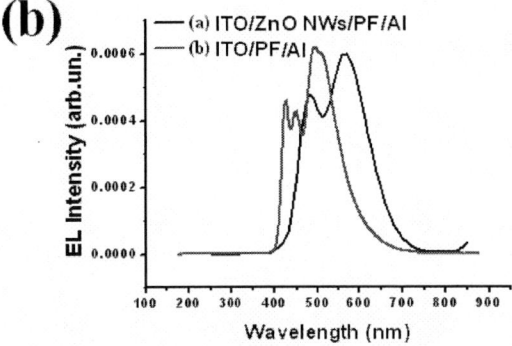

Fig. 2. I-V characteristics (a) and electroluminescence spectra (b) of device A (ITO/ZnO NWs/PF/Al) and device B (ITO/PF/Al)

Stabilization of Temperature Characteristics of Micromirror for Low-Voltage Driving Using Thin Film Torsion Bar of Tensile Poly-Si

Minoru Sasaki[1], Masayuki Fujishima[2], Kazuhiro Hane[2], and Hideo Miura[2]

[1]Dept. of Advanced Science and Technology, Toyota Technological Institute,
Hisakata 2-12-1, Tenpaku-ku, Nagoya 468-8511, Japan
E-mail: mnr-sasaki@toyota-ti.ac.jp
[2]Dept.of Nanomechanics Eng., Tohoku University, Aramaki 6-6-01 Aoba-ku, Sendai, 980-8579, Japan

Abstract

The micromirror with the tense thin film torsion bar can realize the low-voltage driving. The temperature characteristic is improved using polycrystalline (poly-) Si thin film taking advantage of the following features. The large tensile stress is obtained by the crystallization of amorphous (a-) Si film. The doping realizes the electrical connection. The poly-Si has the almost same coefficient of thermal expansion (CTE) with that of Si substrate.

Keywords: temperature characteristics, micromirror, thin film torsion bar, tensile stress, crystallization induced stress

1. INTRODUCTION

The electrostatic driving is preferred for the micromirror due to the advantage of low power consumption. The heat generation and the resultant degradation of the reliability can be avoided. However, the electrostatic force is rather small. The driving voltage of ~100 V is necessary. This is the drawback for adopting the micromirror in the existing equipments. Realizing the large rotation angle with the low driving voltage is ideal. The goal is working with the voltage source of the existing system, for example 5, 12, or 24 V. Limited devices have succeeded to this requirement [1].

The tense thin film torsion bar is proposed in our previous study, and SiN film is used realizing the low-voltage driving [2]. The tensile stress is ~780 MPa. The tension is for the rigidity against the vertical displacement. Since SiN is insulating, the metal layer is indispensable for the electrical connection. Cr/Au metals are used. Figure 1 shows the previous micromirror. The surface roughness is attributed to the attack of HF vapor on SiN film during the sacrificial layer etching. The performance of this device changes drastically depending on the temperature. Figure 2 shows the mirror rotation angle obtained at different temperatures. At 17 °C, the mirror rotates by 4.3° at 5 V. When the temperature increases to 60 °C, the mirror rotation at 5 V decreases to 1.3°. This can be attributed to the materials on the torsion bar having different CTEs.

In this study, the micromirror with the thin film torsion bar is realized using the tense poly-Si film for improving the temperature characteristics. Its CTE is almost same with that of substrate Si. The large tensile stress is obtained by the crystallization of a-Si film.

2. PRINCIPLE

As shown in the inset of Fig. 1, the previous torsion bar includes Cr/Au metal and SiN films. When the thin film consists of two layers having different thermal expansions, the film can bend under the stress balance between two layers. Bending moment M generated by the temperature

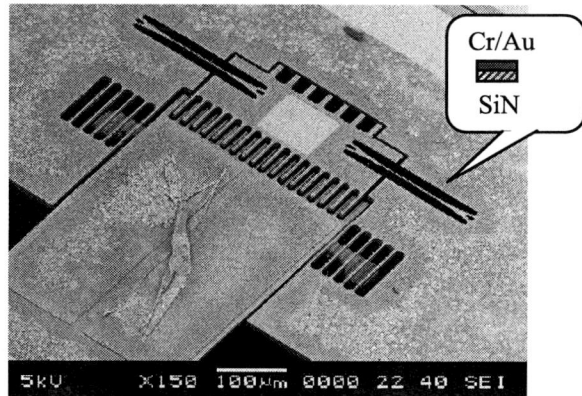

Fig. 1. Previous micromirror. The fixed comb deflect downward realizing the vertical comb.

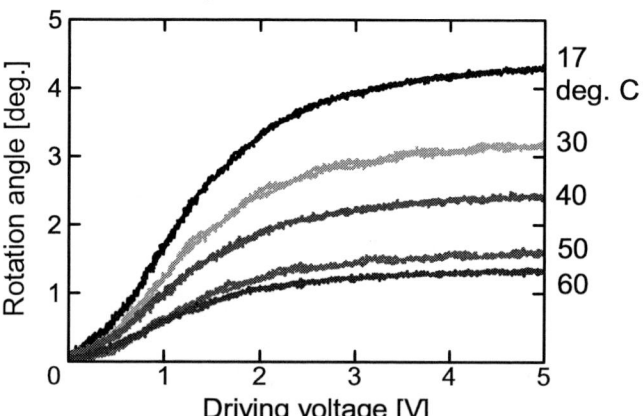

Fig. 2. Temperature dependence of the mirror rotation angle. The torsion bar consists of SiN and metal layers.

change is expressed as follows.

$$M = \frac{t_1 + t_2}{2} \cdot \frac{(\alpha_2 - \alpha_1)\Delta T}{\dfrac{1}{E_1 w t_1} + \dfrac{1}{E_2 w t_2}} \quad (1)$$

978-1-4244-1917-3/08/$25.00 ©2008 IEEE

ΔT is the temperature change. Subscripts 1 and 2 show layer numbers. E is Young's modulus. w and t are width and thickness of the layer, respectively. α is CTE. When the difference of α is large, the structure is strongly affected by ΔT. The thin film torsion bar will deflect destructing the perpendicular relation between the tension and the rotational motion increasing the rotational spring constant [3].

The poly-Si is known to generate the tensile stress. The mechanism generating the tensile stress is the crystallization from the amorphous phase. Nee et al. deposited the film in the mixture of poly- and a-Si phases. When the film is deposited under the pure a-Si phase, the larger tensile stress can be obtained. The tensile stress of >700 MPa is reported [4]. In addition, the doping of poly-Si gives the electrical connection. Using our CVD equipment, the deposition is carried out at 525 °C using SiH_4 gas. After the annealing at >650 °C, the tensile stress of ~600 MPa is obtained [5].

3. RESULTS

Figure 3 shows the fabricated micromirror using the tensile poly-Si film. The design of the comb and the mirror is nearly same. Compared to the previous device in Fig. 1, the device surface is smoother. This can be explained by the chemical inertness against HF vapor used in the sacrificial SiO_2 layer etching. The center area on the mirror is where the poly-Si directly connects to the crystal Si layer. The electrical connection is obtained. The magnified image shows the torsion bar. The design size is 200x4x0.3 μm^3.

Figure 4 shows the round trip data of the mirror rotation angle. The mirror is grounded and the fixed comb is biased. The comb-to-comb height difference d at the initial state is the parameter. When d is 6 μm, mirror rotation angle reaches $3.6°$ at 15V. When d is 12 μm, mirror rotation angles are $5.5°$ at 5V, and $7.6°$ at 10V. Low-voltage driving is realized. The thin film torsion bar is a little stiffer compared to the previous one using SiN film. The thickness of 300 nm is stably obtained.

Figure 5 shows the mirror rotation angle at different temperatures. At 18 °C, this mirror shows the rotation angle of $4.0°$ at 15 V. When the temperature increases up to 120 °C, the fluctuation of the rotation angle is $0.13°$ and 3 % to the full-scale. The rotation angle at 15 V decreases or increases randomly. Temperature characteristic is improved.

This research was supported by a grant-in-aid for scientific research on priority areas (no. 19016003). The facilities used for this research include the micro/nano-machining research and education center, at Tohoku University.

REFERENCES

[1] D. Hah, S. T. Huang, J. Tsai, H. Toshiyoshi, M. C. Wu, J. Microelectromechanical Systems, 13, April (2004) 279-289.
[2] M. Sasaki, S. Yuki, K. Hane, IEEE Photonics Technology Letters, 18, No. 15, (2006) 1573-1575.
[3] M. Sasaki, S. Yuki, K. Hane, IEEE J. Selected Topics in

Fig. 3. Fabricated micromirror. The tensile thin film torsion bar consists of single doped poly-Si layer.

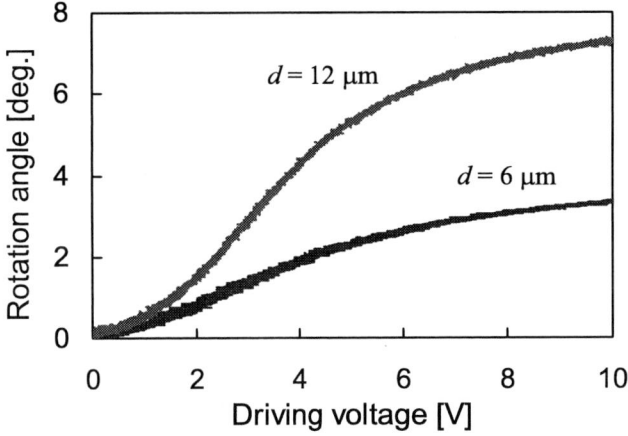

Fig. 4. Mirror rotation as a function of the driving voltage.

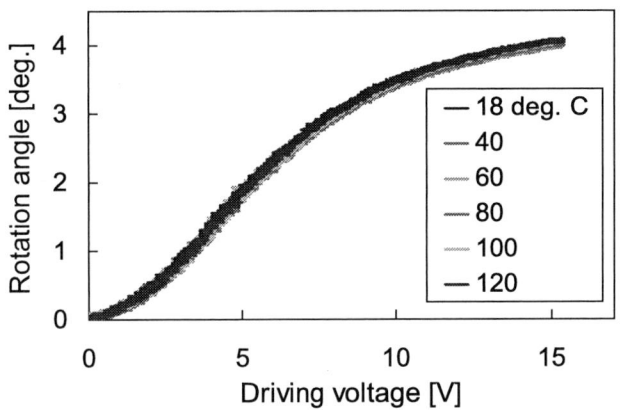

Fig. 5. Mirror rotation angle as a function of driving voltage at different temperatures.

Quantum Electronics, 13, Issue 2 (2007) 290-296.
[4] H. Miura, H. Ohta, N. Okamoto, T. Kaga, Trans. Japan Society Mechanical Eng. A 58, (1992) 1960-1965 (in Japanese).
[5] M. Sasaki, T. Sasaki, K. Hane, H. Miura, Journal of Optics A: Pure and Applied Optics, 10 (2008) 044004 (8pp).

Simulation-based design of a MEMS X-ray optic for X-ray astronomy

M.Koshiishi[1], Y.Ezoe[2], I.Mitsuishi[1], M.Mita[1], K.Mitsuda[1], T.Takano[3], and R.Maeda[3]

[1] Institute of Space and Astronautical Science (ISAS), Japan Aerospace Exploration Agency (JAXA)
[2] Tokyo Metropolitan University
[3] National Institute of Adbanced Industrial Science and Technology (AIST)

ABSTRACT

An ultra light-weight X-ray optic using MEMS technologies was designed for X-ray astronomy. Numerical simulation was utilized to estimate allowable fabrication accuracies. Obtained X-ray images with a fabricated optic were consistent with the design.

1. INTRODUCTION

X-ray focusing systems are widely used in medical science, microanalysis and astronomy. Since X-rays are difficult to focus refractively, grazing-incidence optics are generally utilized. However, the usable grazing-incidence angle is small less than several degrees at 1 keV. Hence, in astronomical satellites, hundreds and thousands of X-ray mirrors are prepared to increase an effective area. Also, a surface roughness of each mirror should be on the order of incoming X-ray wavelength i.e., ~ 1 nm. An innovative light-weight and low-cost X-ray optic is thus needed for future satellite missions.

We have been developing novel light-weight X-ray optics based on MEMS technologies [1-4]. We use smooth Si (111) side walls obtained after anisotropic KOH etching of Si (110) wafers for X-ray mirrors. To date, we have confirmed X-ray reflection [1] and imaging [4]. Here we present a simulation-based design for our test X-ray optic.

2. DESIGN

Figure 1 is a concept of our test X-ray optic. It consists of mirror chips and a optic mount. Incident X-rays are reflected on Si (111) side walls within the mirror chips. A mount has step structures and through-hole to incline chips against parallel incident X-rays. In order to focus X-rays into a point, inclination angles increase toward the edge of the mount. The maximum reflection angle (1.73 degree) is equal to the critical angle of Si at 1 keV. Thus, this optic can be used for soft X-rays less than 1 keV. The focal length is determined by the reflection angle and the optic diameter as 750 mm. Process flows and parameters of the mirror chips and the mount are described in [4]. Below we focus on a detailed design of this optic.

Firstly, we determined two key parameters for the mirror chip: a slit width d and a slit pitch p (fig 1c). To maximize the effective area, the pitch p should be as small as possible and the width d should be $d = t \cdot \tan\theta$, where t is a wafer thickness and θ is a reflection angle. For example, if t is 300 μm and θ is 1 degree, d will be 5 μm. However, we must consider surface tension of pure water in cleaning process after etching. The minimum pitch can be describe as,

$$p \geq \left(\frac{5}{16} \frac{\gamma t l^3}{E d^2} \right)^{\frac{1}{3}}, \qquad (1)$$

where γ is the surface tension of pure water, t is the slit length, and E is the Young's modulus of Si (111) planes. Figure 2 shows the necessary p for the mirror chips.

Secondly, we have to constrain fabrication accuracy (Δ_{mgn} and Δ_{stp} in fig. 1 c and d) and margins between the mirror and the mount (Δ_{mgn}), because possible shift and rotate of the chip in the mount can degrade image quality. To quantitatively evaluate this effect, we constructed numerical simulation of this optic. In this simulation, we consider the optic geometry, X-ray reflectivity depending on the reflection angle, and a position of the focus. Since the reflectivity is also a function of the X-ray energy and the mirror surface roughness, we assumed 1.25 keV and 0 nm, respectively. Free parameters are Δ_{mgn} and Δ_{stp}.

Figure 3 displays simulated images with different Δ_{mgn} and Δ_{stp}. In the ideal case ($\Delta_{\mathrm{mgn}} = 0\mu$m and $\Delta_{\mathrm{stp}} = 0\mu$m) (fig. 3 a), the image size is determined by those of the mirror chips. The FWHM (Full Width at Half Maximum) of the focus is 3.8 mm. If we increase Δ_{mgn} to 50μm (fig. 3 b), the image size does'nt change significantly. On the other hand, Δ_{stp} of 5μm affects the image (fig. 3 c). This is because the Δ_{stp} directly influence the reflection angle. The 5 μm error in Δ_{stp} can change the reflection angle by at most 0.2 degree. Figure 4 summarizes the dependences of the FWHM on Δ_{mgn} and Δ_{stp}. From this result, we constrained $\Delta_{\mathrm{mgn}} < 50\mu$m and $\Delta_{\mathrm{stp}} < 1\mu$m.

To achieve these accuracies, we tuned conditions of a dicing cut and a deep reactive ion etching for the chip and the mount, respectively. The realized accuracies were $\Delta_{\mathrm{mgn}} < 30\mu$m and $\Delta_{\mathrm{stp}} < 1\mu$m. Obtained X-ray images of the optic [4] were consistent with these simulations.

REFERENCES

[1] Y. Ezoe et al., "Micro pore X-ray optics using anisotropic wet etching of (110) silicon wafers", Appl. Opt., 45, pp. 8932-8938, 2006

[2] M. Koshiishi et al., "Design and Fabrication of a MEMS X-ray Optic using Anisotropic Wet Etching of Si wafers," IEEE/LEOS Int. Conf. on Optical MEMS and Their Applications, pp. 84-85, 2006

[3] M. Koshiishi et al., "The first light of a single-stage MEMS X-ray optic", Proc. of SPIE, 6688, 668814, 2007

[4] Y. Ezoe et al., "A micromachined X-ray collector for space astronomy", Sens. & Act. A, in press.

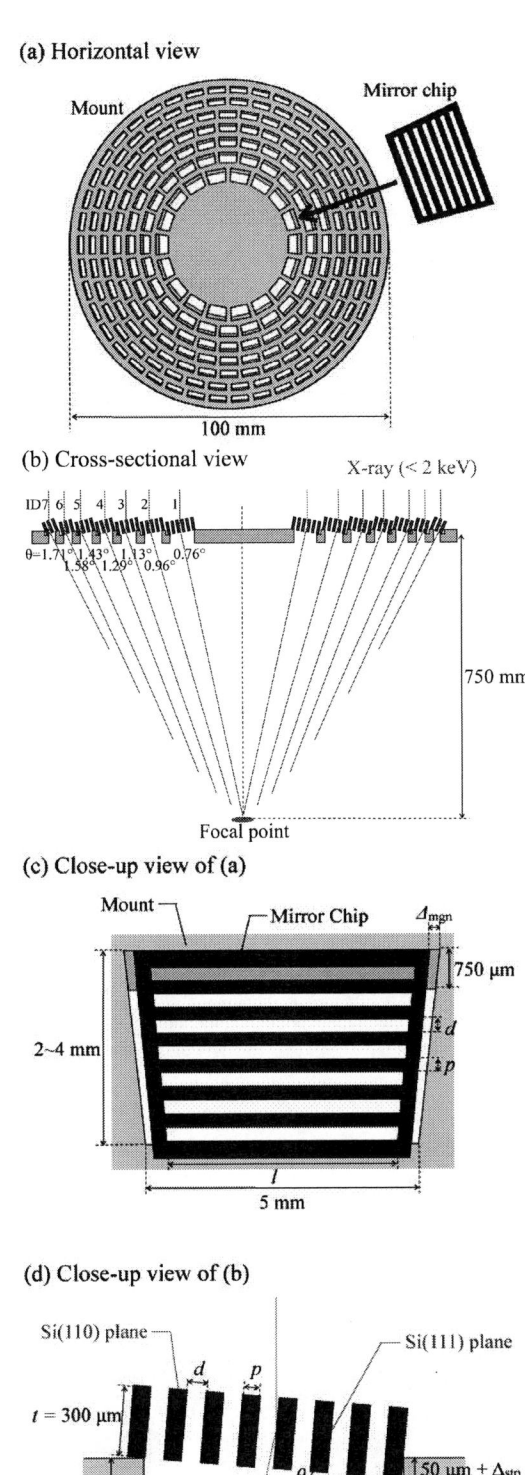

(a) Horizontal view

(b) Cross-sectional view

(c) Close-up view of (a)

(d) Close-up view of (b)

Figure 1: Conceptual views of our MEMS X-ray optic.

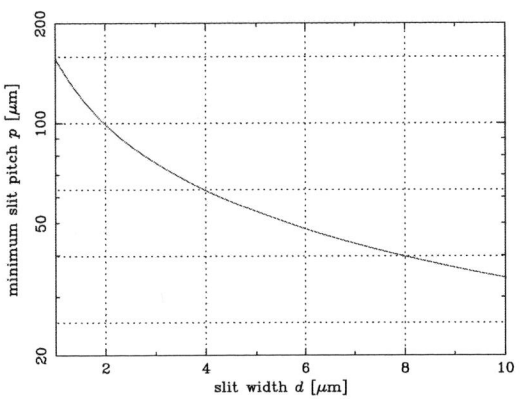

Figure 2: The necessary slit pitch p as a function of the slit width d.

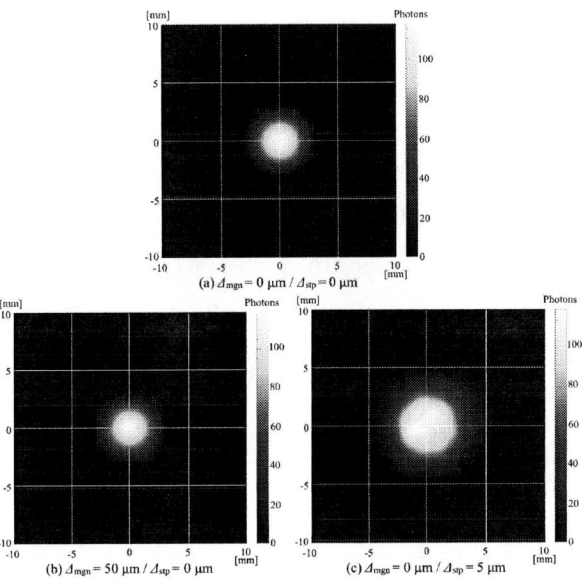

(a) $\Delta_{\mathrm{mgn}} = 0\ \mu\mathrm{m}$ / $\Delta_{\mathrm{stp}} = 0\ \mu\mathrm{m}$

(b) $\Delta_{\mathrm{mgn}} = 50\ \mu\mathrm{m}$ / $\Delta_{\mathrm{stp}} = 0\ \mu\mathrm{m}$

(c) $\Delta_{\mathrm{mgn}} = 0\ \mu\mathrm{m}$ / $\Delta_{\mathrm{stp}} = 5\ \mu\mathrm{m}$

Figure 3: Simulated X-ray images with different fabrication errors of Δ_{mgn} and Δ_{stp}.

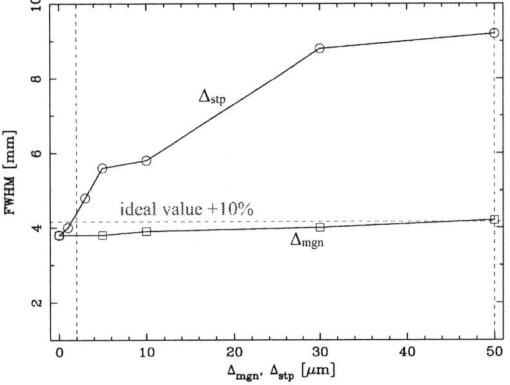

Figure 4: The FWHM image size as a fucntion of Δ_{mgn} and Δ_{stp}.

Fabrication of Sub-micrometer Si Spheres with Atomic-scale Surface Smoothness Using Homogenized KrF Excimer Laser Reformation System

Shih-Che Hung, Shu-Chia Shiu, Cha-Hsin Chao and Ching-Fuh Lin

Graduate Institute of Photonics and Optoelectronics, National Taiwan University,

No. 1, Sec. 4, Roosevelt Road, Taipei, 10617 Taiwan

Recently, microspheres have received increasing attention as optical microcavity resonators due to their high quality morphology dependent resonances (MDRs) [1]. Numerous applications have been realized by using microsphere MDRs. However, most of the research being conducted is based on noncrystalline materials such as glass [2] or polymers [3]. Since silicon is the dominating material in current integrated circuit industry, fabricating microspheres on a Si-based substrate with present electronic devices in a single chip are expected. The study contributes to compact and low-cost integrated devices for mass-market applications. Here, a novel technique using the homogenized KrF excimer laser reformation is presented as an alternative method for fabricating Si spheres.

In our experiment, the silicon spheres were fabricated on a commercial SOI wafer with a 1-um-thick single-crystal silicon film and a 500-nm-thick buried oxide (BOX). The samples were first spin-coated with an electron-beam resist (ZEP520, Zeon Chemical) with a thickness of 360 nm and patterned with electron beam (e-beam) lithography system (Elionix ELS-7500EX). After the resistor defined by e-beam lithography system, a chromium layer with a thickness of 100 nm, working as an etching mask, was deposited by E-Gun evaporation subsequently. After Chromium evaporation, the samples were immersed in a resist remover (ZDMAC, Zeon Chemical) to lift off the undesired chromium. As the etching mask was prepared, reactive-ion-etch (RIE) was then used to dry etch the sub-micrometer / micrometer Si rods. After RIE etching, the Cr layer on top of the ridge waveguide is removed by a remover solution (CR-7, Cyantek). Finally, the samples were placed in a vacuum chamber for laser illumination. Vacuum environment at a pressure of 5 x 10^{-4} Pa was present. This suppresses the surface oxidation and impurity in-diffusion to preserve the quality of illuminated surface. Energy densities imposed on the samples were applied at 2 J/cm^2 in the experiment. The beam power was 14 MW for 25-ns illumination. In addition, normal incidents, namely illumination form the top of rod structures, were conducted here. The process was eventually finished for one pulse of irradiation with the above-mentioned condition.

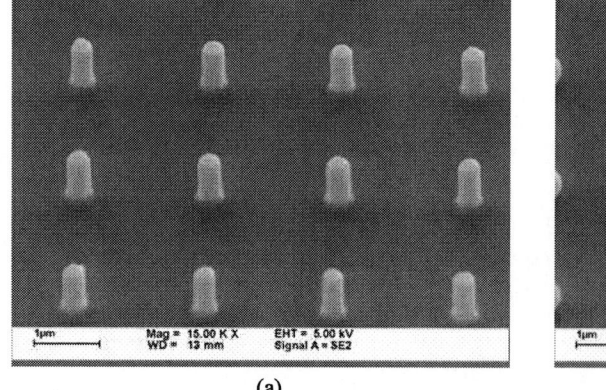
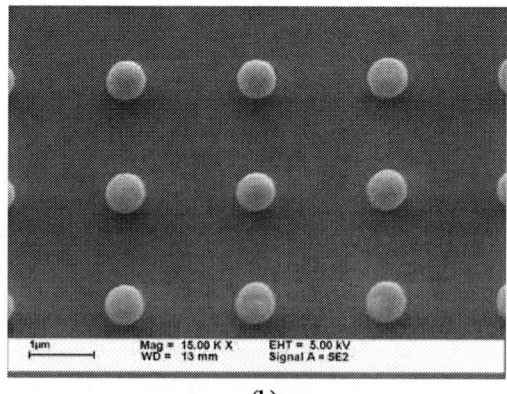

(a) (b)

Fig. 1 30 degree tilted SEM images (a) before illumination (b) after illumination

978-1-4244-1917-3/08/$25.00 ©2008 IEEE

High-resolution scanning electron microscopy (SEM) was used to examine the fabrication results. Fig. 1(a) shows the 30 degree-tilted SEM image of the sample at the process before excimer laser illumination. On the other side, Fig. 1(b) shows 30 degree-tilted SEM image of the sample after excimer laser illumination. The Si rods have heights of 1 um and diameters of 360 nm after RIE etching. The Si profiles transformed from the rods to the spheres after excimer laser illumination. These Si sub-microspheres with greatly smooth surfaces have diameters of about 450 nm. The principle of laser reformation for fabricating Si spheres is to melt the Si device layer by a high energy laser pulse. The Si in the device layer absorbs laser energy and changes the energy into thermal energy. In a very short time, the Si is then melted into liquid. In the liquid phase, the surface tension reforms the Si structure to the spheres since the surface tension forces the surface area to be the minimum.

For a passive sub-micrometer optical confinement device, surface roughness has critical influence on the confinement efficiency. In previous work [4], surface roughness was evaluated by atomic force microscopy. By one shot of laser pulse, the RMS roughness of the laser-reformed surface reduces from 14 nm to 0.28 nm. By 5 shots of laser pulses, the RMS roughness reduces to 0.24 nm. Such a high roughness reduction is due to surface tension which enables the surface area to be the minimum by nature. It is an additional benefit when we fabricate Si spheres using excimer laser reformation. On the other hand, the MWPCD responses for an untreated Si wafer, a furnace-treated Si wafer, and an excimer-laser-illuminated Si wafer were measured. The carrier lifetime of the original wafer is 1818 µs. It becomes 981 µs after illumination with one shot of the KrF excimer laser pulse with an energy density of 1.4 J/cm^2 at normal incidence. In contrast, the carrier lifetime of the furnace-treated wafer reduces to 106 µs. The laser reformation technique results in less damage to surface quality of Si wafers due to two reasons. First, the recrystallized Si is still single crystalline if the melted Si recrystallizes at a sufficiently fast regrowth rate [5]. Second, the pulse duration is so short that the defect diffusion is insignificant and the defect diffusion is further reduced by placing the Si wafer in a vacuum chamber at a base pressure of 5 x 10^{-4} Pa. These properties promise the application of the Si spheres in optical confinement device.

In conclusion, the fabrication process of sub-micrometer Si spheres with atomic-scale surface smoothness using homogenized KrF Excimer laser reformation system is presented. After excimer laser illumination process, the rods with diameters of 360 nm reformed into greatly smooth spheres with diameters of 450 nm.

REFERENCES

[1] Vahala, K.J., "Optical microcavities," World Scientific, Singapore, (2004)

[2] P. Jeuch, J. P. Joly, and J. M. Hode, "P-glass reflow with a tunable CO/sub 2/laser," in *Laser and Electron Beam Interactions with Solids. Proc. Materials Research Society Annual Meeting.* Amsterdam, The Netherlands, 1982, pp. 603–608.

[3] H. Yang, C. Ching-Kong, W. Mau-Kuo, and L. Che-Ping, "High fillfactor microlens array mold insert fabrication using a thermal reflow process," *J. Micromech. Microeng.*, vol. 14, pp. 1197–1204, 2004.

[4] Shih-Che Hung, Eih-Zhe Liang, and Ching-Fuh Lin, "Silicon waveguide sidewall smoothing by KrF excimer laser reformation," accepted by J. Lightwave Technology, March 2008

[5] B.C. Larson, J.Z. Tischler, and D.M. Mills, "Nanosecond resolution time-resolved x-ray study of silicon during pulsed-laser irradiation," Journal of Materials Research 1, 144-154 (1986).

Space Instruments based on MOEMS Technology

B. Guldimann, L. Venancio, K. Wallace, J. Perdigues, Z. Sodnik, B. Furch
ESA-ESTEC
PO-Box 299, 2200 AG Noordwijk,
The Netherlands
Benedikt.Guldimann@esa.int

A summary of current activities of the European Space Agency (ESA) related to optical MEMS technologies with a strong emphasis on the instrument aspects is presented. Future space applications and planned activities where optical MEMS might be an interesting option conclude this paper.

I. INTRODUCTION

Space and particularly planetary exploration missions demand very small, light weight, low power, rugged, flexible and high performance instruments. Micromachined devices might be key components offering exactly these characteristics for a broad range of applications. Optical MEMS technologies, usually also called MOEMS technology, for space instruments are currently being developed or investigated within ESA R&D programmes.

II. SPACE INSTRUMENTS

Telecommunications

An important application of optical MEMS in space telecom payloads is to provide circuit switching and redundancy functionalities. ESA is presently investigating novel payload concepts for space telecom based on photonic technologies, whose core element for switching RF/microwave channels is an optical switch [1]. A promising technology for implementing such an optical switch is MEMS technology. These photonic payloads aim at broader bandwidth, signal transparency, and enhanced routing flexibility at lower mass and size than typical for telecom payloads in full microwave implementation.

Another application of optical MEMS for telecommunications is beam steering in an inter-satellite laser link. Such a link has been established for the first time between the GEO and LEO satellites Artemis and SPOT 4 in 2001 and between two LEO satellites TerraSAR-X and NFIRE in 2008. The steering mirrors use classical electromagnetic or piezoelectric actuation; they are therefore relatively heavy components. An ongoing ESA activity aims at miniaturisation of a high precision optical micro-scanner and tracker. The requirements on the achievable scan angle and tracking precision are very challenging. They depend mainly on the front optics aperture and output scan angle of the laser link instrument.

X-ray Telescope

For ESA's X-Ray Evolving Universe Spectroscopy Mission (XEUS) a new technique for manufacturing large scale x-ray optics is being developed, using commercially available silicon wafers [2]. These wafers already have a surface roughness which is sufficiently low for x-ray reflection (grazing incidence). The silicon wafers are processed to form wedged plates with a rib structure on their non-reflecting side. Each plate is then pre-bent to the required shape, stacked and bonded to the underlying plate to form a monolithic silicon structure with pores on a sub-millimetre scale. These silicon stacks (fig.1) are aligned to form an x-ray optic unit, which forms the elementary part of a large x-ray telescope aperture of diameter 4 m and focal length 35 m. One of the particularly challenging processes is the bonding of 45 structured silicon plates, having a surface area up to 15 cm x 10 cm.

Figure 1: A silicon plate stacks aligned and mounted to form a mirror module. X-rays are focused after 2 reflections from the mirror like surfaces. Courtesy of cosine Research B.V.

Lidar optical circuits

Differential absorption atmospheric remote sensing, altimetry, range-finding and Doppler anemometry are some lidar applications in space. A novel hollow waveguide lidar optical circuit approach has been investigated for a wavelength of 1.5 μm in the frame of an ESA activity. Hollow waveguides, machined into the surface of a dielectric substrate, are used to guide optical beams through a circuit of components. The optical components are simply inserted into alignment slots which are precisely machined into the substrate. A silicon wafer substrate based hollow waveguide circuit, machined using Deep Reactive Ion Etching (DRIE) has also been investigated. Results show that the hollow waveguide circuit shows a significant improvement on the optical component misalignment tolerance compared to free space optical bench circuits and requires only a few fabrication steps.

Spectrometers

Most scientific and earth observation spacecraft, have one or several spectrometers on board. Their miniaturisation is therefore interesting, particularly for planetary probes.

The Programmable Micro Diffraction Grating (PMDG) is an optical MEMS device which could find multiple applications in space spectrometers. From the instrument designer point of view the PMDG can be implemented in the pupil plane or in the image plane of a spectrometer. First results of an ESA study investigating space instrument concepts based on existing PMDG technology show that it could have advantages in the infrared domain when implemented in the pupil plane of a spectrometer. In this case the first order diffraction efficiency of a PMDG could be over 80% using a blazed profile with at least 4 ribbons per period. When implemented in the image plane of a spectrometer, after a dispersive element, it can be used as a wavelength selective spatial modulator in order to select certain wavelengths and reject others. The potential PMDG applicability to ESA missions is under investigation for both discussed implementations. The length of the PMDG ribbons limits the field of view of a spectrometer. Existing PMDG technology seems therefore to be limited to single point source spectrometry, which somewhat narrows its application range for space instruments. Another ESA activity aims to develop a new, European based, PMDG technology.

A slightly different diffractive optical MEMS device for a gas sensing instrument is also currently under development. The primary objective is to build a fully functional demonstrator of a sensor for methane (CH_4) gas concentration measurement. This activity aims for a simple, robust, and potentially low-cost design. The key component of this sensor instrument is a micromechanically controllable two-state optical filter device. The filter device is realized using a Configurable Diffractive Optical Element (CDOE) technology [3].

In traditional spectrometer designs, there is an inherent trade-off between resolution and light throughput. While spectral resolution increases as slit width decreases, the narrowing of the input slit reduces photon throughput and consequently measurement sensitivity. An ESA activity has demonstrated that by using a micromachined multi slit mask SNR improvements of more than a factor of 2.5 have been achieved on absorption measurements of SO_2 for a detector-noise-limited measurement, in accordance with the theory.

III. PLANNED ACTIVITIES INVOLVING MOEMS

The Near Earth Object Micro Explorer (NEOMEx) is an ESA straw man mission to provide a focus application for a micro-system based spacecraft concept. In the frame of a technology development program it is planned to start an activity on a micro-system capable of demonstrating that space optics applications can benefit from microtechnology.

IV. FUTURE MOEMS BASED SPACE INSTRUMENTS

Spectrometer

Spatial Light Modulation (SLM) is a very useful function for both object selection in space as well as spectroscopy. The James Webb Telescope (JWST) has a spectrometer (NIRSpec) on board [4] with a micro-shutter array as key component for object selection. Applying a micromirror array (SLM) differently, namely as an adaptive and programmable aperture/slit mask, could improve the performances of future spaceborne spectrometers whilst maintaining reasonable size and mass.

Telescope primary mirror

Future space telescopes, in particular for science and Earth observation from the Geostationary Orbit (GEO), are foreseen to become larger for increase resolution. The primary mirror's area mass needs to be less than 10 kg/m^2 if diameters of over ten meters shall be attained. MEMS technology could contribute to fulfil this challenging requirement by providing light weight large area wave front correction mechanisms for instance.

Adaptive optics

Adaptive mirrors are key components for future large-aperture space telescopes and high power laser based systems such as Raman spectrometers (ExoMars), lidars (Aladin), altimeters and deep-space inter-satellite communications. For very large space telescopes the required precision for surface shape control and deployment stability cannot be reached without active control. In such cases the telescope will likely have to be corrected by an adaptive mirror using MEMS technology for instance. Optical MEMS based adaptive optics could also be used in high power laser based systems. Thermal stresses in the optical system create optical beam shape instabilities. In the worst case a beam focalises on an optical component and causes damages. An adaptive optical element is therefore very useful to stabilize and control the laser beam shape.

ACKNOWLEDGMENT

The authors thank the industry and research partners for their exciting work performed in the frame of ESA's activities mentioned in this paper.

REFERENCES

[1] B. Benazet, M. Sotom, M. Maignan, J. Perdigues, "Microwave Photonics Cross-connect Repeater for Telecommunication Satellites", Proc. SPIE 6194, 619403 (2006).
[2] M. Bavdaz et al., "The XEUS X-ray telescope", Proc. SPIE 6266, 62661S (2006).
[3] Hakon Sagberg, Thor Bakke, Ib-Rune Johansen, Matthieu Lacolle, and Sigurd T. Moe, "Two-state Optical Filter Based on Micromechanical Diffractive Elements", Proc. IEEE Optical MEMS and Nanophotonics, p. 167-168 (2007).
[4] Robert F. Silverberg et al. "A microshutter-based field selector for JWST's multi-object near infrared spectrograph", Proc. SPIE 6678, 66780Q (2007).

Polymer Biochips with Micro-Optics for LIF Detection

H. Hosseinkhannazer[1], J. N. McMullin[2], L. W. Kostiuk[1]

Departments of Mechanical Engineering [1] and Electrical and Computer Engineering [2]
University of Alberta, Edmonton, Alberta, Canada

Abstract: We present the design, fabrication and experimental testing of polydimethylsiloxane (PDMS) biochips with integrated waveguides that have liquid or UV-cured polyepoxyacrylate (PEA) cores. The use of inexpensive micro/nanofabrication methods and materials results in biochips suitable for laser-induced fluorescence detection applications.

Introduction

Miniaturized fluidic systems developed for diagnostics, usually referred to as lab-on-a-chip (LOAC) or micro-total-analysis systems (μTAS), are becoming important tools in biological analysis [1]. Opto-biochips are LOACs with integrated waveguides that deliver light to the microchannels and have been shown to give comparable or superior results to standard confocal microscopy when used for laser-induced fluorescence (LIF) detection [2]. In this paper, we describe single-layer opto-biochips fabricated by casting heat-curable polydimethylsiloxane (PDMS) on micromachined silicon (Si) masters. Using innovative methods of waveguide integration and new materials, opto-biochips with multiple waveguides for excitation and detection were fabricated. The simultaneous LIF detection of microparticles with two different types of fluorophores is demonstrated experimentally. Inexpensive LOACs fabricated in this way have excellent commercial potential.

Device Fabrication, Design and Characterization

The Si masters were fabricated using deep reactive-ion etching (DRIE), then silanized and used for replicating the micro-features into PDMS. The Si wafer was first coated with a 100-nm aluminum oxide etch mask using pulsed reactive sputtering. Cryo-DRIE was used to etch the Si to a depth of 50μm. PDMS was cast on the silanized silicon master, heat-cured, removed from the master and capped with a thin PDMS film using O_2 plasma bonding. The resulting disposable biochips are suitable for lab-on-a-chip applications.

Each biochip contained three liquid/solid-core waveguides directed at a common point in the middle section of a 50-μm-wide, Y-shaped, microchannel that has one input and two output reservoirs as shown in Fig. 1. Fig. 2 shows the SEM image of the waveguide/channel intersection region in the PDMS layer before capping and filling of the waveguide channels. The 70-μm-wide waveguide channels are filled with glycerol (n = 1.47) for liquid-core waveguides (LCW) and a novel formulation of UV-curable polyepoxyacrylate (PEA - n = 1.54) to make solid core waveguides (SCW). The bulk of the PDMS chip has an index n = 1.43.

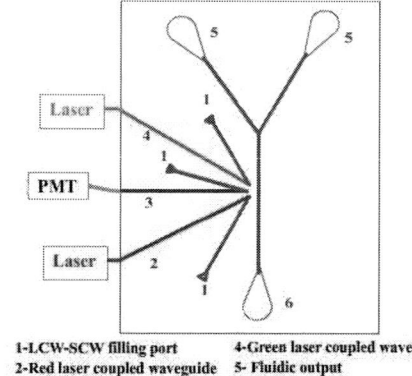

1-LCW-SCW filling port   4-Green laser coupled waveguide
2-Red laser coupled waveguide   5- Fluidic output
3-Pickup waveguide   6- Fluidic input

Fig. 1 Complete biochip layout

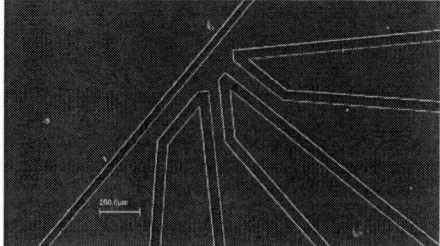

Fig. 2 Close-up of fluid and waveguides microchannels in cast PDMS

The SCWs and LCWs were tested for optical loss by measuring 632-nm laser light scattered through the top surface of the chip as a function of the distance from the input point. The approximate losses were 1.3 dB/cm for the LCWs and 6.3

978-1-4244-1917-3/08/$25.00 ©2008 IEEE   128

dB/cm for the cured PEA SCWs respectively. Although SCWs are more lossy than LCWs, the transmitted power is adequate for LIF detection applications.

Experiments and Results

Two-colour LIF detection experiments were performed using 15-μm fluorescent polystyrene beads excited by either 632 nm or 532 nm light. Using multimode optical fibers, laser light at these wavelengths was coupled into two of the waveguides in the biochip as shown in Fig. 3. The beads were carried through the microchannel by electro-osmotic flow and excited by the lasers at the beam/channel intersection. Fluorescence from the beads was collected in the third waveguide (see Fig. 1) and delivered by optical fiber to an off-chip photomultiplier (PMT) detection system. The laser beams were modulated at two different frequencies allowing the identification of the specific bead type by analyzing the PMT signal with a Fourier transform

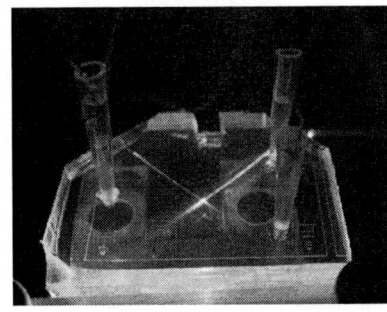

Fig.3 Red and green lasers coupled into the solid-core waveguides and directed to the detection point.

technique [3] as shown in Fig. 4. This system allows for the excitation and detection of multiple types of microparticles using a single PMT.

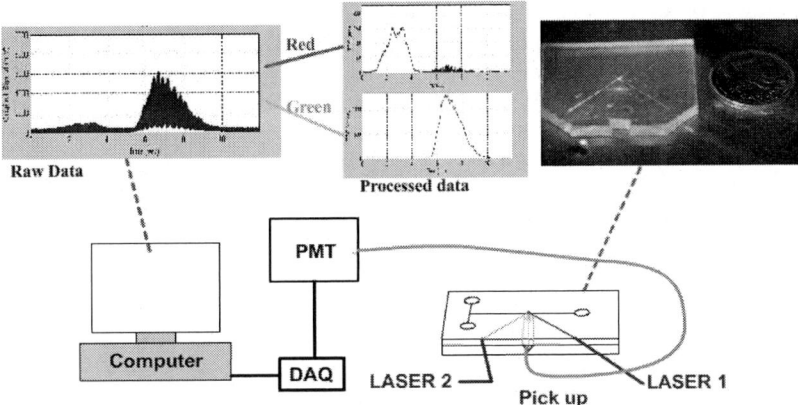

Fig. 4 Experimental setup and data processing. Distinct signals for the two types of fluorescent beads are derived from the raw PMT data.

Conclusions

In this paper, the design, microfabrication and experimental testing of a PDMS opto-biochips with integrated liquid-core and solid-core optical waveguides has been described. As described above, experiments demonstrated the potential usefulness of these devices in lab-on-a-chip applications that use laser-induced fluorescence detection with multiple fluorophores.

References

[1] M. Toner, and D. Irimia, "BLOOD-ON-A-CHIP," Annual Review of Biomedical Engineering, vol. 7, no. 1, pp. 77-103, 2005.

[2] Christopher L. Bliss, James N. McMullin, Christopher J. Backhouse, "Rapid Fabrication of a Microfluidic Device with Integrated Optical Waveguides for DNA Fragment Analysis", , *Lab Chip*, vol. 7, pp. 1280-1287, 2007.

[3] C. L. Bliss, C. J. Backhouse, and J. N. McMullin, "Two-colour microparticle detection in PDMS biochips with integrated optics", in proceedings of SPIE Photonics North conference, Ottawa, 2007, pp. 67960A-6

3D MODELING OF PHOTONIC DEVICES USING DYNAMIC THERMAL ELECTRON QUANTUM MEDIUM FINITE-DIFFERENT TIME-DOMAIN (DTEQM-FDTD) METHOD

E. H. Khoo[1†], S. T. Ho[2], I. Ahmed[1] and E. P Li[1], Y. Y. Huang[2]

[1]Institute of High Performance Computing, SINGAPORE
[2]Northwestern University, EVANSTON, UNITED STATES

[†]Email: khooeh@ihpc.a-star.edu.sg; Tel: (65) 6419-1550; Fax: (65) 6419-1380

This paper reports on the modeling of the semiconductor photonic devices using 3D DTEQM-FDTD. The model includes the physics of Pauli Exclusion principle, Fermi-Dirac thermalization and state filling to describe the complex electron dynamics in semiconductor media to realize the full potential of FDTD [1]. The carrier intraband and interband transition dynamics, energy band filling and thermal equilibrium are demonstrated in the model. The DTEQM-FDTD [2] model includes the essential physics of complex dynamical media and yet computational efficient. It is applicable to a wide range of atomic and molecular media by applying the appropriate rate equations and energy level structure.

Figure 1 shows the structure of the microdisk laser, which is chosen as an example to demonstrate the validity of the 3D DTEQM-FDTD model. The microdisk has a radius of 1 μm and thickness of 200 nm. The microdisk allows only one field mode in the z direction. The refractive index of the microdisk is 3.4 and is surrounded by air medium. An optical pulse is incident on the microdisk to initial simulated lasing emission. To ensure sufficient carrier in the microdisk, an electrical pump source with current density of $30k/cm^2$ is used.

Figure 2 shows the electric field pattern of the microdisk along the z direction using the 3D DTEQM-FDTD model. Fig. 2(a) shows the E_z field distribution at the x-y plane at thickness of 100 nm. For lasing, the resonance in the microdisk is build-up by using whispering gallery modes (WGMs) [3, 4]. For the x-y field pattern distribution, the WGM E_z field has a radial mode number, **n** = 1 and azimuthal mode number **m** = 12. Fig. 2(b) shows the field distribution in the x-z direction at the center of the microdisk.

Figure 3 shows the field amplitude distribution for the microdisk in the x-z plane. The field peak amplitude is near the edge of the microdisk, which demonstrates the presence of whispering gallery resonance modes. Some electromagnetic fields tunnel through the microdisk-air barrier and leaks to the surrounding air medium. It's amplitude decays sharply due to coupling with the radiation unguided mode in the air.

The simulation of the microdisk is set to run for 30000 steps. Fig. 4 shows the E_z field amplitude in the microdisk with the simulation steps. The field energy in the microdisk has buildup to more than 20000 J. In addition, the paper also

simulates a more practical microdisk with a 1 μm Al $_{0.8}$Ga$_{0.2}$As film substrate below. It is observed that the difference between both cases is quite small.

Figure 5 shows the azimuthal mode number of the microdisk at different radius. It is observed that smaller microdisk radius has lower azimuthal mode number. This is because there is less space for the WGM to travel around the microdisk.

Figure 6 shows the lasing spectra of the microdisk laser. The lasing spectra of the microdisk for the experimental measurement and simulated result match very well at the resonance wavelength of 1.56 μm. The Q factor for the experimental and simulation is 3052 and 7658. The reason for this different could be due to the heating of microdisk and optical noise during experimental measurement.

In conclusion, a 3D DTEQM-FDTD computational model is developed for electrodynamics simulation in photonic devices. The model integrates the atomic and quantum physics to the FDTD method and calculates the electromagnetic field properties in photonic devices. The 3D DTEQM-FDTD provides a power and efficient model for the dynamical media at atomic and molecular scale and is able to converge very well with experimental results.

REFERENCES:

[1] K. S. Yee, "Numerical solution of the initial boundary value problems involving Maxwell's equations in isotopic media", *Trans. Antenna Propag.*, vol. 14, pp. 302-307, 1966.

[2] Y. Y. Huang and S. T. Ho, "Computational model of solid-state, molecular, or atomic media for FDTD simulation based on a multi-level multi-electron system governed by Pauli exclusion and Fermi-Dirac thermalization with application to semiconductor photonics", *Opt. Express*, vol. 14, pp. 3569-3587, 2006.

[3] Lord Rayleigh, "The problem of whispering gallery", *Scientific Papers*, vol. 5, pp. 617-620, 1912.

[4] S. L McCall, A. F. J. Levi, R. E. Slusher, S. J. Pearton, and R. A. Logan, "Whispering gallery mode microdisk laser", *App. Phy. Lett.*, vol. 60, pp. 289-291, 1991.

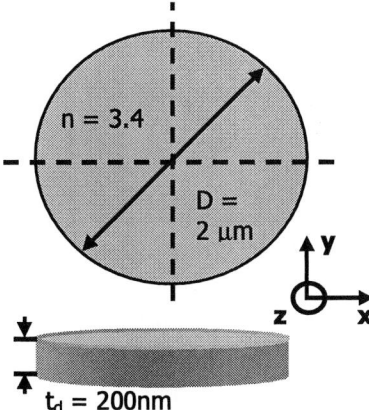

Figure 1. Schematic structure of the microdisk in 3D. The microdisk has a thickness of 200 nm to allow only 1 mode in the z direction.

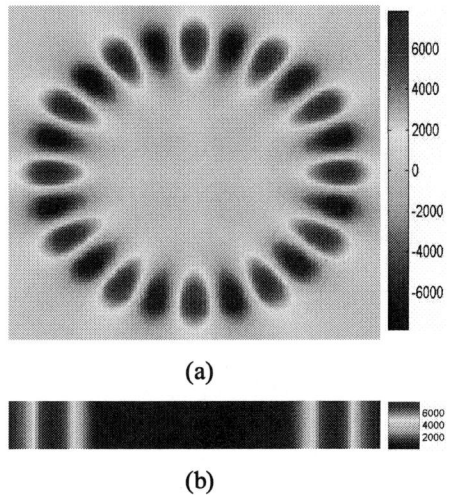

(a)

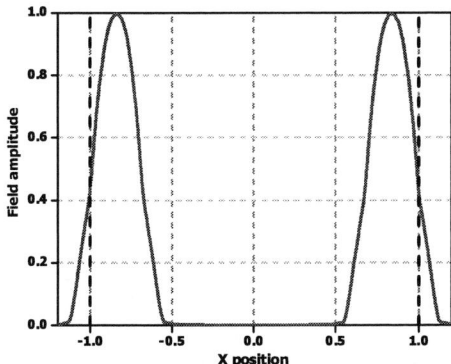

(b)

Figure 2. Field pattern distribution of the microdisk with 3D DTEQM-FDTD method. (a) Field distribution in the x-y plane (b) field distribution in the x-z plane with $n = 1$ and $m = 12$.

Figure 3. E_z field amplitude across the microdisk. Maximum field amplitude of the WGM is obtained near the edge of microdisk, which is represented by the dash line.

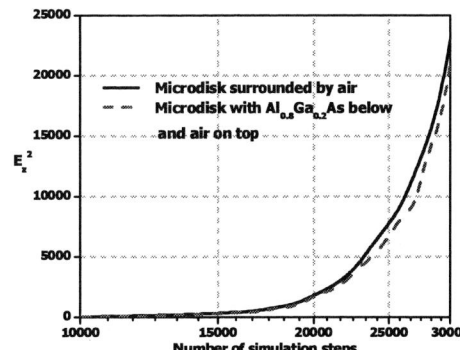

Figure 4. Ez field amplitude for both cases of microdisk simulation. This is small different between both cases.

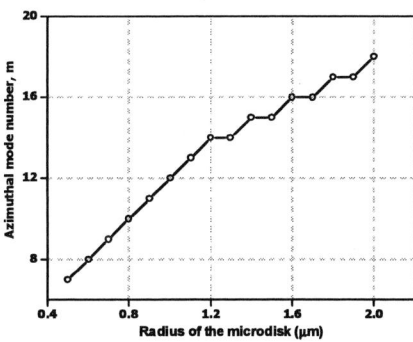

Figure 5. Graph of the azimuthal mode number, **m** vs microdisk radius. The **m** value increase as the radius increases.

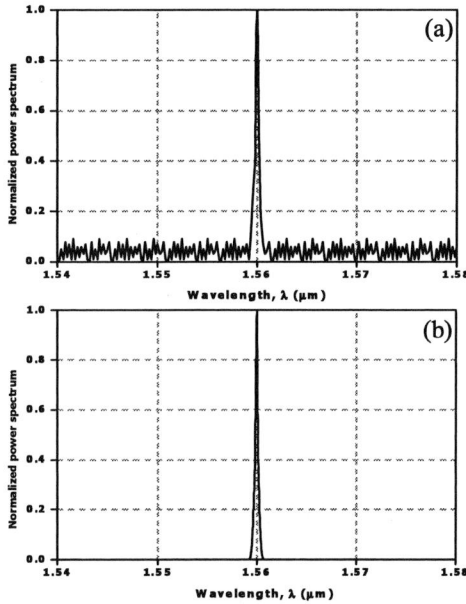

Figure 6. Comparison of the wavelength spectra between the (a) experimental measurement and (b) simulated results. Both graphs show good agreement at $\lambda = 1.56$ µm.

131

High-density Piezoelectric Actuator Array for MEMS Deformable Mirrors Composed of PZT Thin Films

Isaku Kanno, Shogo Tsuda and Hidetoshi Kotera

Micro Engineering, Kyoto University, Kyoto 606-8501, Japan

Abstract

In this paper, we report a piezoelectric deformable mirror composed of high-density actuator array for low-voltage adaptive optics. A piezoelectric Pb(Zr,Ti)O$_3$ (PZT) film was deposited on a Pt-coated silicon on insulator (SOI) substrate, and a diaphragm structure of 15 mm in diameter was fabricated by etching a Si handle wafer. A hexagonal 61-element unimorph actuator array was produced with an Al or Au reflective layer over the backside of the diaphragm. In order to reduce the dead space for the lead lines between the actuators, they were prepared on the polyimide insulating layer. The displacement of each actuator was measure by using a laser Doppler vibrometer and we confirmed a relatively large displacement of more than 1 μm by applying a voltage of 10V$_{pp}$.

Introduction

The development of small size and low-cost deformable mirrors (DMs) has attracted attention for adaptive optics (AO) aiming the industrial and medical applications. To achieve small and low-cost DMs, microelectromechanical systems (MEMS) DMs have successfully been developed and manufactured as commercial products [1],[2]. The MEMS DMs are commonly actuated by an electrostatic force, and the design and fabrication of the microactuator system have already been established. However, several issues still remain unsolved; the large driving voltage and small stroke which is limited by the gap distance of the electrodes. On the other hand, a piezoelectric actuator can generate a large force and its response is very fast. In addition, concave and convex deformation can be easily generated by just changing the polarity of the applied voltage. In previous report, we described the piezoelectric MEMS DMs with 19 actuator array driven by Pb(Zr,Ti)O$_3$ (PZT) thin films [3]. It demonstrated large stroke of 5 μm under low applied voltage of 10V, and we could generated the deformation of low-order Zernike mode up to 4th order. In this study, we designed and fabricated the piezoelectric MEMS DMs for increasing resolution by compensation of higher order Zernike modes. The 61 hexagonal actuator array is fabricated on the continuous mirror membrane composed of PZT/Si layer and the mirror surface is deformed by applying individual input voltage on each actuator. We estimated the displacement by finite element method (FEM) calculation and successively measured actual displacement of the DMs.

Design and Fabrication

Figure 1 shows the structure of the deformable mirror we fabricated. The PZT thin film of 2 μm in thickness was deposited on a Pt/Ti-coated silicon on insulator (SOI) substrate by rf-sputtering. The handle wafer was

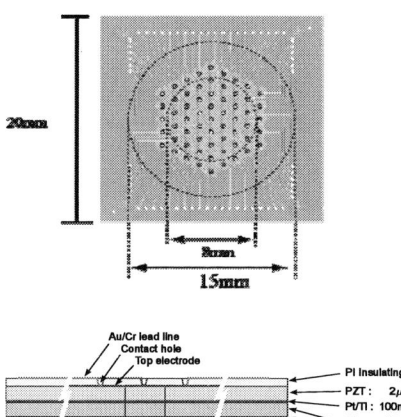

Fig. 1. Surface pattern and cross-sectional structure of piezoelectric MEMS DMs.

etched out to the buried oxide (BOX) layer which acts as an etch-stop layer. The 61 individual Al electrodes were prepared on it, and micro-unimorph actuator array of PZT/Si was accomplished. The electrodes have hexagonal shape with sides of 700μm. Successively polyimide layer with through holes on the electrodes was covered over the PZT film as an insulating layer and lead lines were arranged on it. In this configuration, we need no space for lead lines between the actuators and high density actuator array could be obtained. The diameter of the membrane and the active mirror area are 15mm and 8mm in diameter, respectively. The exposed BOX layer was coated by Al or Au as a mirror surface. The images of the piezoelectric MEMS mirror are shown in Fig. 2. The application of the voltage between Al individual electrode and Pt bottom electrode induces longitudinal strain of the PZT films

978-1-4244-1917-3/08/$25.00 ©2008 IEEE 132

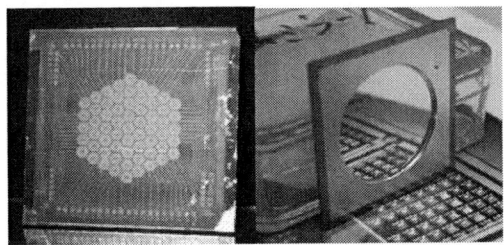

Fig. 2. Pictures of the fabricated MEMS DMs; (a) actuator side, and (b) mirror side.

which causes the vertical deformation of the mirror surface and a variety of the deformation can be generated by control of the applied voltage to each Al electrode.

Results and discussion

A FEM calculation was performed using MARC to estimate displacement profiles. We used the typical bulk material properties of the PZT whose transverse piezoelectric coefficient e_{31} and Young's modulus are -6.5 C/m^2 and 70 GPa, respectively. Figure 3 shows the displacement profiles under the application of 10 V on the center electrode. The calculation indicated that relatively large displacement of 8.5μm would be expected.

The deflection of the mirror was observed by a laser Doppler vibrometer (LDV). The sine wave signal of 10V$_{pp}$ was applied to 5 actuators on different position and the displacement profiles are shown in Fig. 4(a). The 3 actuators near center of the mirror show almost same profiles with peak displacement of about 1.2 μm and 2 outside actuators show relatively low displacement due to the restriction effect of the rim of the mirror. However, the displacement of the center

actuator was lower than that of FEM calculation. It might be attributed to the lower piezoelectric properties of thin-film PZT as well as the internal stress of the membrane.

The hysteresis of the displacement is measured using a fringe count laser interferometer and result is shown in Fig. 4(b). We applied bipolar triangular signals of 20 V$_{pp}$ at 1 Hz to the center electrode. The measurement revealed that the actuator shows the relatively low hysteresis, which is advantageous for simple control of the deformation of the mirror surface.

REFERENCES

[1] J. A. Perreault, et al., "Adaptive optic correction using microelectromechanical deformable mirrors", Opt. Eng. 41, pp. 561-566, (2002)

[2] M. A. Helmbrecht, et al., "Piston-tip-tilt positioning of a segmented MEMS deformable mirror", Proc. SPIE, 6467,p. 64670M (2007)

[3] I. Kanno, et al., "Development of deformable mirror composed of piezoelectric thin films for adaptive optics", IEEE J. Select. Topics Quantum Elect, 13, pp. 155-161 (2007)

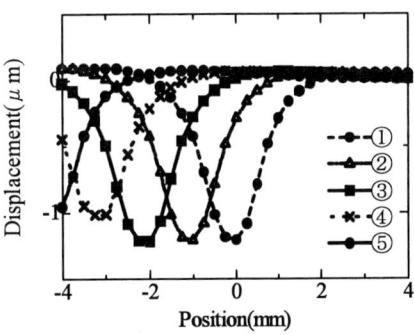

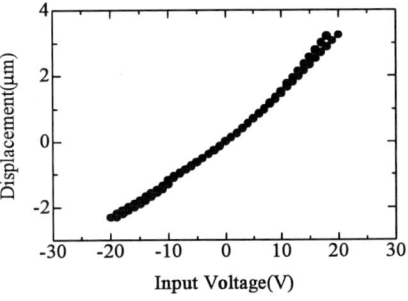

Fig. 4. Displacement of the mirror surface measured by a laser Doppler vibrometer; (a) 2-dimentional displacement profile of each actuator under 10V$_{pp}$, and (b) displacement hysteresis of center actuator.

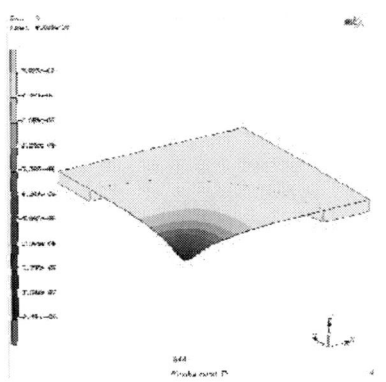

Fig. 3. FEM simulation of displacement profile of piezoelectric MEMS DMs. Application of 10 V on the center actuator generates the displacement of 8.5μm

Monolithic Integration of a Tunable Photodetector Based on InP/air-gap Fabry-Pérot Filters

Thomas Kusserow, Ricardo Zamora, Julian Sonksen, Nethaji Dharmarasu and Hartmut Hillmer
Institute of Nanostructure Technologies and Analytics (INA) and Center for Interdisciplinary Nanostructure
Science and Technology (CINSaT), University of Kassel, D-34132 Kassel, Germany
Tel +49-561-8044315, Fax +49-561-8044488, E-mail kusserow@ina.uni-kassel.de

Tetsuji Nakamura, Tetsuo Hayakawa and Balasubramanian Vengatesan
NanoTech Laboratory, Canare Electric Co. Ltd, 2888-1 Rikka, Kumabari,
Nagakute-cho, Aichi-ken, 480-1101 Japan

Abstract

We present a tunable photodetector MEMS device consisting of an InP/air-gap Fabry-Pérot filter combined with a pin-photodiode structure. Epitaxial growth and advanced surface micromachining are used for the monolithic integration of both devices. Special process steps regarding detector geometry, photo current and signal quality have been investigated to assure high performance.

Key words: InP, GaInAs, photodetector, pin photodiode, near infrared, air-gap DBR, tunable filters.

I. INTRODUCTION

Tunable optical MEMS devices like filters, detectors and emitters are well adapted for telecommunication and sensoric applications. Due to the ability to adjust the wavelength those structures cover a wide spectral range with a single device, e.g. to switch channels in multiplexer systems or for spectroscoptic measurements. To detect wavelengths in the near IR, our device is based on the InP material system, which is suitable for wavelengths from 1 μm on, up to the spectral range over 2 μm. Current integrated tunable optical MEMS devices mostly make use of a hybrid approach, combining detector or emitter structures with dielectric DBR mirrors [1] or using bonding technologies [2]. Also resonant InP/air-gap detectors with only one DBR, and thus having no tunability, have been presented [3]. We present a fully integrated photodetector based on a Fabry-Pérot filter design, which is tunable by electrostatic actuation.

II. DESIGN AND IMPLEMENTATION

The integrated detector device is based on tunable InP/air-gap Fabry-Pérot filters [4]. Its structure is grown on a InP substrate by low-pressure MOCVD starting with the pin-photodiode layers and an intermediate layer as antireflection and protection coating, on which the filter structure is grown (Fig. 1). The filter consists of two DBR mirrors surrounding the cavity to achieve high reflectivity and hence quality of the filter. Due to the refractive index contrast between InP and air of 2.17, a broad stopband and high reflectivity of >99.5% can be obtained by only 2.5 periods at each DBR. As the filter wavelength is defined by the length of the cavity, tuning is possible by electrostatic actuation of the differently doped mirrors when applying a reverse bias.

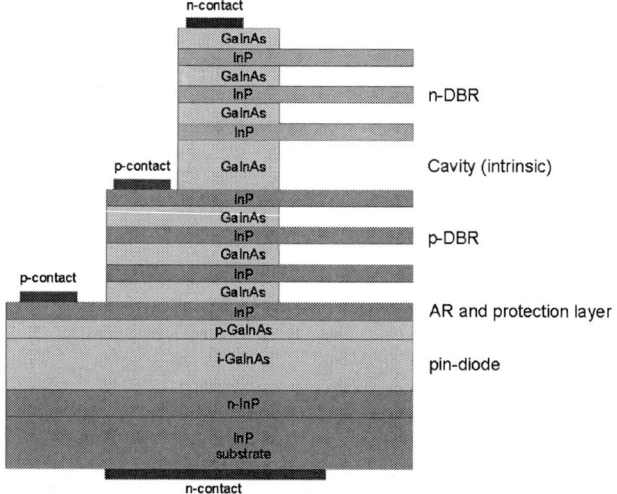

Figure 1: Crossectional view of the detector design. The InP/air-gap DBR mirrors are p- and n-doped, respectively. Ohmic contacts are applied to the filter and the detector for electrostatic tuning and photocurrent read-out.

The lateral design of the filters consists of the filter membranes, which are fixed by four suspensions to the supporting posts and is fabricated by RIE dry etching steps successively to epitaxy. Electrical contacts for tuning and detector operation are applied on the supporting posts and the substrate using electron beam and thermal evaporation. The air-gap structures are fabricated by selective sacrificial wet etching of the GaInAs layers with $FeCl_3$. Since etching is only desired at the filter, the supporting posts and surrounding areas are covered by a SiO_2 etch protection coating. Additionally, the intermediate InP layer between pin-photodiode and filter structure prevents etching of the GaInAs layers of the detector. Also the thickness of this layer was chosen to have antireflective properties for improved coupling efficiency into the detector during operation. The etchant is subsequently replaced by isopropanol, acetone and CO_2, respectively, which is then removed using a critical point dry process to avoid collapsing of the membranes due to capillary forces.

Fully released air-gap structures (Fig. 2) are sensitive to stress in the membrane layers. Hence, special care has to be taken during the design to avoid this effect, e.g. introduced by arsenic carry-over during the epitaxy. For our devices we introduce a compensation layer containing Ga, which leads to very small residual stresses and bending [5]. Fabry-Pérot filters with a wide tuning range, fabricated by the described process have been presented already[6].

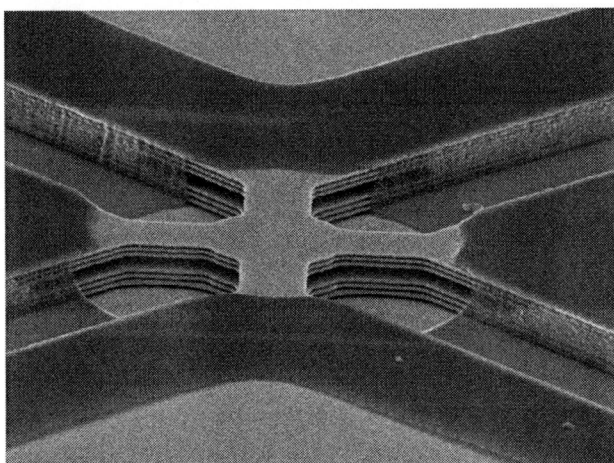

Figure 2: *Fully processed and release InP/air-gap Fabry-Perot filter with λ cavity (top). Gradient stress is well controlled to avoid sticking of the InP membranes (bottom).*

For the monolithic integration of the photodetector new constraints have to be considered to achieve high quality operation. To keep losses of the photocurrent low, improved

ohmic contacts have to be implemented. Especially the contacts on n-type semiconductors, consisting of Ti/Pt/Au, show high resistivity behaviour. Although this is negligible for electrostatic actuation, it becomes important for the detector device as sensitivity is decreased. Thus n-type contacts are replaced by using Zn/Au layers, which have a very low resistivity.

As the lateral geometry of the pin-photodiode is important to obtain high frequency response, single detector devices are separated by a RIE dry etch step, leaving only the minimum photodiode area possible.

Signal quality of the filter device is determined by the ratio between light transmitted by the Fabry- Pérot filter and light reaching the photodetector directly. To minimize this undesired noise signal, a light absorbing layer is deposited on the pin-structure, with transparent areas only at the filters. Two different approaches have been investigated: a metal layer on top of a dielectric layer and a nanoporous dielectric layer with incorporated dye. Both show high absorption values for a wide spectral range, without furtherly affecting the properties of device.

III. CONCLUSION

We present a monolithically integrated, tunable Fabry- Pérot filter for the near IR range. The device consists of InP/air-gap DBR mirrors and cavity with a pin-photodiode structure underneath. Additionally required process steps have been investigated to assure high quality performance.

ACKNOWLEDGMENTS

The financial support by the German Research Foundation DFG, is gratefully acknowledged. The authors thank M. Bartels, S. Ferwana, D. Gutermuth, A. Hasse, S. Irmer, I. Kommallein, F. Römer, A. Tarraf, I. Wensch, M. Wulf for technical support and stimulating discussions.

REFERENCES

[1] Aziz, M. Pfeifer, J. Wohlfarth, M. Luber, C. Wu, S. Meissner, P. , „A new and simple concept of tunable two-chip microcavities for filter applications in WDM systems", *IEEE Phot. Techn. Lett.*, Vol. 12, No. 11, 1522- 1524 (2000)

[2] G. L. Christenson, A. T. T. D. Tran, Z. H. Zhu, Y. H. Lo, M. Hong, J. P. Mannaerts, and R. Bhat: "Long-Wavelength Resonant Vertical-Cavity LED/Photodetector with a 75-nm Tuning Range", *IEEE Phot. Techn. Lett.*, Vol. 9, No. 6 (1997)

[3] Xiamomin Ren, Hui Huang, Yingzhe Chong, Yongqing Huang: "1.57 µm InP-based resonant-cavity-enhanced photodetector with InP/air-gap bragg reflectors", *Microwave and optical Techn. Lett.*, Vol. 42, No. 2 (2004)

[4] A. Spisser, R. Ledantec, C. Seassal, J. L. Leclercq, T. Benyattou, D. Rondi, R. Blondeau, G. Guillot, and P. Viktorovitch, "Highly Selective and Widely Tunable 1.55-µm InP/Air-Gap Micromachined Fabry–Perot Filter for Optical Communications", *IEEE Phot.Technol. Lett.*, VOL. 10, NO. 9, (1998)

[5] T. Kusserow, N. Dharmarasu, H. Hillmer, T. Nakamura, T. Hayakawa and B. Vengatesan, „Tailored Stress in InP/GaInAs Layers for InP/air-gap DBR-Filters with Photonic Crystals", *Proceedings Optical MEMS 2006*, 88-89 (2006)

[6] A. Hasse, S. Irmer, J. Daleiden, N. Dharmarasu, S. Hansmann and H. Hillmer, „Wide continuous tuning range of 221 nm by InP/air-gap vertical-cavity filters.", *Electron. Lett.*, 42, 974 (2006)

Polarization Singularities and Local Field Symmetries in Photonic Crystals

Jeffrey F. Wheeldon and Henry Schriemer

Centre for Research in Photonics at the University of Ottawa
800 King Edward Ave., Ottawa, ON Canada K1N 6N5

Abstract—We show that the local state of polarization of a two-dimensional photonic crystal exhibits distinct polarization singularities that map to the stabilizer groups of field symmetry image points for a particular eigenmode.

I. INTRODUCTION

THE state of polarization of the electromagnetic field is one of its most fundamental properties, quantifying the vectorial character of the optical response. The field solutions of complex systems may not exhibit a uniform polarization state if the direction of the field vector acquires a spatial dependence, thus the local state of polarization is of interest, as it may become singular at specific locations. Three distinct types of polarization singularities exist: pure circular polarization (C point); pure linear polarization (L line); and disclinations (where the field vanishes and the polarization state is asymptotically undefined) [1-3]. We show such polarization singularities in a two-dimensional photonic crystal, noting that the symmetry constrains local states of polarization to subsets of the system's polarization response.

II. THEORY

The point group associated with a two-dimensional photonic crystal of hexagonal dielectric profile is $C_{6v} = \{E, 2C_6, 2C_3, C_2, 3\sigma_x, 3\sigma_y\}$, where C_n denotes n-fold ($2\pi/n$) rotational operations; σ_v indicates mirror reflection operations with respect to particular planes, v; and E is the identity operator. Considering only the TE mode, the polarization of the electric field, $\mathbf{E}_{n,\mathbf{k}}(\mathbf{r})$, lies within the plane, where $\mathbf{k}$ is the wavevector and n is the mode order. The point group associated with a particular wavevector is identified by a subset of symmetry operators that leave the Maxwell operator globally invariant. Since the wavevector must remain invariant under any symmetry operation (to within a reciprocal lattice vector), $\mathbf{E}_{n,\mathbf{k}}(\mathbf{r})$ must be associated with a particular subgroup of C_{6v} for each $\mathbf{k}$. We consider the K point in the Brillouin zone, $\mathbf{k} = (4\pi/3a)\hat{\mathbf{x}}$, where the eigenfunction is associated with the $C_{3v} = \{E, 2C_3, 3\sigma_y\}$ point group [4].

The electric field has symmetry images at positions that are related to each other by the symmetry operators of the eigenfunction's point group. When a position $\mathbf{r}'$ is coincident with its symmetry image, the local field vector (not necessarily merely the mode) transforms as

$$R\mathbf{E}_{n,\mathbf{k}}(\mathbf{r}') = \chi_R \mathbf{E}_{n,\mathbf{k}}(\mathbf{r}'), \quad (1)$$

where the value of the character, χ_R (in general, complex), is set by the local symmetry group and the chosen operator; that is, by the irreducible representation of $\mathbf{E}_{n,\mathbf{k}}(\mathbf{r}')$. Each stable point has a set of operators that form a subgroup of the eigenmodes's point group; this is the stabilizer group.

To examine how the state of local polarization is affected by its stabilizer group, we study the stable points of the electric field along the mirror symmetry planes, and at the centers of rotational symmetry. We express the local electric field as

$$\mathbf{E}(\mathbf{r}') = [\mathbf{P}(\mathbf{r}') + i\mathbf{Q}(\mathbf{r}')] \equiv \Omega(\mathbf{r}') \begin{pmatrix} D(\mathbf{r}') + iB(\mathbf{r}') \\ C(\mathbf{r}') \\ 0 \end{pmatrix}, \quad (2)$$

where $D(\mathbf{r}')$, $B(\mathbf{r}')$, and $C(\mathbf{r}')$ are real values that depend on location, and $\Omega(\mathbf{r}')$ is a phase factor that preserves the form of (2). The field at the K point ($n = 1$) will be analytically determined in the vanishing contrast limit for expediency [5].

III. RESULTS

The photonic crystal may be envisaged as an array of cylindrical rods in the locations of the solid black dots, shown in Fig. 1. The local polarization states, shown by the ellipses in Fig. 1, have been determined by mapping the calculated field over one temporal period across an array of locations within the neighborhood of the unit cell. Polarization singularities of all three types are noted: the disclinations at the positions of the solid black dots, C points at the open black circles (to within the resolution of the ellipse density), and L lines along the heavy black lines. We show that the existence of polarization singularities may be derived by group theoretic arguments directly from the system symmetry.

In a hexagonal lattice, the mirror symmetry planes form equilateral triangles, as denoted by the heavy black lines in Fig. 1. The stabilizer group is $C_{1h} = \{E, \sigma_v\}$ at such positions. Choosing σ_y, for illustration, with $\mathbf{r}'$ any point along a horizontal heavy black line (e.g., x-axis), (1) may be written as

978-1-4244-1917-3/08/$25.00 ©2008 IEEE

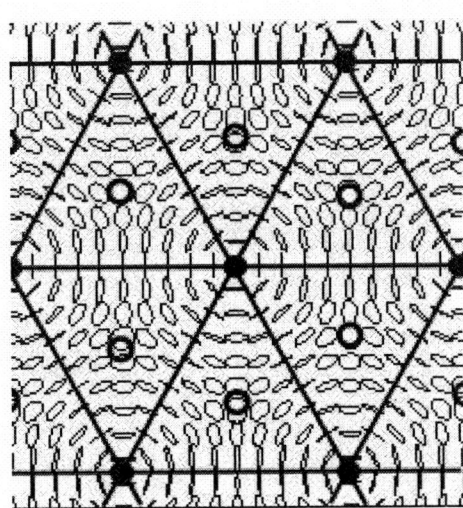

Fig. 1. The local polarization state for the TE electric field mode at the K point of a hexagonal photonic crystal in the vanishing contrast limit. Shown are the polarization ellipses (in blue), and the stable points (black lines – L lines; open black circles – C points; solid black dots – disclinations).

$$\Omega \begin{bmatrix} 1 & 0 & 0 \\ 0 & -1 & 0 \\ 0 & 0 & 1 \end{bmatrix} \begin{pmatrix} D+iB \\ C \\ 0 \end{pmatrix} = \chi_{\sigma_y} \Omega \begin{pmatrix} D+iB \\ C \\ 0 \end{pmatrix}, \quad (3)$$

where the matrix gives the mirror operation and the form of the field vector derives from (2); $\mathbf{r}'$ has been suppressed for convenience. The character for this stabilizer group must be the same as that of the eigenmode under this operation, since the field vectors of (3) transform as the eigenmode. The eigenfunction, $\mathbf{E}_{1,\mathbf{k}}(\mathbf{r})$ itself, transforms as (1), being the partner function of the A_2 irreducible representation of the C_{3v} point group. For $R = \sigma_y$, we have $\chi_{\sigma_y}(C_{3v}) = -1$ [4], and hence $\mathbf{E}_{1,\mathbf{k}}(\mathbf{r}')$ (at any stable point $\mathbf{r}'$ on a mirror plane) must have $\chi_{\sigma_y}(C_{1h}) = -1$ (so it must be the partner vector for the B irreducible representation). Now that the character is known, equating left- and right-hand sides of (3) requires $D(\mathbf{r}') = B(\mathbf{r}') = 0$, and $C(\mathbf{r}')$ be the local field magnitude. Hence, the local states of polarization at such stable points are linear and perpendicular to the mirror symmetry plane. By symmetry, this holds for all three mirror planes.

Let us next consider whether there are any stable points associated with centers of rotational symmetry. Letting $R = C_n$ in (1), and employing (2), we find

$$\Omega \begin{bmatrix} \cos\left(\frac{2\pi}{n}\right) & -\sin\left(\frac{2\pi}{n}\right) & 0 \\ \sin\left(\frac{2\pi}{n}\right) & \cos\left(\frac{2\pi}{n}\right) & 0 \\ 0 & 0 & 1 \end{bmatrix} \begin{pmatrix} D+iB \\ C \\ 0 \end{pmatrix} = \chi_{C_n} \Omega \begin{pmatrix} D+iB \\ C \\ 0 \end{pmatrix}, \quad (4)$$

where the matrix gives the rotational operation for any n; again, $\mathbf{r}'$ has been suppressed for convenience. Inspection of (4) for $n = 1$ and $n = 2$ fails to reveal polarization singularities. By contrast, for $n = 3$, when we equate left- and right-hand sides of (4), upon separating their real and imaginary parts, we find that $\chi_{C_3} = \left(-1 + i\sqrt{3}\right)/2$ with $D(\mathbf{r}') = 0$ and $C(\mathbf{r}') = B(\mathbf{r}')$. Two sets of stable points are associated with this character: one identifies C points and the other, disclinations.

C points exist at locations $\mathbf{r}'$ where $D(\mathbf{r}') = 0$ and $C(\mathbf{r}') = B(\mathbf{r}') \neq 0$, whence it is apparent that the local polarization state must be circular. These locations, associated with purely C_3 symmetry, are given by the open black circles at the centers of the equilateral triangles in Fig. 1. The stabilizer group for the C points is thus the cyclic Abelian group, $C_3 = \left\{E, C_3, C_3^{-1}\right\}$, whose characters in general are given by $\chi = \exp(2\pi i p/3)$ for $p = 1, 2, 3$. Equation (4) constrains the choice of $p = 1$ for $\chi_{C_3}(C_3)$.

Disclinations are identified at a second set of stable points associated with $C(\mathbf{r}') = B(\mathbf{r}') = 0$ and the remaining centers of rotational symmetry. These coincide with the intersections of the mirror symmetry planes, shown by the solid black dots in Fig. 1. The stabilizer group for these field vectors, which transform as (4), is $C_{3v} = \left\{E, 2C_3, 3\sigma_y\right\}$. For these stable points, the character for the C_3 operator must be the same as that of the eigenmode, $\chi_{C_3}(C_{3v}) = 1$, since their groups are the same. Equation (4) will then only be satisfied if the field vanishes. The locally vanishing field, coupled with the continuity requirement as one asymptotically approaches such a stable point along the mirror symmetry planes, dictates that such stable points be disclinations. This is seen in Fig. 1.

I. CONCLUSION

We have shown that a two-dimensional hexagonal photonic crystal may manifest all three types of polarization singularities, L lines, C points and disclinations. We did this for a band 1 TE eigenmode at the K point by presenting the local states of polarization across the unit cell via direct calculation, and independently by exploiting group theoretic arguments, where the stable points were identified and their group representation determined.

REFERENCES

[1] J. V. Hajnal, "Singularities in the transverse fields of electromagnetic waves. I. Theory," *Proc. R. Soc. Lond.*, vol. A 414, pp.433-446, June 1987.
[2] M. V. Berry and M. R. Dennis, "Polarization singularities in isotropic random vector waves," *Proc. R. Soc. Lond.*, vol. A 457, pp. 141-155, August 2001.
[3] J. F. Nye, *Natural focusing and fine structure of light*. Bath, UK: IOP Publishing Ltd, 1999.
[4] K. Sakoda, *Optical Properties of Photonic Crystals*. New York, USA: Springer, 2005.
[5] Jeffrey F. Wheeldon, Trevor Hall and Henry Schriemer, "Symmetry constraints and the existence of Bloch mode vortices in linear photonic crystals," *Opt. Express*, vol. 15, pp.3531-3542, February 2007

Polymer deformable membrane mirrors for focus control using SU-8 2002

Erwin Dunbar, Matthew Leone, Sarah Lukes, David L. Dickensheets

Electrical and Computer Engineering Department, Montana State University, Bozeman, Montana 59717 USA
406 994-7147 (ph) 406 994-5958 (fax) davidd@ee.montana.edu

Abstract

Large stroke deformable membrane mirrors are designed for primary focus control and compensation of focus-induced spherical aberration. SU-8 2002 epoxy membranes 2 μm thick and up to 1 cm diameter, with symmetric aluminum coatings, are described.

Introduction

Deformable membrane mirrors are versatile adaptive optical elements that provide achromatic wavefront control. When designed for mostly parabolic displacement, a deformable membrane mirror can operate as a variable-focus lens. Electrostatic silicon nitride membrane mirrors with only two degrees of freedom and two annular actuating electrodes have been shown capable of diffraction limited focus control in microscopy applications.[1] However, silicon nitride is a stiff material (modulus of elasticity ~260 GPa) and exhibits residual tensile stress between 10 and 100 MPa in a low-stress formulation, requiring relatively large electrostatic force for deformation. For example, 1 μm thick silicon nitride membranes 1 mm in diameter demonstrated approximately 0.01 μm/(V/μm)^2 displacement at membrane center.[2] Practical devices have been limited to less than 5 μm stroke (maximum center deflection of the membrane) for 1 mm diameter mirrors with an air gap of 12 μm and less than 300 volts for actuation.

One application for MEMS focus control is optical microscopy. Especially for *in vivo* confocal microscopy, rapid focus control can be used to achieve arbitrary *x-y-z* scanning and may be used for image stabilization in the presence of sample motion. It has been shown, however, that focus control in a high NA microscope will require deformable membrane mirrors capable of stroke on the order of 30 μm.[3] Current generation silicon nitride membrane mirrors cannot achieve this large displacement.

Recently, investigators have reported deformable membrane mirrors made from SU-8, a permanent epoxy negative photoresist.[4] Advantages of this material include low cost (compared to LPCVD deposited silicon nitride films), ease of use (spin coating and UV patterning) and low elastic modulus (~2 GPa) and residual stress (~5 MPa).[5] It is the mechanical properties in particular that promise greater membrane stroke for a given device geometry, with a Young's modulus that is two orders of magnitude less than silicon nitride, so that 1 μm thick membranes 1 mm in diameter may be expected to deflect on the order of 1 μm/(V/μm)^2. Scaling laws would indicate that a 1 μm thick SU-8 membrane with an air gap of 75 μm and capable of 30 μm displacement could be less than 10 mm in diameter.

To investigate SU-8 membrane mirrors for focus control we adopted a simple device structure illustrated in Figure 1. Membranes are formed on one silicon wafer, and drive electrodes and a spacer are formed on a second wafer. An aligned bonding step results in the structure illustrated. This two-wafer process allows flexibility in choice of membrane material, diameter and thickness, spacer thickness and electrode patterns.

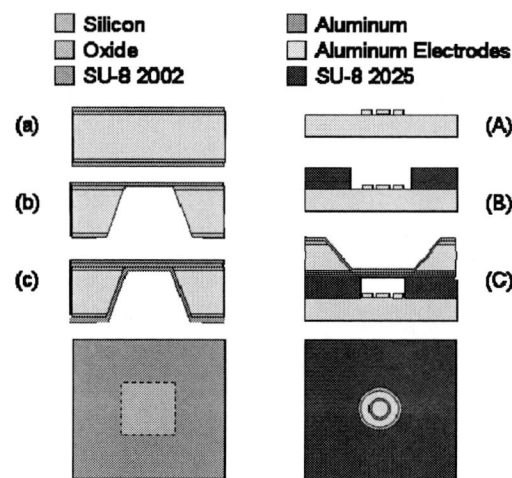

Figure 1 Steps (a)-(c) form the metal coated free standing membrane, suspended in a silicon frame. Steps (A)-(B) form the electrodes and spacer, and the final device (C) with the membrane wafer bonded to the electrode wafer. The bottom drawings show the membrane circular shape and the shape of the annular electrodes.

Fabrication Methods

For the substrate and membrane wafers, we are using single-side-polished silicon <100> wafers. The first step in creating the electrode wafer is aluminum evaporation to 200 nm thickness and patterning for the electrodes. For the test devices reported here the electrode is not patterned. After electrode deposition and patterning, SU-8-2025 is spun on at a thickness of 25 μm to serve as a spacer layer between the electrode and the membrane grounded layer. The spacer layer is then patterned by exposing for 8 seconds at 19.4 mW/cm^2 and then developed. A hard bake is performed, ramping from 95° to 200°C and held for 60 minutes (to further crosslink the SU-8).

The membrane wafer was made by first growing thermal oxide to a thickness of approximately 620 nm. Membrane regions were defined by patterning the back (rough) side of the wafer and etching the oxide in buffered HF to open etch windows on the back and remove all of the oxide from the front of the wafer. After stripping the photoresist, SU-8-2002 was spun on the front-side of the wafer at a thickness of 2 μm. This serves as the flexible mirror membrane. The SU-8 is blanket exposed and hard baked as well (following the same process used for the SU-8-2025). Following the hard bake, the wafer is etched in TMAH at 75°C until the wafer has been etched completely through from the back-side, stopping at the SU-8 layer on the front side. After a DI water rinse, Aluminum is sputtered to a thickness of 100 nm on both the front side (to serve as the ground plane) and on the backside (which serves as the reflective surface). Symmetric metallization is used to minimize stress gradients through the membrane structure to promote initial flatness.

Eventually we will use wafer level bonding of the membrane and electrode wafers. The prototype devices were diced first and rinsed prior to bonding. A simple CCD camera jig was used for bonding alignment of the separate dies. An epoxy joins the two wafer layers together. An overhang is left to allow for electrical contact to the membrane and actuating electrode.

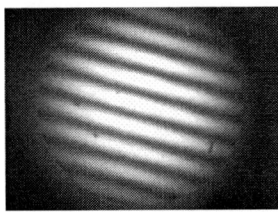

Figure 2 Interferometric image (λ=670nm) of central 1.5 mm region of a 3 mm square membrane, illustrating initial flatness to better than 50 nm.

Results

Free standing, metalized membranes as large as 1 cm have been produced by this method. The membranes withstand routine handling, including DI rinse and gentle N2 blow dry. Roughness of both the optical surface (wafer side) as well as the TMAH exposed side of the membrane were below the 10 nm noise floor of our optical profilometer. Initial flatness of a 3 mm square membrane, shown in Figure 2 imaged interferometrically at 670 nm wavelength, is better than 50 nm peak-to-valley over the central 1.5 mm in the field-of-view of our phase-shift profilometer.

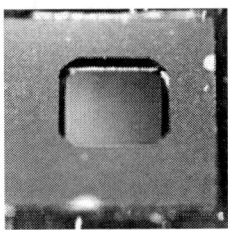

0 volts   250 volts

Figure 3 Assembled membrane device for initial testing. 5 mm square membrane suspended over a 3 mm circular opening in the spacer.

Figure 3 shows a photograph of an assembled membrane device. Assembly was done after scribing and breaking the wafers. Although the die were rinsed in isopropanol to remove particles created during the dicing

procedure, our initial bonded devices show some wedge separation. Figure 4 is a series of interferograms taken with different voltage applied to a membrane suspended over a 3 mm diameter circular opening in the 25 μm thick spacer layer. Figure 4a, with 0V applied, shows the initial wedge separation, with the membrane in contact with the wafer only at the left edge of the spacer opening. Figure 4b, with 96V applied, shows the membrane now making contact along the full perimeter, but the initial deformation leads to some astigmatism of the membrane in this position. Figure 4c shows the membrane with 270V applied and exhibiting 5 μm of deflection at the center, relative to the height at the edge of the spacer opening. We observed stable deflection for this device in excess of 17 μm relative to the edge of the spacer layer, which is 68% of the spacer thickness. Actual gap thickness may be greater than 25 μm, as a result of the bonding process. Center displacement is plotted vs. applied voltage in Figure 5.

References

[1] Himmer *et al.*, Optical MEMS 2005, p185-186 (2005).
[2] Himmer *et al.*, Opt. Letters **26**(16) 1280-1282 (2001).
[3] Dickensheets, J. Micro/Nanolith., MEMS, MOEMS v.7, 021008 (2008).
[4] Friese, *et al.*, JMEMS **17**(1), p11-19 (2008).
[5] SU-8 2002 Datasheet, Microchem, Inc.

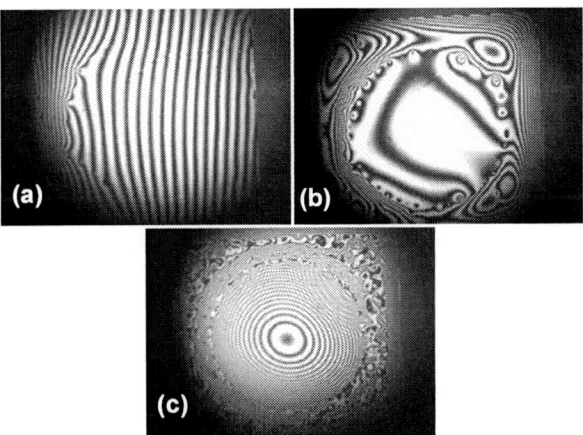

Figure 4 Interference images of assembled membrane device. (a) rest condition, with membrane mostly not in contact with spacer layer; (b) 96 volts applied, and membrane in contact with spacer perimeter; (c) 270 volts applied, and membrane displaced 5 μm at center.

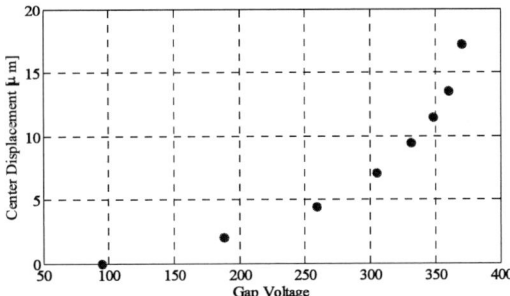

Figure 5 Displacement of 3 mm diameter membrane at center, as a function of applied voltage.

Mechanically Coupled Comb Drive
MEMS Stages

Aslıhan Arslan, Çağlar Ataman, Sven Holmstrom, Karin Hedsten*, H. Rahmi Seren, Hakan Urey, Peter Enoksson*

Koç University, Electrical Engineering and Optical Microsystems Laboratory, Istanbul, TURKEY
* Chalmers University of Technology, Department of Microtechnology and Nanoscience, Gothenburg, SWEDEN
Phone: +90-212-338 1772, E-mail: asarslan@ku.edu.tr, hurey@ku.edu.tr

Abstract- An electrostatic large clear-aperture in-plane scanner with a novel actuation principle is presented for fast and large stroke scanning applications. 9 µm resonant deflection at 11.51 KHz with 100 Vpp excitation is observed.

Keywords: Mechanical coupling, MEMS stage, electrostatic actuation.

Introduction
Damping power losses can be quite large in comb-actuated structures resulting in small deflections particularly at high frequencies. In this article, a novel MEMS actuation technique is proposed to overcome this limitation. Analytical and numerical modeling of the technique is performed and confirmed with experimental results.

An in-plane MEMS stage is fabricated and tested during this study. Efforts are under way to integrate the SOI fabrication process for the MEMS stage with a monolithic microlens array process to fabricate microlens arrays on MEMS stages. The main application is beam steering for endoscopic imaging applications that require deflections on the order of the diameter of a single microlens in the microlens array for full-angle beam steering [1].

Principle of Mechanical Coupling
Figure 1(a) illustrates the basic structure that includes an additional outer frame around the inner frame. The inner frame is connected to the outer frame with four flexures. Microscope picture of the fabricated device is shown in figure1 (b). Clear aperture of the device is 1.5 mm x 1.5 mm and the overall size is 3.8 mm x 4 mm.

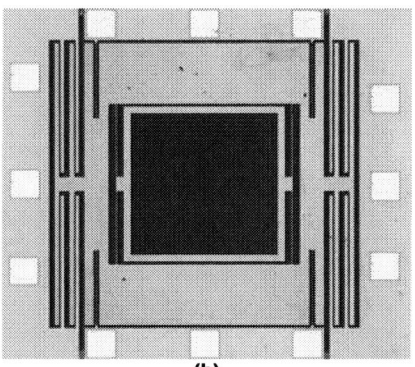

(b)
Fig 1. (a) 3D illustration of the design. (b) Microscope picture of the fabricated MEMS device.

Electrostatic comb drive actuators are only present on the outer frame. As illustrated in figure 2, in a lower vibration mode (joint mode), determined mainly by the outer frame flexures, the inner and outer frames move in-phase and deflect by nearly the same amount. In a higher vibration mode (coupled mode), determined mainly by the inner frame flexures, the inner and the outer frames move out-of-phase and the motion of the outer frame is transferred to the inner frame mechanically with a mechanical gain factor [2].

Main advantages of the proposed scheme are that the inner frame has no electrical connections and has low damping. The outer frame deflections are smaller and therefore the damping losses at the comb fingers can be kept low.

Numerical Analysis
Mechanical coupling efficiency is defined as the ratio of the inner frame deflection to the outer frame deflection and is dependent on vibration frequency. For the coupled-resonant mode where two frames move out-of-phase, a simple

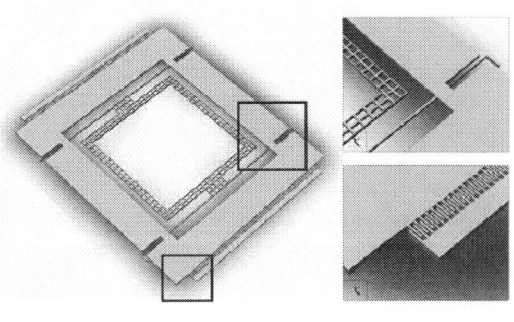

(a)

approximate equation can be obtained for the coupling factor M:

$$M = \frac{k_1}{\sqrt{(k_1 - m_1\omega)^2 + b_1^2}} \qquad (1)$$

where k_1, m_1 and b_1 are the stiffness, mass, and damping coefficient associated with the inner frame and its flexures and ω is the oscillation frequency of the frames.

The maximum oscillation amplitude for the inner frame is obtained at the coupled-resonance frequency denoted by ω_c. Using some simplifying assumptions, it can be shown that the coupled resonance frequency is given by:

$$\omega_c^2 = \frac{k_1}{2m_2} + \frac{\omega_2^2}{2} + \omega_1^2 \qquad (2)$$

Where ω_1 and ω_2 are the natural frequencies of the inner frame and the outer frame obtained using $\sqrt{k_1/m_1}$ and $\sqrt{k_2/(m_1 + m_2)}$ ratios. k_2 and m_2 are the stiffness and the mass of the outer frame.

Figure 2 presents the frequency response analysis results obtained with MATLAB assuming harmonic excitation.

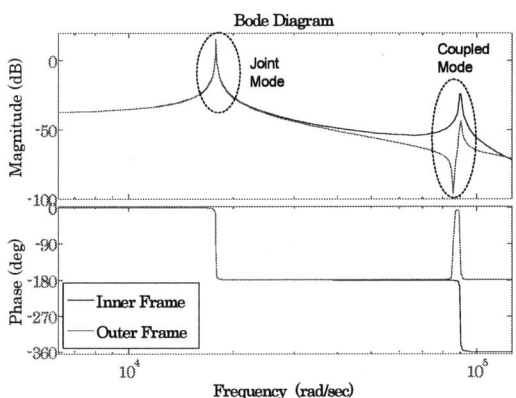

Fig 2. Numerical simulation results for the coupled equation of motion illustrating the inner and outer frame deflections

Experimental Results
Devices are fabricated on 20μm SOI wafer using a simple fabrication process.

Comb drive scanner designs are tested to validate the mechanical coupling mechanism. Complementary biased sinusoids are applied to comb fingers on either side of the outer frame for differential driving to ensure harmonic excitation. According to FEM analysis, in plane sliding mode resonance frequency of the scanner is 11.44 KHz while experimental result shown in figure 3

is 11.51 KHz. At this frequency, 1μm deflection is obtained with 15 Vpp excitation. The coupling efficiency is found to be 5. The maximum voltage applied to the scanner was 100 V and 9 μm deflection is obtained at 11.51 KHz. For the joint mode at 2.5 KHz, 30 V p-p input results in 15 μm deflection.

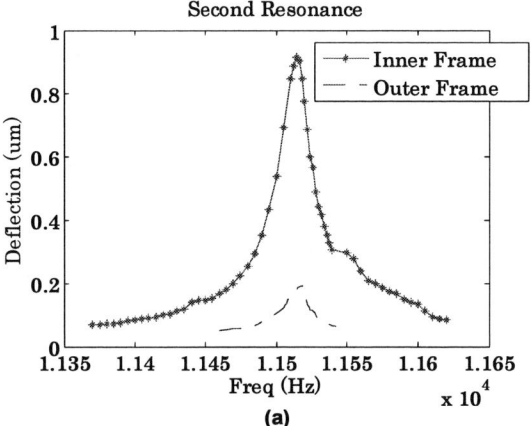

Fig 3. (a) Experimental result of frequency response around coupled resonance frequency of the scanner.

Conclusion
A new actuation technique is developed based on the mechanical coupling principle. At the coupled mode, small deflections of the outer frame are transferred to the inner frame with a mechanical gain. Closed form gain factor and coupled resonance frequency formulas are derived analytically.

Stages will be integrated with microlens arrays for beam steering for endoscopic imaging applications.

Partial funding from MC2 access, FP6 program and TUBITAK (Turkey) Grant No: 106E068 are gratefully acknowledged.

References
[1]   A. Akatay, C. Ataman, H. Urey," Design and optimization of microlens array based high resolution beam steering system," Opt. Lett, Vol. 15, pp. 4523-4529, 2007

[2] A.D.Yalcinkaya, H. Urey, D. Brown, T. Montague, and R. Sprague, "Two-Axis Electromagnetic Microscanner for High
Resolution Displays," J. MEMS, Vol. 15, No. 4, August 2006

Fano resonance phenomenon utilizing Photonic Crystal Rods for tunable filter applications

XiongYeu Chew[1], Guangya Zhou*[1], Hongbin Yu[1] and Fook Siong Chau[1]

[1]Micro and Nano Systems Initiative, Dept. of Mechanical Engineering, National University of Singapore, 10 Kent Ridge Crescent, 119260, Singapore
Email mpezgy@nus.edu.sg

Abstract

A novel approach of utilizing the fano resonances of photonic crystal rods to achieve high-Q filter capabilities. Here we present the optimization of peaks attenuation and Q-factor properties by permutation of rod dimensions.

I Introduction

An interesting phenomenon known as the fano resonances are recently proposed for photonic crystal (PC) slabs. This property manifests itself when the normal incidence light interacts with the inherent guided modes of photonic crystals. According to numerical and theoretical approaches[1][2], when the guided modes decouples from the incoming light due to mode mismatch, fano resonances are formed. The transmission properties of a PC slab exhibits a fabry-perot like varying curve that is an inherent transmission characteristic of a uniform dielectric slab superimposed with fano resonances peaks. Some of the fano resonance peak sensitivities are also dependent on the incident angle of illumination as shown by Crozier et. al. [3]. Various applications are proposed by utilizing this interesting phenomenon for out-of-plane vertical light propagation. W. Suh et. al.[4] proposed using this phenomenon as a flat-top filter where a wide band of attenuation is observed. They also proposed integrating two PC slabs mechanically to utilize it as a displacement sensors[5]. Rosenberg et. al.[6] demonstrated experimentally showing the existence of fano resonances even in asymmetrically configured PC slabs showing the robustness of its characteristics. Interestingly, all of the theoretical and design approaches were focused on PC slab due to its ease in fabrication and integration. In this paper, we enabled the design of a new class of tunable vertical photonic devices by utilizing the fano resonances in a free-standing PC rods structures embedded on a layer of PDMS. In such design, the PDMS can easily be bonded[7] with mechanical actuators to enable mechanical tunable filtering device- see Fig. 1. The study first shows the different characteristics of fano resonances in rods structures comparatively to slab structures. Then we demonstrate the optimization of attenuation and Q-factor of fano resonance by altering rod dimensions. These results can be utilized in future for the design of miniaturized vertical incidence tunable filters depending on the requirement of Q-factor and attenuation rate.

II Modeling methodology

The PC structures rods we studied is a hexagonal lattice with a lattice constant a=930nm, with a radius for all rods of r/a=186nm. The band analysis is numerically shown in Fig. 2, utilizing a 3D plane-wave-expansion method. Studying the band diagram we can approximate the frequencies of fano resonances by locating mode matching doubly degeneracy modes. We showed that the condition of fano resonances excitation in PC rods similarly requires a doubly degenerated mode that consists of a matched mode distribution with the corresponding incident light source. The transmission results of the PC rods are then shown numerically in Fig. 3 utilizing a 3-D finite-difference time-domain (FDTD) approach with sufficiently fine mesh. We've found that the theoretical approach[1] of non-degeneracy modes corresponds well to PC rods. However, we observed from the transmission results that PC rods do not have the varying fabry-perot curves due to decoupling of the dielectric media. Instead we observe a a good transmission with a sharp reflective peak at near communication bandwidth of λ=1.55µm corresponding approximately well to frequencies at $\omega \approx 0.59 \times (2\pi c/a)$ shown in Fig. 2. Likewise we also observe other fano resonance at λ=1.43µm and 1.24µm. The fano resonance at higher order modes are not very useful as a wavelength selective filter as they are in close proximity to each other causing an inconsistent transmission on the side lobes of the spectra. Hence only the corresponding lowest order fano resonance mode is selected for optimization.

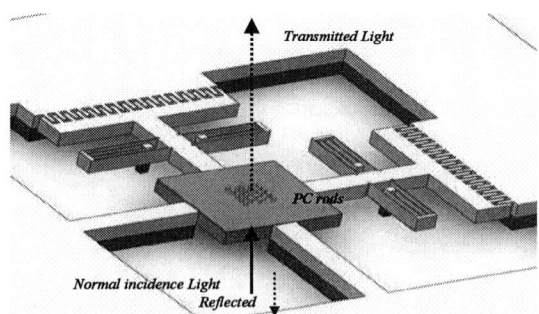

Fig. 1. Illustration of a proposed filter/reflector based on PC rods bonded on top of PDMS.

978-1-4244-1917-3/08/$25.00 ©2008 IEEE

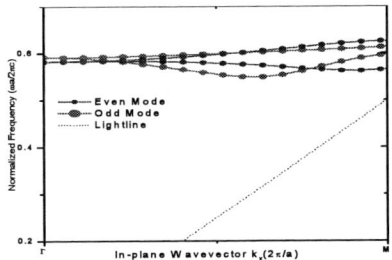

Fig. 2. Band diagram of a Hexagonal PC rods. We see doubly degeneracy at odd modes at $\omega \approx 0.59 \times (2\pi c/a)$. Higher order modes are not shown in this band diagram.

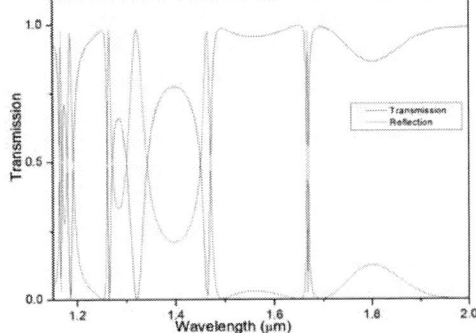

Fig. 3. Transmission plot of PC rods showing multiple fano resonance peaks. The significant lorentzian like fano resonance peak that we found useful is at $\lambda \approx 1.55 \mu m$.

III Results and Discussion

We show numerically that the Q-factor and shape properties of the corresponding fano resonance are highly dependent with the height dimensions of the PC rods. By linearly increasing the rod height in constant steps, we observe an exponential increase in the Q factor. 13 sets of curves are shown in Fig. 4a demonstrating the transformation of the fano resonance at a 10% incremental of PC rod height. It can be observed that the fano resonances at $h \approx 0.74 \mu m$ are relatively broad and slowly sharpening into a vey fine peak when the PCrod height is increased to $h \approx 1.1 \mu m$. Beyond $h \approx 1.2 \mu m$, the peak broadens out again. Fig. 4b shows the magnified perspective with a 1% increment of PC rods ranging from $h \approx 1.116 \sim 1.2 \mu m$. We can see that at this range, the attenuation of fano resonance is associated with the increase of Q-factor. At approximately $h \approx 1.15 \mu m$

the fano resonance is reduced to an extremely small and sharp fano resonance peak with a Q-factor of 275×10^3. Fig. 5 quantitatively shows the Q-factor of the corresponding peaks observed in the previous figures. We see a symmetrical increase and decrease in q-factor in variation of the PC rod height. By utilizing this relation, we can devise a optical filter/reflector with an optimum PC rod height to achieve necessary Q-factor and required attenuation ratio. Further optimization of a proposed device will be presented in the full paper.

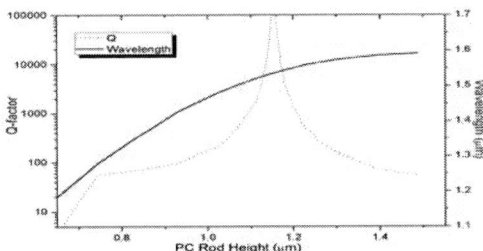

Fig. 5. Shows the relations of PC rod height versus Q-factor of fano resonance. The corresponding wavelength when PCrod height is adjusted is also shown.

IV References

1. T.Ochiai, K. Sakoda, "Dispersion Relation and optical transmittance of a hexagonal photonic crystal slab", Phys. Rev. B, **63** 125107 (2001)
2. S. Fan, J.D., Joannopoulos, "Analysis of guided resonances in photonic crystal slabs, "*Phy. Rev. B.* **65**, 235112 (2002)
3. K.B. Crozier, V. Lousse, O. Killic, S. Kim, S. Fan and O. Solgaard, "Air-bridged photonic crystal slabs at visible and near-infrared wavelength," Phys. Rev. B **73** 115126 (2006)
4. W. Suh & S.Fan, "All-pass transmission or flattop reflection filters using a single photonic crystal slab", Appl. Phys. Let., **84** 24, 4905 (2004)
5. W. Suh, M.F. Yanik, O. Solgaard & S. Fan, "Displacement sensitive photonic crystal structures based on guided resoannce in photonic crystal slabs", Appl. Phys. Lett. **82,** 13 (2003)
6. A. Rosenberg, Michael W. Carter, J. A. Casey, Mijin Kim, Ronald T. Holm, Richard L. Henry, Charles R. Eddy, V. A. Shamamian, K. Bussmann Shouyuan Shi, Dennis W. Prather, "Guided resonances in symmetrical GaN photonic crystal slabs observed in the visible spectrum," Opt. Express **13**, 6564 (2005)
7. E. Iwase, H. Onoe, K. Matsumoto, I. Shimoyama, "Hidden Vertical comb-drive actuator on pdms", MEMS 08,**116**,(2007)

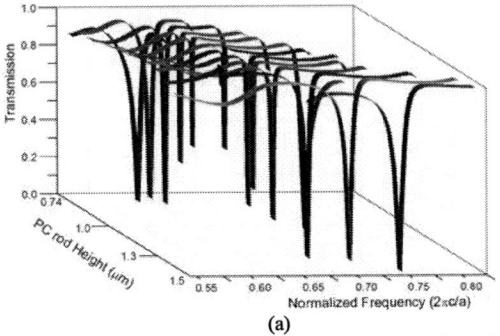
(a)

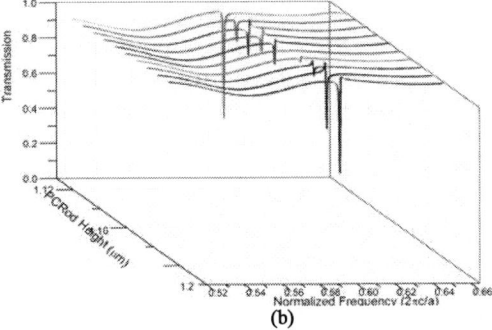
(b)

Fig. 4. (a) and (b)Shows the transmission peaks at varying PC rod height range at 0.7~1.5μm and at 1.12~1.2μm respectively.

143

A Glass Cantilever Beam Sensor Combined with a Spherical Reflecting Mirror for Sensitivity Enhancement

Chun-da Liao and Jui-che Tsai

Graduate Institute of Photonics and Optoelectronics and Department of Electrical Engineering
National Taiwan University
No. 1, Sec. 4, Roosevelt Rd., Taipei 10617, Taiwan
Tel: +886-2-3366-3700 Ext. 247, Fax: +886-2-3366-3686, E-mail: jctsai@cc.ee.ntu.edu.tw

Abstract

The incorporation of a micro spherical reflecting mirror (MSRM) onto a cantilever beam sensor magnifies the optical deflection angle and therefore enhances the sensitivity of an optical sensing system. The curvature of the spherical reflecting mirror determines the angle magnification. Given a certain bending amount and a fixed distance between the cantilever and position sensing detector, the cantilever integrated with a micro spherical mirror with a smaller radius of curvature produces a larger optical deflection. Currently at an intermediate stage, our optical sensing system includes a traditional cantilever and a separate spherical mirror with a radius of curvature of 10 cm, which experimentally simulates a system with a MSRM-integrated cantilever beam. The experimental results show that the sensitivity is enhanced by 70%.

Keywords: Cantilever beam sensor, spherical reflecting mirror, MEMS sensor

1. INTRODUCTION

MEMS cantilevers have been frequently seen in a wide range of research instruments. One of the most well-known examples is the atomic force microscope (AFM), where interactive forces between the tip and the sample lead to the cantilever deflection [1]. The resolution of the image depends on the sharpness of the tip and it can reach to the atomic level. MEMS micro cantilevers are also used as biomedical sensors; for example, they can be a means of DNA detection [2], [3]. When bio-molecules adhere to the bottom of the cantilever, the induced stress difference between the top and bottom surfaces leads to a bending, resulting in an angular deflection of the reflected laser beam from the cantilever's top surface. This optical deflection can be measured by a position-sensing detector (PSD) to determine the amount of bending, which further provides the information on the bio-molecule adhesion process. The sensitivity of a sensing system is determined by the minimum optical deflection that the system can detect [4]. It strongly depends on the characteristic of the PSD. Generally the resolution of a PSD is roughly sub-μm. However, external circuits and stray light in the environment produce extra noises in the signal readout, which degrades the system sensitivity. In the case of a slight optical deflection, the PSD output change can be overwhelmed by these noises and cannot be observed. Stretching the distance between the cantilever and PSD is one of the simplest ways to increase the laser beam displacement on the PSD under the same optical angular deflection, therefore enhancing the system sensitivity [Fig. 1(a)]. Nevertheless, this leads to a bulky system and an expanded laser beam that may be truncated the PSD aperture.

In this paper, we propose a cantilever structure combined with a micro spherical reflecting mirror (MSRM). The schematic is shown in the Fig. 1(b). Given a certain bending amount and a fixed distance between the cantilever

and position sensing detector, the cantilever with a MSRM magnifies the deflection angle so the laser beam displacement on the PSD is increased. In other words, the sensitivity of the optical sensing system is enhanced while maintaining the system compactness.

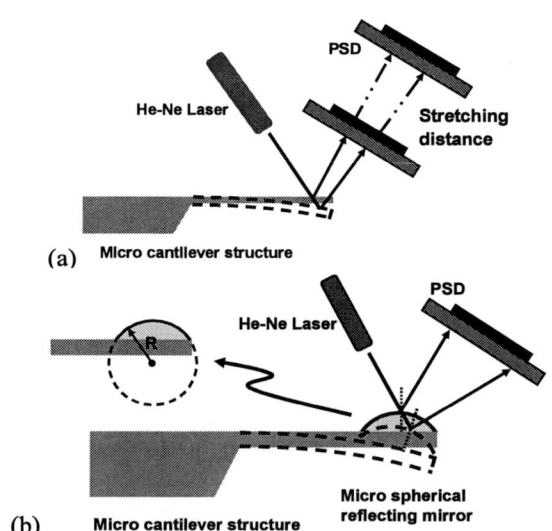

Figure 1. The optical sensing systems with (a) a traditional micro cantilever and (b) with a MSRM-integrated cantilever, respectively.

2. THEORETICAL ANALYSIS

The theoretical calculation results are shown in Fig. 2. The square, triangular, and circular symbols are for the radii of curvature (R) of 8mm, 10mm, and 12mm, respectively. It can be seen that a smaller radius of curvature of the MSRM leads to a larger deflection angle of the laser beam, i.e. a larger laser beam displacement on the PSD. The crisscross symbols are for the traditional cantilever structure. With

978-1-4244-1917-3/08/$25.00 ©2008 IEEE

R=8mm, the MSRM-integrated cantilever produces an optical angular deflection which is 2.5 times as large as that of the traditional approach.

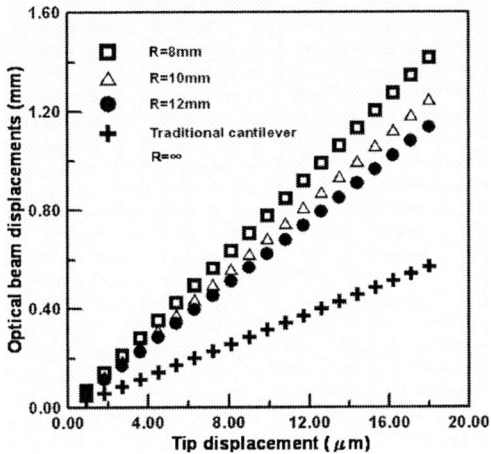

Figure 2. Calculation results: the optical beam displacement on the PSD vs. the cantilever tip movement. The sensitivity is improved by the magnification of the optical deflection.

3. EXPERIMEMT

Currently at an intermediate stage, our optical sensing system includes a traditional cantilever and a separate convex mirror with a radius of curvature of 10 cm, as shown in Fig. 3. This experimentally simulates a system with MSRM-integrated cantilever beam. The convex mirror is suspended over the glass cantilever. The use of glass as the cantilever material offers the advantage of high biocompatibility. It also has a relatively low Young's module (65-90 GPa) compared with other common materials such as silicon and silicon nitride. Furthermore, it provides the potential of fabricating an optically-transmitting-type device, in contrary to the current reflecting architecture.

When the cantilever is un-bent, the incident angle of the He-Ne laser beam is 60^0 and the reflected light from the cantilever is aimed at the center of the suspended convex mirror. The light pathlength from the cantilever to the PSD is totally 10 cm. The cantilever bends while its tip is being pressed by a PZT-controlled probe. At the maximum tip movement 18 μm, the optical displacement of the laser beam on the PSD is 0.98 mm for the system with a separate convex mirror and it is 0.56 mm for the traditional architecture (without the convex mirror). Therefore, the sensitivity of the sensing system with a convex mirror is enhanced by more than 70% compared with that obtained by the traditional approach. The complete experimental results are shown in Fig. 4. The dash lines are the results of numerical calculations for comparison.

4. CONCLUSIONS

We have proposed a glass cantilever beam sensor combined with a spherical reflecting mirror for sensitivity enhancement. Enhancement of more than 70% is demonstrated experimentally. The fabrication of a

MSRM-integrated cantilever sensor and its experimental results will later be reported elsewhere.

ACKNOWLEDGEMENTS

This work was supported by National Science Council of Taiwan under grants NSC 95-2221-E-002-053 and NSC 96-2221-E-002-198-MY2, and Excellent Research Projects of National Taiwan University, 95R0062-AE00-06.

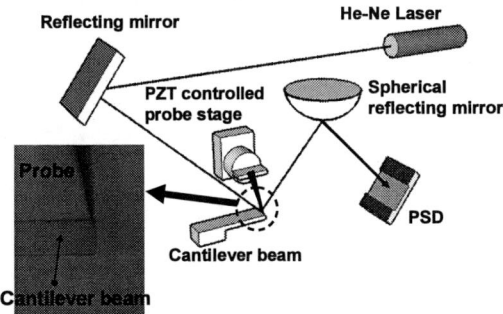

Figure 3. The experimental setup. The inset shows the optical microscope image of the glass cantilever.

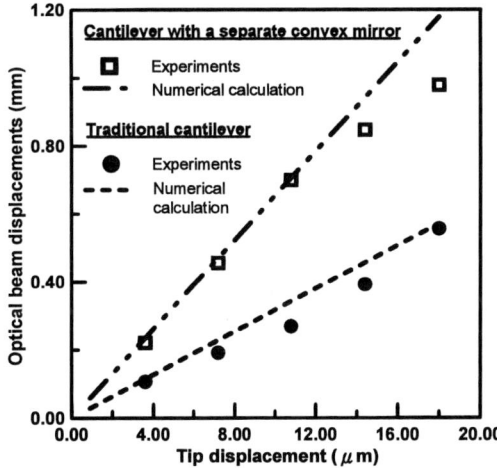

Figure 4. Experimental results. The circular and square symbols represent the experimental data whereas the dash lines are the numerical calculation results.

REFERENCES

[1] R. Wiesendanger, "Scanning probe microscopy and spectroscopy: methods and applications," *Cambridge University Press*, 1998.

[2] T. P. Burg et al., "Weighing of biomolecules, single cells and single nanoparticles in fluid," *Nature*, Vol. 446, pp. 1066-1069, 2007.

[3] J. Fritz et al., "Translating biomolecular recognition into nanomechanics," *Science*, Vol.288, pp. 316-318, 2000.

[4] D. Klaitabtim et al., "Design consideration and finite element modeling of MEMS cantilever for nano-biosensor applications," *5th IEEE Conference on Nanotechnology*, vol. 1, pp. 311-314, 2005

A Micromirror Scanner with Vertical Combs Tilted by Assembly Process

Min-Ho Jun[1], Sungsik Yun[1], Man Geun Kim[1], Sung-Kil Lee[1] and Jong-Hyun Lee[1,2]

[1]MEMS Lab., School of Information and Mechatronics, [2]School of Medical System Engineering,
Gwangju Institute of Science and Technology (GIST)
261 Cheomdan-gwagiro (Oryong-dong), Gwangju, Republic of Korea
(TEL) +82-62-970-2395, (FAX) +82-62-970-2384, e-mail: jonghyun@gist.ac.kr

Abstract

We have developed a simple assembly technology to realize the tilted vertical combs for electrostatic micromirror scanners. The in-plane vertical comb electrodes are easily transformed into out-of-plane tilted comb by asymmetrical pushing down the levers of the spring that is attached to the micro mirror. The fabricated mirror scanner showed the optical scan angle of up to 1.3° at 60V and the resonant frequency of 3.15 kHz.

Keywords: mirror scanner, vertical combs, tilting, assembly, alignment pillar

1. INTRODUCTION

Since optical microscanner was developed in early 1980's, it has been applied to wide variety of applications, such as display, communication, and biomedical imaging [1, 2, 3], etc. Electrostatic actuators are typically preferred over other actuation mechanisms owing to lots of advantages in the device performances such as low power consumption and simple fabrication, etc.

To drive the mirror in large scanning angle, the comb actuator needs a vertical offset between a movable comb and a fixed comb. The unleveled combs, which can be fabricated by two step silicon etching, were used to realize the offset in the comb actuator [4]. It is, however, difficult to fabricate the two combs in different levels due to multi-layered etch masks. Another method is to use the angled vertical comb, which requires additional process such as reflow [4] and Joule heating [6] though it allows larger scanning angle.

In this paper, we have proposed the new fabrication method to enable the movable vertical comb to tilt during an assembly with no additional process. The tilted vertical comb is fabricated to demonstrate its application possibility through the experiment for the scanner.

2. DEVICE CONFIGURATION

Fig. 1 shows the schematic of the proposed micromirror scanner. The scanner mainly consists of a device part and tilting part. When two alignment pillars are inserted into the alignment holes, the device and the tilting parts are aligned. Simultaneously the comb electrodes with a mirror are tilted by pushing down the levers of the spring that is attached to the micro mirror. The tilted combs allow the mirror to scan in large angle with no additional fabrication process.

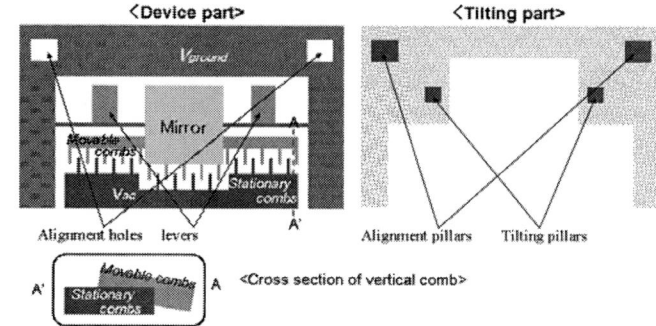

Fig. 1 Schematic of the electrostatic mirror scanner with vertical combs tilted by assembly process with a tilting part

3. FABIRCATION

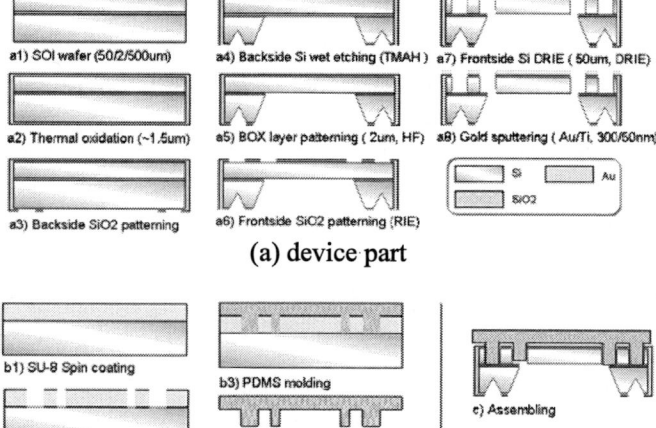

(a) device part

(b) tilting part (c) assembly

Fig. 2 Fabrication sequence of a micromachined mirror scanner with vertical combs tilted by assembly process

Fig. 2 illustrates the fabrication sequence of the mirror scanner with the tilting part. Fig. 2(a) and Fig. 2(b) represent the fabrication sequence of the device part using DRIE process and the tilting part using molding process, respectively. Fig. 2(c) shows assembly process of two parts to tilt the vertical combs.

The device part is fabricated using SOI (silicon-on-insulator) wafer (Fig. 2a1). Firstly, the SOI wafer is thermally oxidized (Fig. 2a2) and the backside is patterned for etch mask (Fig. 2a3). Then, the backside silicon is etched by TMAH solution, followed by HF etching of the buried oxide (Fig. 2a4, Fig. 2a5). The device pattern is realized on the SiO_2 layer as a etch mask by standard photolithography process, and transferred to the top silicon of the SOI wafer (Fig. 2a6). For the final process of the device part, the top silicon is etched by DRIE (Fig. 2a7).

To fabricate the tilting part, SU-8 is coated on a (100) silicon wafer by spin coater (Fig. 2b1). A SU-8 cast is built by photolithography process (Fig. 2b2), and the PDMS is poured into the SU-8 cast (Fig. 2b3). When the tilting part covers the device part, the movable combs are tilted, because the tilting pillars push down the levers of the spring.

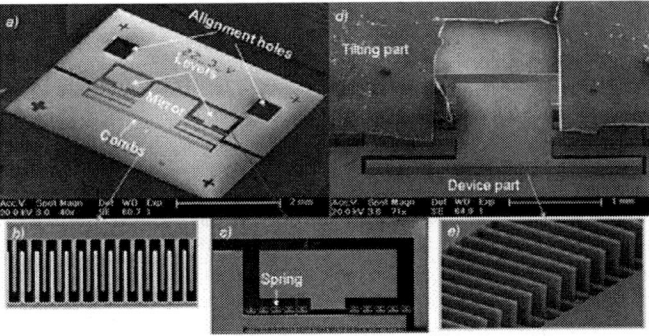

Fig. 3 SEM image of the electrostatic mirror scanner

Fig. 3 is SEM image of the fabricated mirror scanner. The Fig. 3a represents the device part as fabricated, and Fig. 3b and Fig. 3c show the in-plane combs and lever for tilting the combs in detail, respectively. Fig. 3d represents the mirror scanner assembled with tilting part, and Fig. 3e shows the vertical combs actually tilted during assembly process.

4. EXPERIMENTS

The fabricated mirror scanner was evaluated in terms of resonant frequency and optical scan angle. Firstly, the resonant frequency of the scanner was measured by laser vibrometer to characterize the maximum scan speed. The optical scan angle of the fabricated scanner was measured by He-Ne laser and a scaled screen.

As experimental results, the fabricated scanner shows the resonant frequency of 3.15 kHz, and optical scan angle of up to 1.3° at 60V as indicated in the Fig. 4 and Fig. 5,

respectively. The resonant frequency of the fabricated scanner, designed as 2.67 kHz, was shifted to 3.15 kHz, which is due to the fabrication error of the spring. Considering the dimensional error, the compensated theoretical scan angle was compared with the experimental one, and confirmed they matches well each other, as shown in Fig. 5.

Meanwhile, the roughness of the mirror surface should be as small as possible to minimize the scattering loss. The experimental roughness of the fabricated mirror surface is 9nm, which would cause negligible scattering loss (0.02 dB or less) on the condition that the wavelength used in between 800nm to 1500nm and the incident angle is 45°.

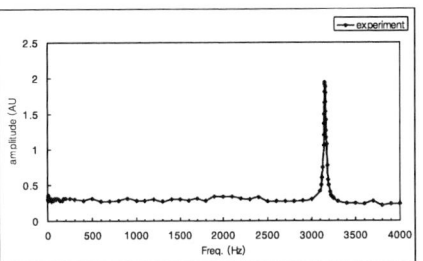

Fig. 4 Experimental result for the frequency response of the fabricated scanner

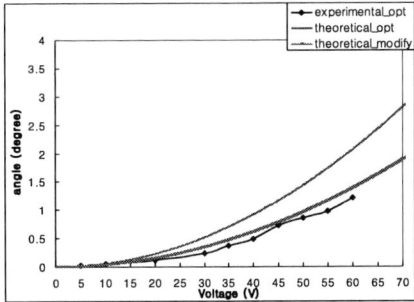

Fig. 5 Comparison of the theoretical and experimental scan angle of the scanner

ACKNOWLEDGEMENT

The research was supported by a grant from the institute of Medical System Engineering (iMSE) in the GIST, Korea.

REFERENCES

[1] H. Urey, et al., in Proc. SPIE, MOEMS Miniaturized Systems, 2000, vol. 4178, pp. 176–185
[2] M. Last, et al., in Proc. of IEEE ISCAS Conf. 2003, Bangkok, Thailand, 2003
[3] J Singhm et al., JMM, vol. 18, 2008, pp. 1-9
[4] D. Hah, et al., JMM, vol. 14, 2004, pp. 1148–1156
[5] Aaron D. Aguirre, et al., in Proc. SPIE, Vol. 5692, 2005, Bellingham, WA, pp. 227-282
[6] J. Kim, et al., JMM, vol. 15, 2005, pp. 1777-1785

Design and fabrication of compact etched diffraction grating demultiplexers based on α-Si nanowire technology

Jun Song and Ning Zhu

Abstract: Silicon nanowire waveguides and related etched diffraction grating (EDG) demultiplexers are studied by α-Si-on-SiO₂ technology. Compact EDG demultiplexers with 10 nm spacing for both echelle and total-internal-reflection (TIR) facets have been fabricated and characterized.

Summary:

Due to the compatibility of the fabrication technology with micro-electronics, silicon photonics has attracted a lot of interests. Both arrayed waveguide gratings (AWGs) and etched diffraction gratings (EDGs) based on Si have been studied in the recent years. However, to maintain single mode propagation along such a silicon nanowire waveguide the cross section of the waveguide needs to be very small. The performance of such sub-micrometer sized devices strongly depends on the fabrication accuracy. For the EDG application, the thickness of waveguides h=250nm is fixed here, and the width of waveguides w=500nm is chosen, which lies in the single mode region.

As examples, two EDGs with the same parameters will be designed: the central wavelength is 1552.5nm; the refractive indexes of silica buffer layer and α-Si:H core laye are 1.46 and 3.58, respectively; the incident angle is 45 degrees; the diffraction order is 6; channel interval is 10 nm; and the internal of output waveguides is 5μm. The only difference between the two devices is the grating type. One is based on echelle facets and the other uses the total-internal reflection (TIR) facets.

Fig. 2 shows some pictures of the fabricated grating structures. The roughness of the sidewall (i.e., the side facet of the waveguides and gratings) is ~10nm, which is directly measured from the scanning electron microscopy (SEM) pictures.

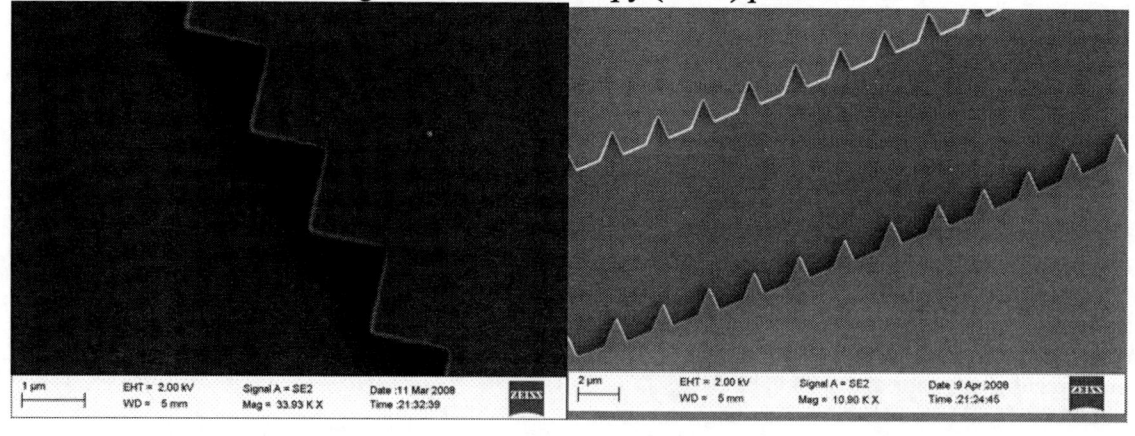

(a)   (b)

978-1-4244-1917-3/08/$25.00 ©2008 IEEE   148

Fig. 2 Pictures of enlarged structures of the fabricated echelle (a) and TIR (b) gratings using Si nanowires technology.

Fig. 3 shows the spectral responses of the fabricated EDGs for the TE polarization. The channel spacings are matched with the design. The large insertion loss is mainly due to the large scattering loss of a-Si nanowire waveguides. We can find that the crosstalks are relatively high here for practical wavelength division multiplexer (WDM) applications. We attribute this to the phase error induced in the grating etching due to the grating-facet variation in short range (sidewall roughness) and long range (stability of the E-beam during exposure). In addition, sidewall roughness also results in most of loss for the sample. In addition, from this figure, one also sees that the loss of EDG using TIR facets will be 3-5 dB lower than that using echelle facet. This is because we have not coated a metal at the backside of the echelle grating in the sample.

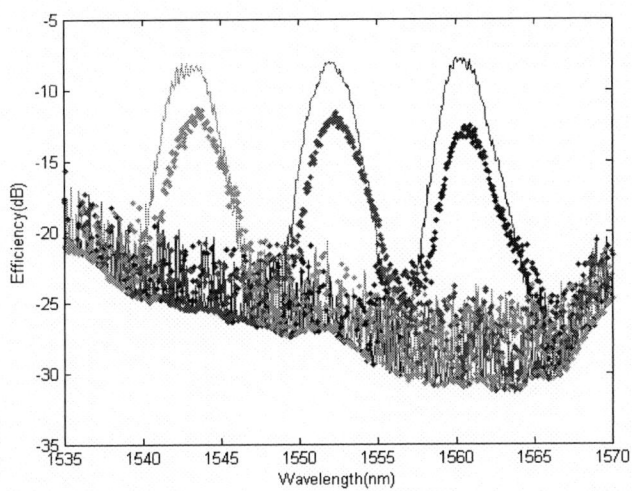

Fig. 5.7 Measured spectral responses of the fabricated EDG for the TE polarization. Solid lines are for TIR facets. Dotted lines are for echelle facets.

Authors' affiliations:
J. Song and N. Zhu are with Centre for Optical and Electromagnetic Research, State Key Laboratory for Modern Optical Instrumentation, East Building No.5, Zijingang Campus, Zhejiang University, Hangzhou 310058, China; and Division of Electromagnetic Theory, Royal Institute of Technology, S-100 44 Stockholm, Sweden. (email: juns@kth.se).

Planar Centering mechanism for dielectrically liquid lens

C. Gary Tsai[1*], L. S. Chen[2], C. L. Peng[1] and J. Andrew Yeh[1, 2]

[1]Institute of Nanoengineering and Microsystem, National Tsing Hua University, Hsinchu, Taiwan, ROC

[2]Department of Power Mechanical Engineering, National Tsing Hua University, Hsinchu, Taiwan, ROC

[*](Tel: +886-3-5715131 Ext.33730; E-mail: d939205@oz.nthu.edu.tw)

Abstract: Two types of planar centering mechanism with shape edge were fabricated and demonstrated their performance in this paper, which effectively reduce the tilt angle of dielectrically liquid lens to 0.2°.

Keywords: edge effect, liquid lens, centering mechanism

I. Introduction

Recently, liquid lens technologies have attracted intensive interests due to its potential to replace the function of the motor in the adaptive optical system. Compared with conventional motor system, the liquid lens is reliable, compactable and has less power consumption because of no moving components. The curvy liquid-liquid interface of two immiscible liquids is the simplest structure of the liquid lens [1]. The driving force can be realized by using either the electrowetting force [2-4] or the dielectric force [5-6]. Until new, liquid lens technology is still in its infant stage and many problems are waiting to be solved. Centering mechanism for holding the optical axis is one of the key issues for liquid lens package. Berge in 2000 firstly proposed a non-uniform electric filed produced by gradient dielectric thickness as the centering mechanism [2]. In 2005, a 3-D geometry taken as the centering mechanism was further proposed and compared with the type of the gradient dielectric thickness in detail [3]. In 2005, two-immiscible liquids directly filled into a cylinder with electrodes on the sidewall in Kuiper's design and the wall as the centering mechanism was used to confine the interface [4]. However, above mentioned centering mechanism are 3-D structures and hard to fabricate in batch process. Herein, we proposed two planar centering mechanisms for the dielectrically liquid lens and investigate their centering performance.

II. Theory and Design

Centering error shown in Fig. 1 is an importance factor to evaluate the optical quality of a liquid lens, leading to off-axis optical aberrations, such as coma and astigmatism. The image quality of an optical system become worse as the optical axis of a lens does not coincide with its geometric axis. For liquid lens, particular centering mechanism is necessary because the solid surface usually is modified to low-friction, which is to low the slipping energy between solid and liquid. But, the droplet will leave its location easily.

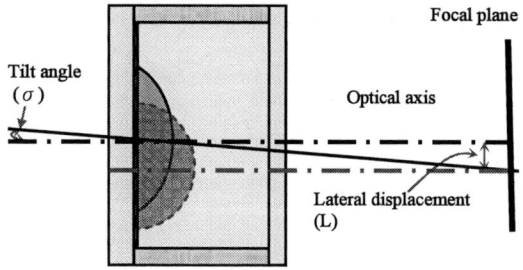

Figure 1 dielectrically liquid lens and centering error

The planar centering mechanisms proposed in this paper were based on the edge effect that strongly resists the spreading of a droplet by the shape edge [7]. Fig. 2 shows a droplet on a PMMA platform to be effectively resisted by the edge effect with a contact angle (θ) was 140°, which was far larger than the instinctive contact angle (θ_0) of water on the PMMA surface, ~60°. The contact angle on the solid edge can be expressed by the Gibbs' inequality equation:

$$\theta_0 < \theta < (180° - \varphi) + \theta_0$$

where φ is the solid edge angle. The optical axis remains unchanged in different liquid volumes as long as edge starts to resist the spreading of liquid. It offers the potential to fabricate liquid lens with different focal lengths in the same package.

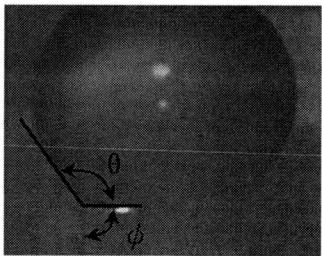

Figure 2 the photograph of the spreading of a droplet resisted by sharp edge

Two types of structures containing the circular disk structure and the circular ring structures were fabricated on the electrodes to hold the optical axis, shown in Fig. 3. SU8 2010 was chosen as the material of centering mechanism because it is the negative photoresistance and easy to lithograph the shape edge. The height of centering mechanisms was 10μm and the radius was 7mm, respectively. The substrate of dielectrically liquid lens can be fabricated by the standard semiconductor fabrication process, a batch fabrication. Therefore, the can be massive fabrication in low cost.

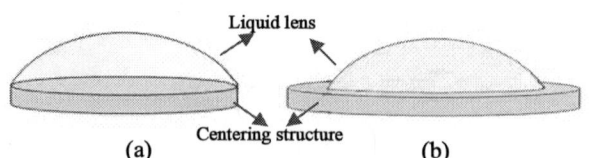

Figure 3 centering mechanisms by the shape edge (a) circular disk structure (b) circular ring structure

978-1-4244-1917-3/08/$25.00 ©2008 IEEE

III. Result and Discussion

The tilt angle of optical axis was measured by the transparent optical system, shown in Fig. 4. A beam profiler (Nanoscan[@], Photon Inc.) was used to measure the focal length and the lateral displacement of optical axis. The tilt angle was calculated by the focal length and the lateral displacement.

$$\varphi = \tan^{-1}(\frac{L}{f_o}) \sim \frac{L}{f_o}$$

Before measuring, green laser has to be aligned to normal incident into the liquid lens to diminish the measurement error. We try to eliminate the mechanical package error for focusing on the performance of the planar centering mechanism. Therefore, the geometric center of the concentric electrodes was taken as the reference axis.

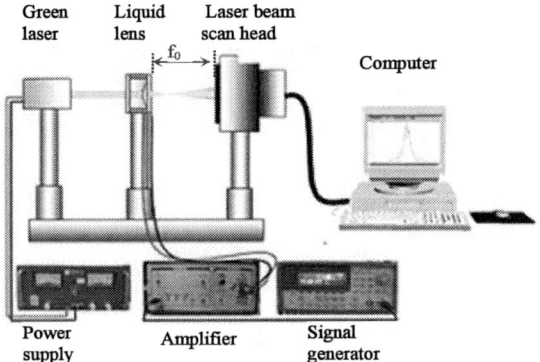

Figure 4 the schematic of optical axis measurement system

The result of centering measurement was illustrated in the Fig. 5. For the dielectrically liquid lens, the concentric electrodes will generate a circular symmetric dielectric force toward to the center of electrode [5]. As increasing the applying voltage, the driving force becomes stronger. The stronger circular symmetric force manipulates the center of liquid lens toward to the centering of electrodes, seeing the line of planar in Fig. 5. Therefore, the planar centering mechanism was deigned to confine the location of liquid lens during no applying voltage. The result showed two centering mechanism can reduce the initial centering error greatly, from 0.8 to 0.2.

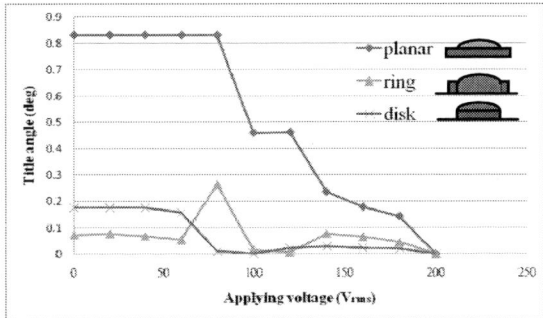

Figure 5 centering performance of two planar centering mechanisms

Comparing the centering performance of two planar centering mechanisms, the ring structure had smaller centering error in the original stage. But a large centering error peak happened at the beginning of liquid deformation. Although the shape edge of the ring can stop the spreading of liquid, it also is a barrier for shrinking the liquid to leave the edge, shown in Fig. 6. The deformation of liquid will be influenced at the actuation beginning and then come back to the small centering error after leaving the edge. Therefore, the circular disk would be the better deign for the centering mechanism because we hope to keep the small centering error during actuation.

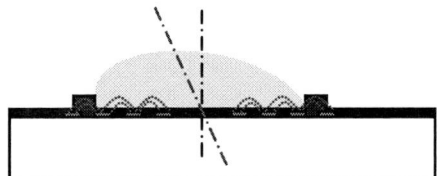

Figure 6 the illustration of ring structure influence on the liquid deformation in the beginning of applying voltage.

VI. Conclusion

Centering mechanism is an essential and important package design needed to be taken into account for liquid lens. We introduced two planar centering mechanisms in this paper and investigated their centering performance, which reduced the centering error greatly. The circular disk structure had better centering characteristic in two of them.

Reference

1. T. Krupenkin, S. Tang and P. Mach, Appl. Phys. Lett., 82, 3, 2003.
2. B. Berge and J. Peseux, *Eur. Phys. J. E*, 3, 2000.
3. B. Berge, *MEMS conference*, 2005
4. S. Kuiper and B. H. W. Hendriks, Appl. Phys. Lett., 85, 7, 2005.
5. C. Cheng and J. A. Yeh, *Opt. Express*, 14, 9, 2006.
6. C. Cheng and J. A. Yeh, *Opt. Express*, 15, 12, 2007.
7. C. G. Tsai, C. M. Hsieh and J. A. Yeh, *Sens. Actuators, A*, .139, 2007.

The two-axis magnetostatic-drive single-crystal-Si micro scanner driven by back-side electroplating Ni film

Chia-Pao Hsu[1], Wen-Chien Chen[1], Tsung-Lin Tang[1], Ming-Chuen Yip[1], Weiluen Fang[1,2]

[1]Power Mechanical Eng. Department, [2]MEMS Inst., National Tsing Hua University, Hsinchu, Taiwan

Abstract- **A simple 2-mask process to realize a two-axis micro scanning mirror is presented. The scanning mirror is made of single-crystal-silicon, and Ni is electroplated at the backside of scanner to induce the magnetostatic driving force.**

I. INTRODUCTION

The two-axis micromachined scanning mirror is a key enabling component for optical application such as display system [1]. The two-poly electrostatic drive two-axis scanner has been reported in [2]. A 9-mask process is required to fabricate the device. Eye-type electrostatic scanning mirror has been implemented in [3]. However the process, which consists of deep RIE, CMP, anodic bonding is not straightforward. A electromagnetic two-axis scanner driven using radial magnetic field has been demonstrated in [4]. A special spray coated process is required.

The two-axis Ni micromirror driven using the coil-less megnatostatic force is reported in [5]. The process requires only 2-mask. However, the mechanical properties of Ni structure need to be considered. This paper reports a simple back-side electroplating process to implement a two-axis magnetostatic-drive micro scanner on single-crystal-silicon.

II. DESIGN CONCEPT

Fig. 1 shows the design of the present two-axis micro scanner. The scanner, consisted of springs, supporting frame, and mirror, is made of single crystal silicon. Fig. 1a shows the front side of the scanner. The whole front side has its original polished mirror surface. In addition, the back-side of the mirror is electroplated with thick Ni film using the present process, as shown in Fig. 1b. The electromagnetic force will be induced by the Ni film, so as to drive the scanner. According to the present scanner design, it is not necessary to pattern the Ni film. Moreover, the optical performance of the scanner will not be influenced by the roughness of Ni film. The whole process needs only two masks using the standard photolithography technique.

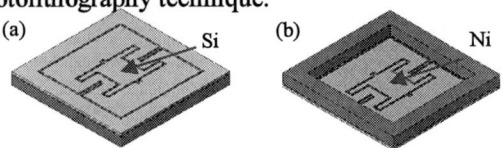

Fig. 1. Design of the two-axis magnetostatic drive scanning mirror.

III. FABRICATION

The process steps are shown in Fig.2. The fabrication process started with a SOI-wafer with device layer thickness of 50μm and handling Si layer thickness of 350μm. As shown in Fig. 2a, a 1-μm-thick thermal oxide was grown on the SOI wafer. After pattern the oxide on the backside of the wafer, the handling Si layer was etched using DRIE and the BOX (buried oxide) layer was removed by BOE (buffered oxide etcher), as shown in Fig. 2b. Similarly, the front-side oxide layer was defined by the second mask, and the device Si layer was patterned using DRIE, as shown in Fig. 2c. Thus, the single-crystal-silicon scanning mirror with its original mirror-polished surface was realized. As shown in Fig. 2d, the Ti adhesion layer and the Cu seed layer were deposited onto backside of the device Si layer. Finally, a Ni thick film was electroplated on the back side of the scanning mirror to enhance the induced magnetostatic driving force, as shown in Fig. 2e.

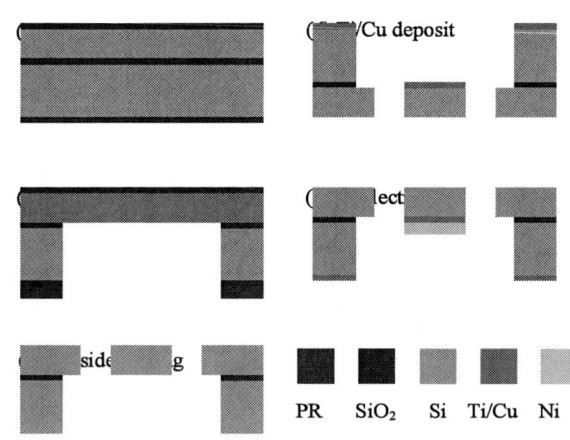

Fig. 2. Fabrication processes of the scanning mirror.

The typical fabrication results are shown in Fig. 3. The SEM micrographs in Fig. 3a show the front-side Si surface of the scanning mirror. The zoom-in micrographs show the torsional springs for two different axes. The SEM micrographs in Fig. 3b show the electroplated Ni at the backside of the mirror.

978-1-4244-1917-3/08/$25.00 ©2008 IEEE

(a)

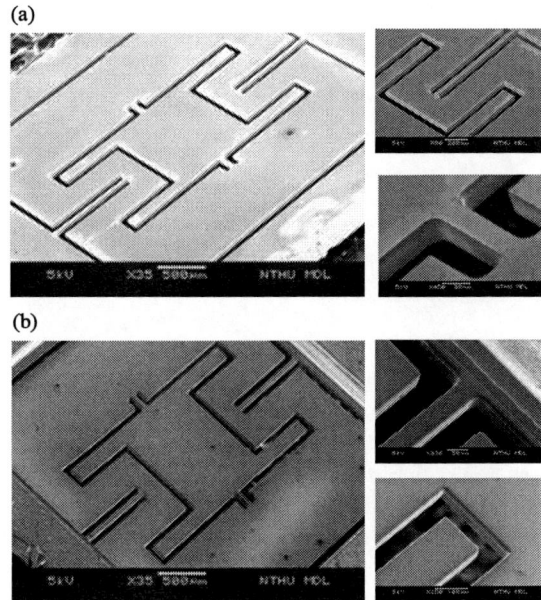

(b)

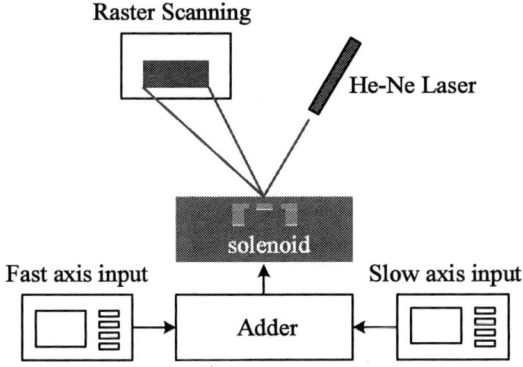

Fig. 3. SEM photos of typical fabricated scanning mirror, (a) front side Si surface, and (b) backside electroplated Ni.

IV. EXPERIMENT AND RESULT

Fig.4 shows the experiment setup to test the fabricated scanning mirror. As shown in Fig. 4, the solenoid was driven by two function generators, so that it can provide the ac magnetic field with multiple excitation frequencies. Thus, more than one resonant modes of scanner can be excited by the present driving mechanism. Fig. 5 shows the typical measured frequency response of two-axis scanner. The two orthogonal resonant scanning modes were 13.831 kHz (torsional mode of fast axis) and 1.317 kHz (torsional mode of slow axis), respectively. The typical optical scan angle was 8° at slow axis and 2.6° at fast axis.

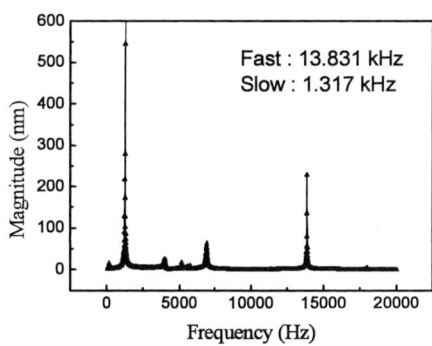

Fig. 5. Typical measured frequency responses for two-axis scanner.

These two orthogonal scanning modes were excited simultaneously when the solenoid was driven by dual ac signals. Thus, the two-axis scanning patterns are achieved, as indicated in Fig. 6.

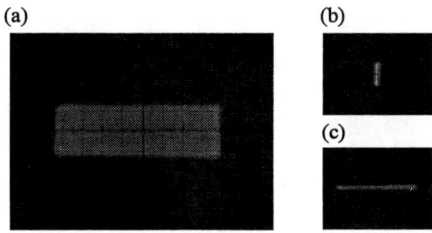

Fig. 6. Typical scanning patterns of the present two-axis scanner, (a) raster scanning pattern, (b) fast axis scanning pattern, and (c) slow axis scanning pattern.

V. CONCLUSION

A simple 2-mask process to realize a two-axis magnetostatic-drive micro scanning mirror is presented. The scanning mirror is made of single-crystal-silicon for its superior mechanical properties. Meanwhile, the thick Ni is electroplated at the backside of mirror-plate to induce the magnetostatic driving force. A maximum optical scan angle of 8° for the 1.317 kHz at slow axis and 2.6° for the 13.831 kHz at slow axis are reported at atmospheric.

ACKNOWLEDGEMENT

This project was(partially)supported by Nation Science Council, Taiwan, under contract no. NSC 96-2628-E-007-007-MY3. The authors would like to appreciate the Nation Science Council Central Regional MEMsS Research Center (Taiwan).

Fig. 4. Schematic illustration of the experiment setup for mirror driving test.

REFERENCE

[1]   H. Urey, et.al., *Proc. SPIE*, p. 106, 2003
[2]   M. Wu, et.al., *IEEE J. Photo. Tech. Lett.*, 18, p.2111, 2006
[3]   Y. Ko, et.al., *Sens. Actuator A*, 126, p.218, 2006
[4]   C. Ji, et.al, *Microsys. Techn.*, 16, p.989, 2007
[5]   H. Yang, et.al., *IEEE/ASME J. of MEMS.*, 16, p.511, 2007

All-Optical Ultra-Compact Photonic Crystal Controllable Logic Gate

P. Andalib and N. Granpayeh

Faculty of Electrical Engineering, K. N. Toosi University of Technology, Tehran, Iran.

Emails: p.andalib@ee.kntu.ac.ir; granpayeh@kntu.ac.ir

Abstract- **In this paper we propose an all-optical photonic crystal controllable logic gate based on nonlinear ring resonator. Simulation and analysis have been done by finite difference time domain and plane wave expansion method.**

I. INTRODUCTION

Nowadays ultra-compact all optical integrated circuits become the most attractive appliance for real time information communication. This amount of compactness obtained by application of photonic crystal (PC) devices which have many other advantages such as low loss. Ultra-fast response of bistable switches, optical limiters , optical logic gates realized with nonlinear Mach-Zehnder interferometers (MZI) and other devices made by nonlinear photonic crystals, makes them the best candidate for high speed data communication.

The two most commonly used memory element types are latches and flip-flops. A latch is a memory element whose excitation input signals control the state of the device hence they are not appropriate for use in synchronous sequential optical logic circuits. In synchronous sequential circuits, output signals from the memory elements are the input signals to the combinational logic, and vice versa. When its enable is active, a latch acts like a combinational circuit, too! Thus we have the possibility of two cascade combinational circuits feeding each other, generating oscillations and unstable transient behavior. This problem is solved in flip flops by using a special timing control signal called clock to restrict the time at which the states of the memory elements may change. [1]. In this paper we proposed a device for implementing this clock signal.

II. NUMERICAL METHODS

We have used numerical two dimensional (2D) finite difference time domain (FDTD) method to analyze the light guiding, resonance and transmission dropping in our photonic crystal device. Perfectly matched layers (PMLs) are used in all around of the simulation area for modeling open space regions and Yee-cell technique used for discretizing fields in the space and time.

For analyzing wave behavior, group velocity, band gaps range and defect modes we obtain dispersion diagram by plane wave expansion method (PWEM).

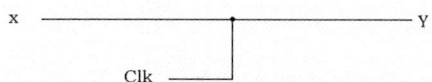

Fig. 1. Schematic of a controllable gate

III. NONLINEARITY

We used nonlinear kerr properties of Si nanocrystal in this device. Semiconductor nanocrystals broad range optical properties, controllable by their size make them interesting and considerable materials. Furthermore the technological importance of silicon caused Si nanocrystals become more interested than other semiconductor nanocrystals. Si nanocrystals show strong nonlinearity in third communication window. High nonlinearity and compatibility with Si technology persuaded us to choose Si nanocrystals as nonlinear material for nonlinear part of our device embedded in SiO2.

Intensity dependent behavior of Si-nc is because of third-order nonlinear effect. Dependency of this material to intensity could be refractive or because of two photon absorption. Two nonlinear parameters which describes for these two effects are, respectively n_2 and β which are related, respectively to real and imaginary part of third order nonlinear susceptibility $\chi^{(3)}$.

Since the real part of $\chi^{(3)}$ is two orders of magnitude higher than its imaginary part, nonlinearity of this material is refractive and linear and nonlinear two photon absorption of this material is negligible in this wavelength [2].

Nonlinear optical refractive index of this material has been measured for plasma-enhanced-chemical-vapour-deposition prepared type at 1552nm by Prakash et al. [2]. Linear refractive index of this nonlinear optical material is 1.5 and its Kerr coefficient is 10^{-16} m^2/W [3].

IV. PHOTONIC CRYSTAL

We have designed this novel resonator with square lattice photonic crystal. Background is BSC glass and the rods are GaAs, the refractive indices of which are 1.507 and 3.59, respectively [4]. We embedded Si nanocrystal rods in SiO2 background, as nonlinear ones in the cavity. Refractive index of this material is about 1.5 in third telecommunication window.

Dispersion diagram of this lattice, obtained by PWE method, shows that this lattice has only TM gap. Therefore, device has been designed for TM mode. The lattice constant of the photonic crystal is selected to be a=455nm, designing photonic crystal device in the third communication window. The radius of the dielectric rods are r=0.2a for increasing transmissibility of the barrier in order to increase coupling.

V. RESULTS AND DISCUSSION

Controllable gate contains a nonlinear add/drop filter consists of a photonic crystal ring resonator, placed between two open

978-1-4244-1917-3/08/$25.00 ©2008 IEEE   154

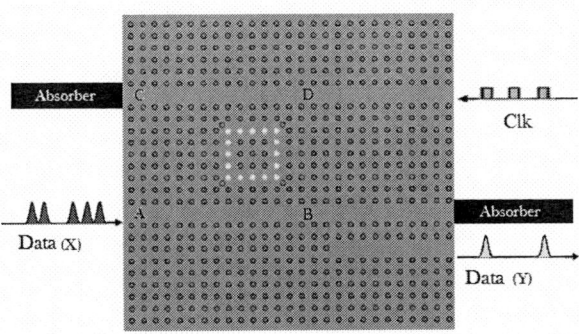

Fig. 3. All-optical PC controllable gate.

open ended waveguides which is demonstrated in Fig. 2. The scatterers in corners of the quasi-square ring resonator are placed for improving transmission spectrum and drop efficiency [5]. This resonator has been formed by replacing GaAs rods with nonlinear Si nanocrystal rods with the same radius. These rods in confronting with low power electromagnetic field have approximately the same linear refractive index as the background, but by confronting with higher power field of the pump laser, high intensity localized electromagnetic field which caused by resonant phenomena inside the cavity changes nonlinear rods refractive index. This relative effective refractive index enhancement changes resonant wavelength of the resonator to higher wavelength. This wavelength shift amount depends to input power, Kerr coefficient and quality factor (Q) of the resonator which represents light confinement in resonator. Higher quality factor can be achieved by increasing coupling section periods between the W1 waveguide and the resonator. This action enhances field confinement inside the resonator, hence enhances external quality factor. We placed two rows of rods between the ring resonator and each open ended waveguide. This structure has 80% drop efficiency and high quality factor, however, with one period distance, very higher drop efficiency can be obtained but it would have lower quality factor. Drop occurs at 1550nm, resonance wavelength of this filter. We select this wavelength as the operational wavelength of this device. For higher input power, high Q of this resonator causes high variation of nonlinear inner rods refractive index and shifts resonance wavelength. Therefore in operational wavelength filter would be in "through" state. Transmissions to each port of the nonlinear add/drop filter are demonstrated in Fig. 3 for different input powers. This diagram shows that for pump power P=330 W/μm and higher the device is in "through" state and for P=33 W/μm and lower it is in "drop" state.

By applying data with power equal to 33 W/μm from port A, the device will be placed in "drop" state and data exit from port C. But, if simultaneously a power equal to 330 W/μm flows into port D as the clock signal, the data will exit from port B. A coupler is added beside port B for adjusting the logic level of the device. This coupler couples 20% of flowing power in waveguide terminated to port D, to the adjacent branch which is the output port, with suitable power level. Data transmissions in presence and absence of clock signal are demonstrated in table I.

TABLE I
DATA TRANSMISSION IN PRESENCE AND ABSENCE OF CLOCK SIGNAL

X		Clk		Y	
Power (W/μm)	Logic level	Power (W/μm)	Logic Level	Power (W/μm)	Logic level
33	1	330	⎍	47.855	1
0	0	330	⎍	14.72	0
33	1	0	⎎	1.87	0
0	0	0	⎎	0	0

In binary number system two power levels can be represented as the two digits 0 and 1, hence Signals in digital optics are set to have two distinct power levels. Our optical logic gate defines 0 to 16W/μm as logic 0 and 30 to 50W/μm as logic 1. Fig. 4 shows this system with the built-in tolerances for variations in the power. A valid digital signal, as demonstrated in this diagram should be within either of the two shaded areas [6]. From these results, this device can function as an instrument for applying clock signal in optical sequential devices.

VI. CONCLUSION

In this paper, we proposed an all-optical photonic crystal controllable gate. The structure of which is a nonlinear ring resonator, placed between two high group velocity waveguides. This device constructed by placing Si nanocrystal rods in a resonator which causes high localized electromagnetic field intensity because of resonance phenomena. By applying clock signal, data will cross to the output port, without any variation in logic level which in the absence of clock signal data will not transfer to output.

REFERENCES

[1] V.P. Nelson, H. Troy Nagel, B.D. Carroll, J.D. Irwin, *Digital Logic Circuit Analysis and Design*, New Jersey: Prentice Hall, 1995.

[2] G.V. Prakash, M. Cazzanelli, Z. Gaburro, L. Pavesi, F. Iacona, G. Franzò, and F. Priolo, "Linear and nonlinear optical properties of plasma-enhanced chemical-vapour deposition grown silicon nanocrystals," *J. Mod. Opt.*, vol. 49, no. 5, pp. 719 – 730, 2002.

[3] C.A. Barrios, "High-performance all-optical silicon microswitch," *Elect. Lett.* vol. 40, no. 14, July 2004.

[4] R.W. Boyd, *Nonlinear Optics*, 2nd ed., USA: Academic press, Elsevier Science, 2003.

[5] Z. Qiang, W. Zhou, R.A. Soref, "Optical add-drop filters based on photonic crystal ring resonators," *Opt. Express*, vol. 15, no. 4, pp.1823-1831, Feb. 2007.

[6] *The 8051 Microcontroller and Embedded Systems*, M. Mazidi and J.G. Mazidi , New Jersey: Prentice-Hall, Pearson Education, 2000.

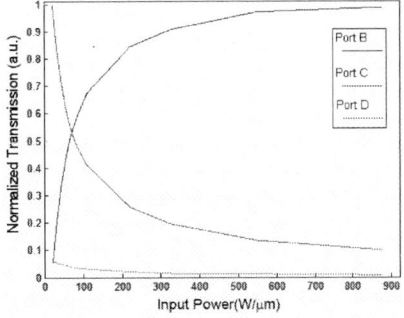

Fig. 3. Transmission to each port of nonlinear add/drop filter

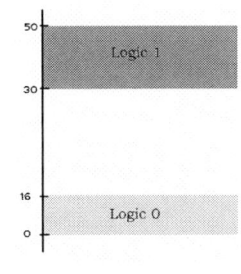

Fig. 4. Power built-in valid tolerance.

Conductive pattern forming method on vertical wall using spray coating and angled exposure technologies

Hiroki Morii[1], Fumikazu Oohira[1], Minoru Sasaki[2], Toshihiko Ochi[3], Asumi Yuzuriha[3]

Tel +81-87-864-2341, Fax +81-87-864-2341, E-mail: s08g525@stmail.eng.kagawa-u.ac.jp

[1]Kagawa University, Faculty of Engineering, 2217-20, Hayashi-cho, Takamatsu-shi, Kagawa, Japan

[2]Toyota Technological Institute, [3]AOI Electronics co. LTD

Abstract

A novel conductive pattern forming method on the vertical wall using a resist spray coating and an angled exposure technologies is proposed. This method makes it possible to decrease the chip package size. The spray coating and the angled exposure technologies enable the uniform resist coating and the patterning the resist on the vertical wall of 600 μm height. Then a conductive pattern layers are sputtered and electroless plated. As the result, the conductive patterns of 200 μm width and 300 μm spacing is successfully formed on the vertical walls.

Keywords: packaging, spray coating, angled exposure, MEMS

1. Introduction

A high density packaging technology is indispensable for the mobile information apparatus equipped with the optical and electronic devices.

This paper describes the conductive pattern forming method on the vertical wall to realize the high density packaging of the devices. For this purpose, a resist spray coating and an angled exposure technologies are examined, and the principle confirmation has been made.

2. Chip On Glass (COG)

Figure 1 shows one of the examples of the chip packaging; Chip On Glass (COG). The chip is mounted on the glass substrate in the COG. In this packaging structure, the optical device is mounted on the glass substrate side to receive the light. Then, when the COG is mounted on the PCB, it must be mounted upside down and the hole must be formed on the PCB, and high density packaging is impossible because the conductive patters are formed outside of the chip as shown in fig.2.

This paper proposes a novel packaging structure for the high density packaging, and develops the conductive pattern forming method for this structure.

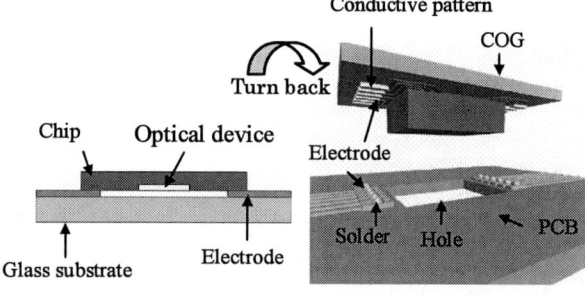

Figure 1 COG structure   Figure 2 Conventional packaging method

3. Proposed packaging structure

Figure 3 shows the newly proposed high density COG structure. The optical device is embedded in the plastic material, and the conductive patterns are formed on the side walls as well as the upper side of the plastic material. Figure 4 shows the schematic view of the packaging on the PCB based on this structure. This structure does not need the hole in the PCB and enables the high density packaging because the conductive patterns are formed on the vertical side wall and upper side of the plastic material.

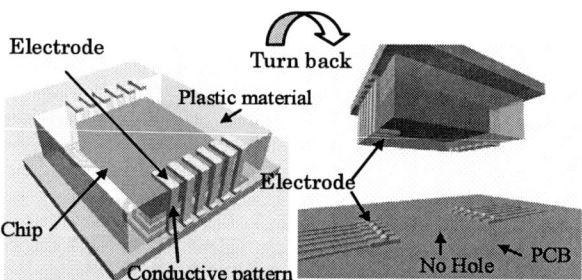

Figure 3 Proposed high density COG structure   Figure 4 Proposed packaging method

4. Conductive pattern forming process

4.1 Fabrication process

In order to form the conductive patterns on the vertical side wall, the resist spray coating and the angled exposure technologies are examined. The spray coating technology and the angled exposure technologies are expected to realize the uniform resist coating and the patterning the resist on the vertical wall of 600 μm height.

Figure 5 shows the fabrication process of the conductive patterns. The photo-resist is spray coated on the glass substrate and the plastic material in which the optical device is imbedded.(1,2) The angled exposure and development are done.(3,4) The seed layer is deposited.(5) Then, the pattern is formed by the lift-off process.(6) Finally, Ni electroless plating is made for the good conductivity.(7)

978-1-4244-1917-3/08/$25.00 ©2008 IEEE   156

(1) Form the plastic material

Plastic material

Glass substrate

(2) Resist spray coating

Resist

(3) Angled exposure

UV

mask

(4) Development

(5) Deposit seed layer

Ti,Pd

(6) Lift-off

(7) Ni electroless plating

Ni

Figure 5 Fabrication process of the conductive pattern

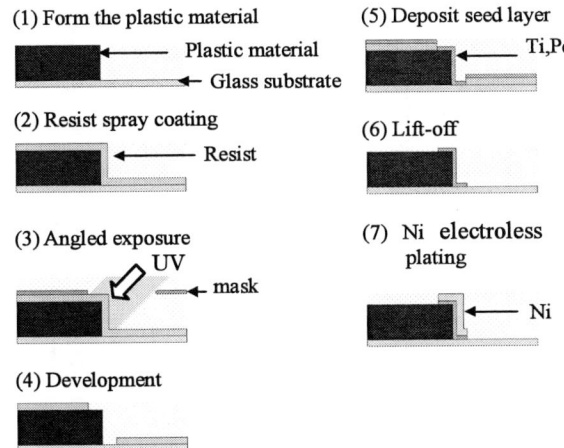

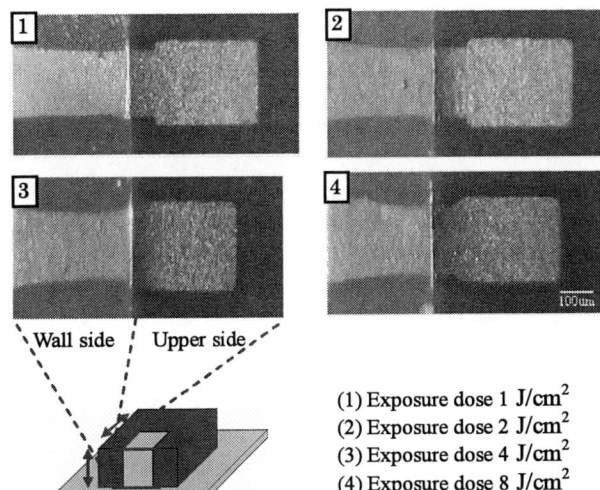

Wall side / Upper side

(1) Exposure dose 1 J/cm^2
(2) Exposure dose 2 J/cm^2
(3) Exposure dose 4 J/cm^2
(4) Exposure dose 8 J/cm^2

Figure 7 Changes of pattern width at different exposure doses

4.2 Spray coating technology

The coating conditions are examined in this spray coating process[1], with the sample temperature, and sample scanning speed, and supplying rate of diluted resist supply. Figure 6 shows the example of the spray coated and patterned chip, with the condition of temperature of 105 ℃, stage speed of 400 mm/s, and supplying rate of 100 gr/h. As shown in fig.6, uniform spray coating can be realized.

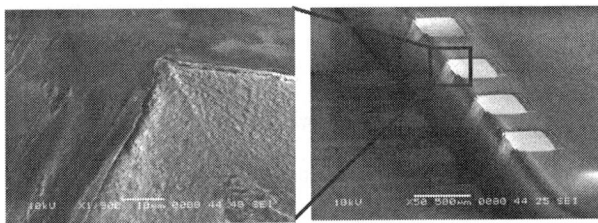

Figure 6 Example of spray coated and patterned chip

4.3 Angled exposure technology

In the exposure process, the exposure condition changes because of the gap between the upper side and the lower side of the plastic material. Therefore, the angled exposure technology is examined to realize the uniform exposure.[2] Moreover, the change of the pattern size with the exposure dose is examined. Here, the height gap between the upper and lower side is 600 μm, the pattern width is 200 μm, and the pattern spacing is 300 μm. The angled exposure dose conditions are 1, 2, 4, and 8 J/cm^2. Figure 7 shows the pictures after the patterning at 45° angled incidence, metal deposition and lift-off process (step 6, in Fig.5). The pattern width increases by 20 μm at 1 J/cm^2, 50μm at 2 J/cm^2, 70 μm at 4 J/cm^2, and 90 μm at 8 J/cm^2 doses. Nevertheless, there is no pattern overlap problem in this case, because the spacing between the patterns is 300 μm.

5. Results

The patterning, seed layer deposition, lift-off, and Ni electroless plating processes have been successfully passed using the sprayed resist. Figure 8 shows the fabricated chip through the proposed process. The Ni conductive layer of 5 μm thickness is electroless plated over the seed layer realizing the resistance lower than a few ohms.

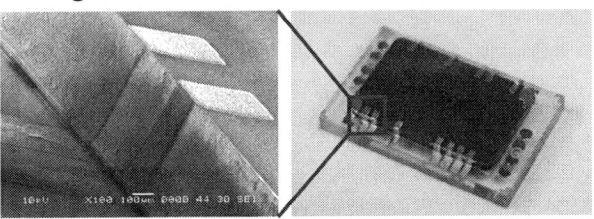

Figure 8 Demonstrated electrical wiring over the vertical side wall using angled exposure technique

6. Conclusions

A novel packaging method for the COG has been proposed to realize the high density packaging. For this purpose, a conductive pattern forming method on the vertical wall using a resist spray coating and an angled exposure technologies has been developed. The spray coating technology and the angled exposure technologies enabled the uniform resist coating and the patterning the resist on the vertical wall of 600 μm height. Then a conductive pattern layers are sputtered and electroless plated. As the result, the conductive patterns of 200 μm width and 300 μm spacing could be successfully formed on the vertical walls. The conductive layer of 5 μm thickness is electroless plated over the seed layer, and the resistance is lower than a few ohms which is sufficient for the normal conductive pattern.

References

[1] Hayato Izumi, Yohei Matsumoto, Seiji Aoyagi, Yusaku Harada, Shoso Shingubara, Minoru Sasaki, Kazuhiro Hane, Hiroshi Tokunaga, IEEJ Trans. SM, Vol. 128-E, No. 3, (2008), pp.102-107.
[2] Vijay Kumar Singh, Minoru Sasaki, and Kazuhiro Hane, Japanese Journal of Applied Physics, Vol. 46, No. 9B, (2007), pp. 6449-6453

A Novel Lens Formation Technology to Implement a 3D Spherical Polymer Lens

Chih-Chun Lee[1], Sheng-Yi Hsiao[1], and Weileun Fang[1,2]
[1]Power Mechanical Engineering, [2]MEMS Inst., National Tsing Hua University, HsinChu, Taiwan

Abstract-This study presents a novel lens formation technology to implement a 3D polymer lens in buffer liquid. The integration of such 3D polymer lens on SiOB and the application for barcode scanner are also demonstrated.

Keywords: micro sphere, micro ball lens, silicon optical bench

I. Introduction

Silicon optical bench has been demonstrated more than ten years [1]. Various optical components have complicated three-dimensional shape, and which will lead to challenge to the existing micromachining processes. Most of the existing technologies for micro optical components, such as LIGA [2], inject [3], thin film optics [4], etc., are mainly planar fabrication processes. As a result, the lens components are usually applied in the out-of-plane optical axis. The approaches to assembly the lens in the in-plane optical axis are reported in [5, 6]. However, the assembly processes are not straightforward. Moreover, the planar processes can only fabricate lens with planar symmetric shape.

In this study, the polymer optical components dispensed in the buffer liquid is designed and implemented. The fully symmetric spherical lens is successfully formed in the buffer liquid. The ball lens can be used in both in-plane and out-of-plane directions. In addition, the fabrication and integration of the micro ball lens to form a SiOB is also demonstrated.

II. Design Concept

It is challenge to realize in-plane micro optical components using the existing micromachining technologies, especially the 3D ball lens. Liquid formation approaches, such as PR-reflowed [7], ink-jet [3], and polymer dispensing [8], are widely used in MEMS process to realize the 3D lens. The surface tension formed lens will have good optical quality [9]. However, the lens fabricated using these planar processes is radial symmetric only in the in-plane direction.

Fig. 1 shows the concept of the present study. The polymer is dispensed in buffer liquid instead of air to form the ball lens. The density of the buffer liquid has the same density as the polymer. The surface tension of polymer dispensed in buffer liquid minimizes its energy state. In addition, the influence of gravity force is balanced by floatage. As a result, the liquid lens material is force free externally in the buffer liquid so as to form a fully symmetric spherical lens. The ball lens can be applied in both in-plane and out-of-plane directions.

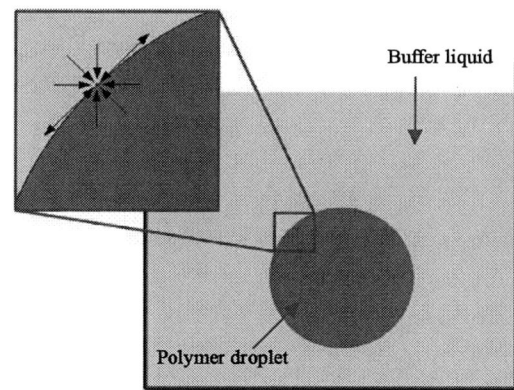

Fig.1 Droplet in buffer liquid is force equilibrium. Spherical lens is formed to satisfy the minimum energy principle.

III. Fabrication and Integration

Fig.2 shows the fabrication process for the polymer lens. Fig.2a shows the dispensed droplet in the buffer liquid, thus the discrete polymer ball lens is realized. The diameter of the ball lens is controlled by the volume of the dispensed polymer. Fig.2b shows the integration of polymer ball lens with thin film lens holder on Si substrate. Thus, the integration of polymer lens to form a SiOB can be achieved.

IV. Result and Measurement

The typical fabrication results are demonstrated in Fig3. Fig.3a shows the formation of micro ball lenses inside the buffer liquid after polymer dispensing. Fig.3b demonstrates a typical micro ball lenses with a diameter of near 400μm. Fig.3d shows the integration of the polymer ball lens (approximately 600μm in diameter) with the silicon nitride lens holder shown in Fig.3c. The diameter of micro ball lens is controlled using a commercial hydraulic polymer dispensing system. The diameter of micro ball lens prepared using the process in Fig. 2a ranges from 200μm to 600μm.

(a) 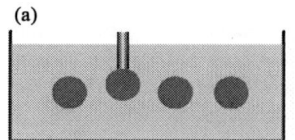   (b)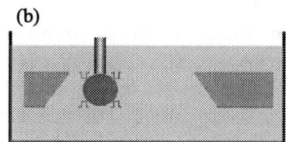

Fig.2 (a) Fabrication of the ball lens using the dispensed polymer droplet in buffer liquid, and (b) fabrication and integration of the dispensed polymer lens with lens holder.

978-1-4244-1917-3/08/$25.00 ©2008 IEEE

As the measurements shown in Fig.4, the variation of ball lens diameter for dispensing processes is near 2%. In addition, the 580μm ball lens has a focal lens of 403μm. The surface roughness of the ball lens measured by optical interferometer is about 10nm, as shown in Fig.5.

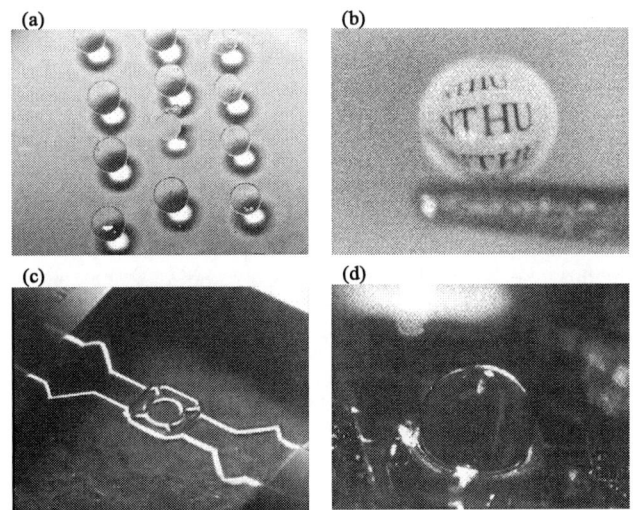

Fig.3 Typical fabrication results for spherical polymer lens, (a) Formation of polymer micro spherical lenses in the buffer liquid, (b) the text "NTHU" image through a polymer lens of 400μm in diameter, (c) Silicon nitride lens holder, and (d) integration of the polymer lens with lens holder.

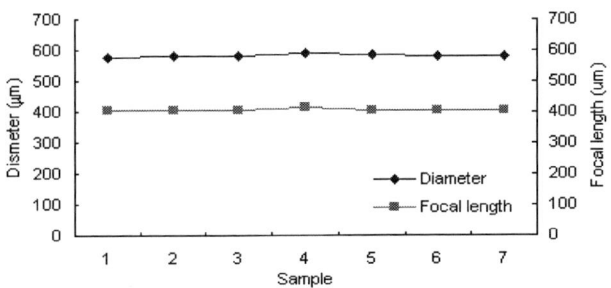

Fig.4 Diameter deviation of the dispensed polymer lenses.

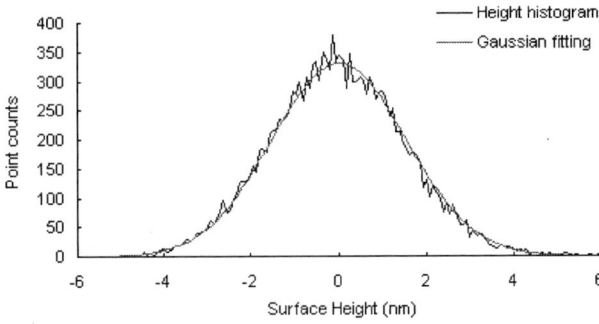

Fig.5 Surface height histogram of the polymer lens (curvature removed). 95% of data is between 6.09nm (peak to valley). 80% of data is between 3.88nm.

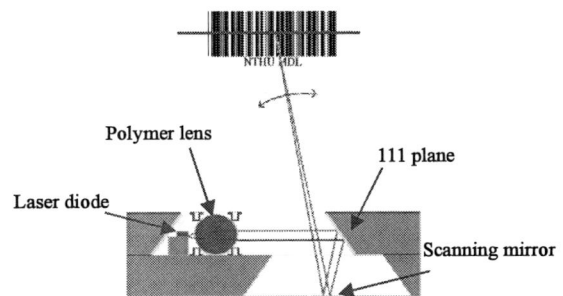

Fig.6 Integration of the polymer lens with SiOB for barcode scanner application.

V. Conclusions

This study presents a novel lens formation technology to implement a 3D polymer spherical lens in buffer liquid. The diameter of the ball lens is controlled by the volume of the dispensed polymer. The typical ball lens diameter ranges from 200μm to 600μm, and its surface roughness is near 10nm. In summary, the polymer lens fabrication process provides a quick forming method for micro ball lens, with perfect surface roughness and self alignment. Moreover, the integration of such 3D polymer lens on SiOB is also demonstrated. Thus, the integration of ball lens on SiOB can be applied for various optical applications, for instance, the barcode scanner illustrated in Fig.6.

ACKNOWLEDGMENT

This work was supported by the Ministry of Economic Affairs, Taiwan, under contract no. 96-EC-17-A-07-S1-011 and by the Nation Science Council, Taiwan, under contract NSC-96-2628-E-007-008-MY3.

References

[1] O. Solgaard, M. Daneman, N. C. Tien, A. Friedberger, R. S. Muller, and K. Y. Lau, *Photonics Technology Letters, IEEE*, **vol. 7**, pp. 41-43, 1995.

[2] S. K. Lee, K. C. Lee, and S. S. Lee, *Journal of Micromechanics and Microengineering*, **vol. 12**, pp. 334-340, 2002.

[3] D. L. MacFarlane, V. Narayan, J. A. Tatum, W. R. Cox, T. Chen, and D. J. Hayes, *Photonics Technology Letters, IEEE*, **vol. 6**, pp. 1112-1114, 1994.

[4] J. Y. Chang, C. M. Wang, C. C. Lee, H. F. Shih, and M. L. Wu, *Photonics Technology Letters, IEEE*, **vol. 17**, pp. 214-216, 2005.

[5] L. Y. Lin, S. S. Lee, K. S. J. Pister, and M. C. Wu, *Applied Physics Letters*, vol. 66, p. 2946, 1995.

[6] J. Hsieh, S. Y. Hsiao, C. F. Lai, and W. Fang, *Journal of Micromechanics and Microengineering*, **vol. 17**, pp. 1703-1709, 2007.

[7] P. Heremans, J. Genoe, M. Kujik, R. Vounckx, G. Borghs, and L. Imec, *Photonics Technology Letters, IEEE*, **vol. 9**, pp. 1367-1369, 1997.

[8] M. Wu, S. Y. Hsiao, C. Y. Peng, and W. Fang, *Journal of Optics A: Pure and Applied Optics*, **vol. 8**, pp. S323-S329, 2006.

[9] M. He, X. Yuan, N. Q. Ngo, W. C. Cheong, and J. Bu, *Appl. Opt*, **vol. 42**, pp. 7174-7178, 2003.

Fabrication of 2D and 3D Photonic Quasi-crystals with High Rotation Symmetry by Holographic Lithography Technique

Chia Chen Hsu, Ngoc Diep Lai, and Jian Hung Lin

Department of Physics, National Chung Cheng University,
168 University Road, Ming Hsiung, Chiayi 621, Taiwan
Tel +886-5-242-8173, Fax +886-5-272-0587, E-mail cchsu@phy.ccu.edu.tw

Abstract

We demonstrate using a double-exposure three-beam and/or three-beam-plus-one interference technique, one can easily and efficiently fabricate 2D and/or 3D quasi-periodic structures. Using the multi-surface prism, our fabrication technique becomes a compact and robust way to produce different kinds of structures, periodic or quasi-periodic.

Keywords: Holography; Photolithography; Optical design and fabrication; Photonic quasi-crystals.

1. INTRODUCTION

Recently there has been considerable interest in the fabrication of photonic quasi-crystals (PQCs), which consist of non-periodic structures [1]. The high degree of symmetry of PQCs affects optical property of structures and allows much weaker direction-dependent property. Accordingly, one can obtain a more isotropic photonic bandgap (PBG) in PQCs [2,3]. In addition, the contrast in refractive index required to create a PBG can be significantly reduced, and thus also the loss. Holographic lithography (HL) is a very promising and inexpensive technique to fabricate PQCs. In this work, we demonstrate both theoretically and experimentally that the use of multi-exposure of three-beam interference technique can be a simple, compact, and efficient way to fabricate not only 2D but also 3D PQCs with 12-fold symmetry. [4,5]

2. EXPERIMENTS AND RESULTS

Figure 1 shows the experimental setup used to fabricate both 2D and 3D quasi-periodic structures. A laser beam emitted from a He-Cd laser at 442 nm was spatially cleaned and extended to a diameter about 2 cm. Then, a mask with four irises was used to select four laser beams with the same profile (5mm in diameter) and same intensity and each of these four laser beams was incident upon each surface of the prism noted as O, A_1, A_2, and A_3 (Fig. 1(a)). After passing through the prism, the beam O (center beam) continued propagating along the z axis, while the other three beams A_1, A_2, and A_3 (ambient beams) changed their directions and overlapped with the center beam at

one point in the z axis (see Fig. 1(b)). The prism was designed to arrange the ambient beams after passing through this prism making an angle θ = 15.7° with respect to the center beam (the z axis). This resulted in a period of 1.1 μm in the xy plan of the periodic 2D structure. A positive photoresist AZ-4620 (Clariant Corp.) was chosen to fabricate quasi-periodic structures. The sample film either with 1μm thickness (for 2D structures) or 12 μm (for 3D structures) was spin-coated on a glass substrate and soft baked at 65° C for 2 minutes and then 95° C for 3 minutes to remove the solvent. The power of each laser beam was fixed at 2 mW and the exposure-time was 2 second for one exposure. After double exposure, the sample was developed for 7 minutes in AZ developer and rinsed by DI water.

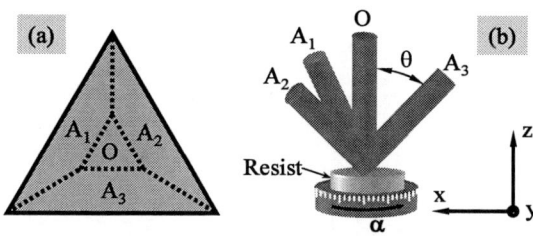

Fig.1 (a) Projection of a multi-surface prism along the z axis. (b) Three ambient beams after passing through the prism changed their directions, and overlapped with the center beam at one point in the z axis in which a photoresist sample was placed for the fabrication. The sample can be rotated by angle α for multi-exposure.

Figure 2(a) shows the experimental result of a 2D

quasi-periodic structure obtained with a double-exposure of the three-beam (A_1, A_2, A_3) interference pattern at $\alpha = 0°$ and $90°$. The structure is quite uniform in a very large area (5 mm × 5 mm, corresponding to the size of the iris). Figure 2(b) shows a zoom on an area of the quasi-periodic structure in which one can clearly see a twelve-fold symmetry structure. Moreover, for different samples obtained in different times of fabrication we found that even quasi-periodic structures changed their forms but still remained a twelve-fold symmetry level. A picture of another 2D quasi-periodic structure is shown in Fig. 2(c) for comparison with that in Fig. 2(a). In both cases, the diffraction pattern of the quasi-periodic structures contains a series of circles constituting twelve bright spots around the zero-order diffraction spot, as shown in Fig. 2(d). Moreover, by using the 2D approximation finite element method to simulate the transmission spectra of these structures for different incident angles, we proved that their photonic gap is quite isotropic. Figure 2(e) shows transmission spectra at $0°$, $6°$, $12°$, $18°$, and $24°$ incident angles of a 2D quasi structure with a low refractive index material (index=2) (for a structure possesses a twelve-fold symmetry level, it is enough to calculate the transmission spectra only for the angles ranging from $0°$ to $30°$). Two isotropic gaps are clearly observed.

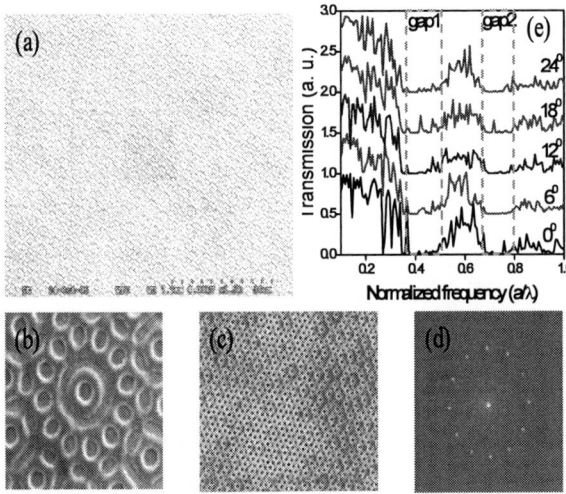

Fig.2 SEM images of 2D quasi-periodic structures: (a) top view; (b) zoom on a particular area, (c) a 2D quasi-periodic structure obtained with another arbitrary rotation center; (d) the diffraction pattern of these structures. (e) A simulation result of transmission spectra of 2D quasi-periodic structures obtained with different incident angles.

To fabricate 3D quasi-periodic structures, we exposed the three-beam-plus-one (A_1, A_2, A_3 and O) interference pattern two times; at $\alpha = 0°$ and $90°$. Figure 3(a) shows the top view of 3D quasi-periodic structure in which we can find a twelve-fold symmetry

level in the xy plan. Figure 3(b) shows a 3D view (tilt angle $\approx 12°$) of an area containing different layers (different z positions). A beautiful diffraction pattern with very bright spots was also obtained showing a twelve-fold symmetry level in the xy plan of the fabricated structure (Fig. 3 (b)). To prove the symmetry level in the xz or the yz plan, it is worth to mention that our samples are not thick enough to easily sending a laser beam to one side of the samples. However, the diffraction pattern shown in inset of Fig. 3(b) is more complicated than that of the 2D quasi-periodic structures shown in Fig. 3 and should reflect an interesting property of these 3D quasi-periodic structures.

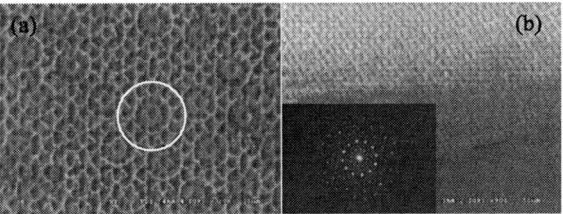

Fig. 3. SEM images of 3D quasi-periodic structures. (a) Top view. The white circle indicates the twelve-fold symmetry in the xy plan of structure. (b) A 3D view showing different layers in the z direction of 3D quasi-periodic structures. Inset of (b) shows the diffraction pattern of the 3D quasi-periodic structures

3. CONCLUSION

We have demonstrated both theoretically and experimentally that using a double-exposure three-beam and/or three-beam-plus-one interference technique, one can easily and efficiently fabricate 2D and/or 3D quasi-periodic structures. With two-exposure of a three-beam interference pattern at $\alpha = 0°$ and $90°$, we obtained a beautiful twelve-fold symmetry 2D quasi-periodic structure. These structures possess an isotropic gap as demonstrated by a simple calculation of transmission spectra at different incident angles. With two-exposure of a three-beam-plus-one interference pattern at $\alpha = 0°$ and $90°$, a 3D quasi-periodic structure was obtained for the first time.

References

[1] P. J. Steinhardt and S. Ostlund, *The Physics of Quasicrystals* (World Scientific, Singapore 1987).
[2] Y. S. Chan, C. T. Chan, and Z. Y. Liu, Phys. Rev. Lett. 80, 956-959, 1998.
[3] M. E. Zoorob, M. D. B. Charlton, G. J. Parker, J. J. Baumberg, and M. C. Netti, Nature 404, 740-743, 2000.
[4] N. D. Lai, J. H. Lin, Y. Y. Huang, and C. C. Hsu, Opt. Express 14, 10746-10752, 2006.
[5] N. D. Lai, J. H. Lin, and C. C. Hsu, Appl. Opt. 5645-5648, 2007.

Fabrication of wall-coated Cs vapor cells for a chip-scale atomic clock

M. Hasegawa[*], P. Dziuban, L. Nieradko, A. Douahi, C. Gorecki, V. Giordano

*Institut FEMTO–ST, UMR 6174 CNRS –Université de Franche–Comté, 16 route de Gray, 25030
Besançon Cedex, FRANCE*

Abstract — Cesium vapor microcells incorporating a Cs dispenser were fabricated. An organosilane monolayer was successfully applied to the microcells' walls as an anti-relaxation coating to improve the relaxation time of Cs atoms.

Fabrication of a chip-scale atomic clock using the principle of coherent population trapping (CPT) have been investigated intensively because it does not require the implementation of a microwave cavity to probe the atomic resonance. In this technology, the fabrication of a micro-scale alkali-vapor cell is a key issue. By incorporating a cesium dispenser, we are able to generate the required cesium vapor after the cell is sealed, and have demonstrated that this process is effective for the fabrication of a microcell [1]. However, some problems remain to be addressed for the practical application of the cells to an atomic clock. The most serious is the disadvantageous broadening of the clock transition associated with collisions of Cs atoms with cell's walls. To ameliorate this problem, we have studied the application of an anti-relaxation coating to the micro-cell walls. Organosilanes, including octadecyltrimethoxysilane (ODS) [2,3] and dimethyldichlorosilane [3,4], have been reported to exhibit the anti-relaxation effect with alkali-atoms, and they are known to react with OH groups on a substrate resulting in the formation of a self-assembled monolayer (SAM). Moreover, organosilanes are fortunately compatible with our design of micro-cell: they react with the native oxides on both Si and glass (SiO_2) surfaces to form an extremely uniform layer, even on 3-dimensional structures with intricate surface topologies. In addition, the layers are known to exhibit sufficient thermal stability to tolerate subsequent processing. Therefore, we focus our investigation on organosilane-SAMs as candidates for an advantageous anti-relaxation coating. In this paper, we investigate the application of organosilane-SAMs to the microcell's wall.

Microcell structure consists of borosilicate glass (Borofloat 33, SCHOTT®) and Si(111) wafers. The cell contains two cavities (Fig. 1): one for the transmission for CPT detection (Cavity A), and the other for storing a Cs dispenser (Cavity B). These two cavities are connected through filtration channels, which avoid the contamination from the Cs dispenser during its activation process. The Cs dispenser employed here is commercially available (SAES®), and is chemically stable at temperatures lower than 700 °C, which is much higher than the temperature of the subsequent anodic bonding (450 °C).

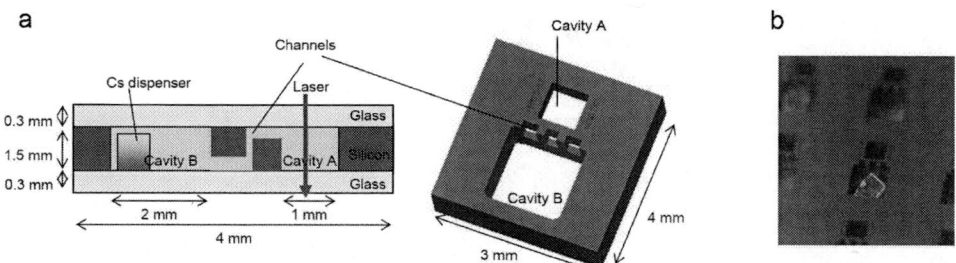

Figure 1. Schematic illustrations a microcell (a) and a photograph of microcells on a wafer (b).

Microcells with and without an anti-relaxation coating were fabricated and compared. The cavities and filter channels were fabricated on a Si wafer using photolithography and the double-sided DRIE process. Then, the Si wafer was bonded with a glass wafer by anodic bonding. A Cs dispenser was placed in the cavity B, followed by the packaging of the cells by anodic bonding in inert atmosphere. After this, the Cs dispenser was locally heated up to 700°C by CW Raman fiber layer as a pump laser source (wavelength, 1455 nm, and line width, 2 nm). Figure 1b shows

978-1-4244-1917-3/08/$25.00 ©2008 IEEE

the image of the microcells fabricated on a Si wafer. Finally, these cells were cut with a dicing saw machine into single-cell chips. For the microcells coated with an anti-relaxation coating, organosilane-SAMs were deposited after the first anodic bonding. SAMs were formed onto both the substrate with microcell-cavities and the glass substrate, which is used for the subsequent anodic bonding. The precursor reagent for SAMs in this study was octadecyltrimethoxysilane (ODS), dissolved in anhydrous toluene. ODS-SAMs were deposited from this solution at 60°C onto the substrates, which were cleaned beforehand with "piranha" solution and rinsed with deionised water.

Since defects are likely to compromise the coating's anti-relaxation effect, we have studied in detail the quality of the ODS-SAM. For the experiments, Si or glass plate substrates were used. Figure 2 shows the cross-sectional images of a water drop on a Si substrate before (Fig. 2a) and after (Fig. 2b) coating of ODS-SAM. Before deposition of the SAM (Fig. 2a), the surface of the substrate was hydrophilic because of the surface oxide layer on the Si substrate. The contact angle increased significantly after immersion of the substrate in organosilane solution (Fig. 2b). The increase in contact angle indicates that the surface property of the substrate changed from hydrophilic to hydrophobic, which suggests that the OH groups on the surface of a substrate have indeed reacted with ODS molecules and the surface has become covered with the alkyl groups of the ODS-SAM. The water contact angle for the sample with an ODS-SAM was 105°. This value agrees well with one reported previously [7]. For the experiment with glass substrates, similar results were obtained. The thicknesses of our layers were measured with a spectroscopic ellipsometer, performed at a nominal angle of incidence (and thus reflection) of 70°, and ranging in wavelength from 275 nm to 700 nm. The thickness of an ODS-SAM was 2.0-2.1 nm. This measurement agrees (within experimental uncertainties) to the value of 1.9 nm reported in the literature [6] also for an ODS-SAM. These results confirm that an ODS-SAM can be successfully formed on both Si and glass substrates.

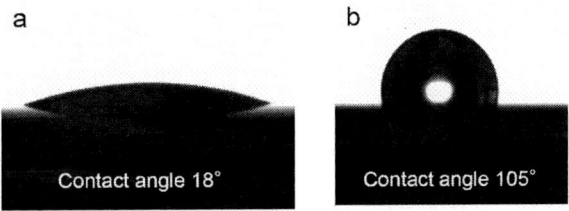

Figure 2. Cross-sectional images of a water drop on a Si substrate without (a) and with (b) an ODS-SAM.

The ODS-SAMs were applied to the micro-cells. In spite of the presence of an ODS-SAM on the surface of the substrates, the cells were successfully sealed by anodic bonding. This result suggests that ODS-SAMs are compatible with our process whilst affording an effective anti-relaxation coating for micro Cs-vapor cells. The optical characterization of the ODS-SAM coated microcells will be discussed in the presentation.

Acknowledgements
The work was supported by Centre National d'Etudes Spatiales (CNES) and by Delegation Generale de l'Armement (GDA) France.

[1] A. Douahi, L. Nieradko, J.C. Beugnot, J. Dziuban, H. Maillote, S. Guerandel, C. Gorecki, V. Giordano, *Electron. Lett*, **43**,279 (2007).
[2] H.N.De Freitas, A.F.A.Da Rocha, M. Chevrollier, M. Oria, *Appl. Phys. B*, **76**, 661 (2003).
[3] D.R. Swenson and L.W. Anderson, *Nucl. Instrum. Method B*, **29**, 627 (1988).
[4] J. Vanier and C. Audoin, *"The Quantum Physics of Atomic Frequency Standards"*, Vol. 1, pp. 388-389, Adam Hilger, Bristol and Philadelphia (1989).
[5] I. Choi, Y. Kim, S.K. Kong, J. Lee, and J. Yi, *Langmuir*, **22**, 4885 (2006).
[6] H. Sugimura, K. Ushiyama, A. Hozumi, and O. Takai, *Langmuir*, **16**, 885 (2000).

The Torque-Enhancement Design for Magnetostatic Scanner

Tsung-Lin Tang[1], Wen-Chien Chen[1], Rongshun Chen[1,2], and Weileun Fang[1,2]

[1]Power Mechanical Engineering, [2]MEMS Inst., National Tsing Hua University, HsinChu, Taiwan

Abstract-This study presents two designs to enhance the magnetostatic torque to drive the scanner, (1) the lever arm, and (2) the ferromagnetic material pattern with higher length to width ratio.

1. INTRODUCTION

The micromachined scanner is a core device for optical applications [1,2]. The design considerations for scanner include large scan angle, high frequency response, and low driving voltages. Electromagnetic force is a promising approach to drive micro device [3]. The non-contact electromagnetic actuator can deliver a large displacement. Magnetic force is introduced to drive the scanner. The coil-less scanner is reported in [4-6] to prevent the complicated fabrication and packaging due to electrical routing.

2. DESIGN CONCEPT

The design concept of this study is illustrated in Fig.1. The scanner is single crystal silicon structure with perfect reflective mirror plate, high quality factor, and high reliability. The driving force can be applied on the scanner through the ferromagnetic material under the magnetic field. The magnetization of the ferromagnetic material is influenced by its planar geometry shape[7]. This study designs the "lever arm" indicated in Fig. 1 and covered with ferromagnetic material to enhance the torque applied on the mirror. The rib structure in Fig. 1 is designed to reduce the mass/stiffness ratio of the mirror plate [8].

The slender (higher length to width ratio, L/W) ferromagnetic materials patterns have larger magnetization. Thus, this study also varying the geometric pattern of the ferromagnetic material covered on the "lever arm" to further increase the magnetostatic force. As a comparison, three different ferromagnetic material patterns shown in Fig.2 are investigated. The double slender patterns (type2, L/W = 11) will have larger torque than the single one (type1, L/W = 5). In type3 design, the ratio L/W is further increased to 18.

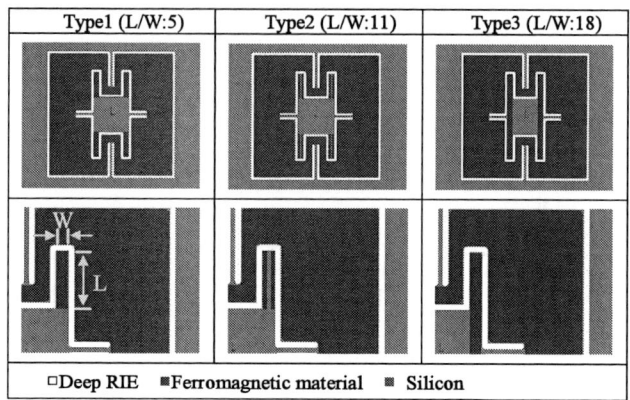

Fig.2 Three different ferromagnetic material patterns.

3. FABRICATION

Fig.3 shows the process to realize this scanner. In Fig.3a-b, this study employed the SOI wafer to precise define the spring thickness using device Si layer. Al was deposited and patterned as the backside hardmask. In Fig.3c, the Au/Ti layers were deposited for electroplating seed layer. Fig.3d shows photolithography and Ni electroplating/molding processes, thus, the pattern of the ferromagnetic material was easily defined. After strip PR, Al was deposited and patterned to define the planar shape of moving structure, as in Fig.3e. After that the Au/Ti layers were etched to exposure the silicon. In Fig.3f-g, DRIE was employed to form the rib structure on the backside of the mirror, and the buried oxide was removed by RIE. Finally, the device layer was etched by DRIE to release the structure in Fig.3h.

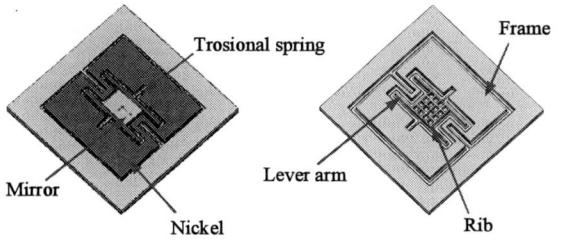

Fig.1 Design of the present scanner, top view (left), and bottom view (right).

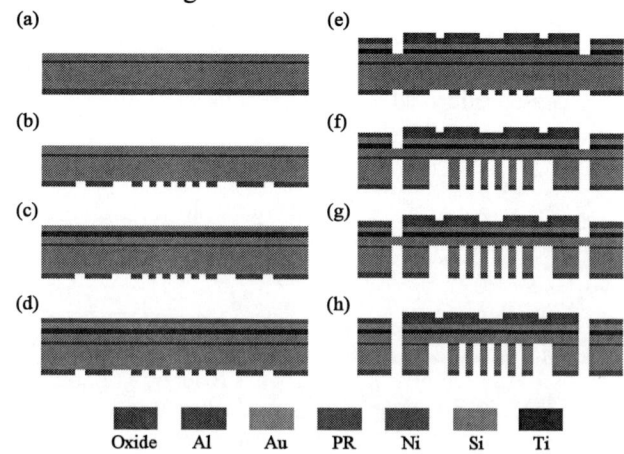

Fig. 3 Fabrication process steps of the present scanner.

978-1-4244-1917-3/08/$25.00 ©2008 IEEE

4. RESULT AND MEASUREMENT

The SEM micrographs in Fig.4 show typical fabrication results of the present 2-axis scanner. The torque-enhancement lever arm is demonstrated in Fig.4a. The backside view micrograph in Fig.4b shows the stiffness-enhancement ribs. The single and double slender patterns electroplated on the lever arm are clearly observed in Fig.4c-d. Fig.5a shows the driving stage for scanning test. The size of driving stage is 11mm×11mm×5mm and it is small enough to integrate into the portable electronic products. Fig.5b shows a typical 2D scanning pattern from the scanner in Fig. 5a. The scanner was driven after magnetizing the ferromagnetic material. Fig. 6 shows a typical frequency response of he scanner.

The measurement results in Fig. 7 show the variation of torque applied on mirror with different ferromagnetic material patterns. According to the measurements, the torque of the double slender pattern (type 2, L/W = 11) is large than that of the single one (type 1, L/W = 5). Since type 1 and type 2 has the same magnetic material volume, the torque difference is mainly resulted from the pattern design (i.e. L/W). The results agree with the prediction of this study. The torque induced from type 3 (L/W = 18) design is much larger than others.

(a)   (b)

(c)   (d)

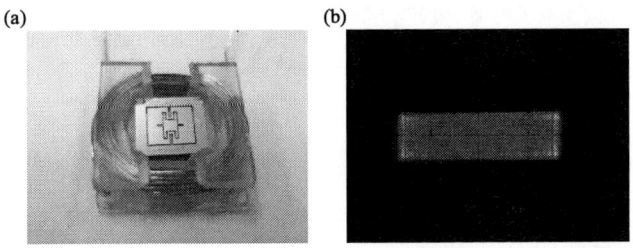

Fig.4 The SEM photos of a typical fabricated 2-axis scanner, (a) Front view with lever arm, (b) the rib at the backside of mirror, (c-d) single slender pattern and double slender pattern on the lever arm.

(a)   (b)

Fig.5 (a) The driving stage, and (b) the typical scanning pattern of 2-axis scanner.

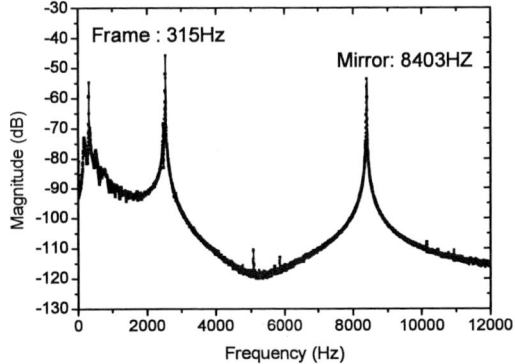

Fig.6 Typical frequency response of the present scanner.

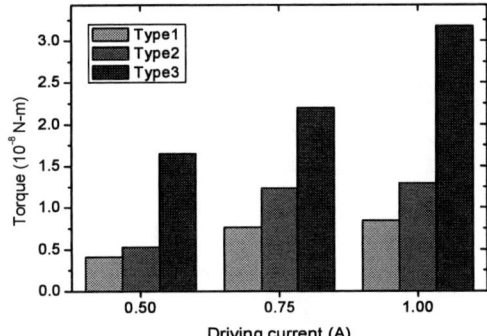

Fig.7 Torque vs different ferromagnetic material patterns.

5. CONCLUSION

This study presents two designs to enhance the torque to drive the scanner, (1) the lever arm, and (2) the ferromagnetic material pattern with higher L/W. This study has successfully designed and fabricated a two-axis scanner to demonstrate the present concept. Thus the present approaches can be employed to enhance the magnetostatic force as well as torque.

ACKNOWLEDGMENTS

This project was(partially) supported by National Science Council, Taiwan, under contract no. NSC 95-2221-E-007-MY3, and Ministry of Economic Affairs, Taiwan, under contract no. 95-EC-17-A-07-S1-011. The authors would like to appreciate the National Science Council Central Regional MEMS Research Center (Taiwan).

REFERENCES

[1] H. Schenk, P. Dürr, D. Kunze, H. Lakner, and H. Kück, Sensors and Actuators A, vol.89, pp. 104-111, 2001.

[2] Y. C. Ko, et al., Sensors and Actuators A, vol.126, pp. 218-226, 2006.

[3] A. D. Yalcinkaya, H. Urey, D. Brown, T. Montague, and R. Sprague, JMEMS, vol.15, pp. 786-794, 2006.

[4] H. A. Yang, and W. Fang , IEEE MEMS'06, Istanbul, Turkey, January, 2006, pp. 774-777.

[5] H. A. Yang, T. L. Tang, S. T. Lee, and W. Fang, JMEMS, vol.16, pp. 511-520, 2007.

[6] A. D. Yalcinkaya, H. Urey, and S. Holmstrom, Photonic Technology Letters, vol.19, pp. 330-332, 2007.

[7] J. W. Judy, and R. S. Muller, JMEMS, vol.4, pp. 162-169, 1997.

[8] M. C. Wu, and W. Fang , JMM, vol.15, pp.1-8, 2005.

SIX PORT WAVEGUIDE FILTER BASED ON CIRCULAR PHOTONIC CRYSTAL

E. H. Khoo[1†], A. Q. Liu[2] and E. P. Li[1]

[1]Institute of High Performance Computing, SINGAPORE
[2]School of Electrical & Electronic Engineering
Nanyang Technological University, SINGAPORE

†Email: khooeh@ihpc.a-star.edu.sg; Tel: (65) 6419-1550; Fax: (65) 6419-1380

This paper present a six port PCWG channel created from a circular photonic crystal array (CPC) with filtering functions. PC based waveguide (PCWG) [1, 2] created along the radial direction of CPC enables the transmission of lightwaves in subwavelength width structures. Additional filter rods are added to the PCWG to server as filtering function so that only allowed lightwaves with designated frequency can pass through the rods set. The other frequencies will be reflected. This PC device allow the distribution of light for different frequencies to different output port channel, which is very useful for multiplexing optical circuit and system.

The waveguide filter consists of several "filter" rods with selected radii as shown in Fig. 1. A filter waveguide is designed based on a rectangular PC structure which consists of rods with $n_r = 3.45$ in air medium to demonstrate the working mechanism of the filter rods. The radius of the rods is $0.2a$ where a and $1.05a$ are the lattice constants in x and z direction. It has a photonic bandgap (PBG) in the frequency of 0.298-0.408 (a/λ) in TM polarization.

Figs. 2(a) and (b) show the band structures for each rod radius in PCWG. The band structures are obtained by using plane waves expansion [3] on column unit cell. When placed next to each other as in Fig. 1, the combined set has a combined band structure shown in Fig. 2(c). There is a narrow frequency range in Fig 2(c). Only frequencies that fall within this range are allowed to pass through the combined filter rods. The other lightwaves frequencies are reflected back. If the condition of phase matching is implemented, only the frequency range of 0.364 ± 0.003 (a/λ).

A varying frequency light source is launched on the left. Fig. 3 shows the transmission spectrum of the waveguide filter. There is a transmission peak at 0.364 (a/λ) which coincides with the narrow filter range in the combined band structure. This demonstrates the working mechanism and the validity of the filter rods design.

Fig 4 (a) and (b) show the field distribution for the waveguide filter at operating frequency 0.364 and 0.37 (a/λ). It is observed clearly that the set of filter rods only allows the frequency at 0.364 to propagate through while lightwaves at 0.37 (a/λ) is reflected back.

Figure 5 shows the CPC six port waveguide filter. The circular photonic crystal (CPC) structure has a concentric period of $1.05a$ and radial period of a [4]. Six PCWG is created in CPC structure by removing rods at the polar angle axis of $\pi/3$ with respect to the x axis as shown in Fig. 5. In each PCWG, different sets of rods with selected radii are added to function as filtering system for different frequencies as indicated near each waveguide output. The filtering function is obtained by matching the band structure of different rods from band engineering and combining them to allow selected frequency transmission in each PCWG. An embedded light source is launched in the centre of six port CPCWG filter.

The light source has frequency range from 0.32 – 0.4 (a/λ) and Fig. 6 shows the transmission spectrum. There are six different peaks with filtering efficiency of more than 96% for each of the waveguide port. It is observed that the bandwidth for each of the transmission peak is very narrow Q-factor more than 1×10^5.

In conclusion, a six port CPC waveguide filter is report in this paper. The waveguide filter consists of rods to provide the filtering function. It has an overall size of 15 μm x 15 μm. It is believe that this can serve as a basic element for building large scale optical circuit in communication system.

REFERENCES:

[1] A. Mekis, J. C. Chen, I. Kurland, S. Fan, P. R. Villeneuve, and J. D. Joannopoulos, Phys. Rev. Lett. 77, 3787 (1996).
[2] E. H. Khoo, T. H. Cheng, A. Q. Liu , J. Li, and D. Pinjala, Appl. Phys. Lett. 91, 171109 (2007).
[3] S. Guo, and S. Albin, Opt. Express 11, 167-181 (2003).
[4] S. Xiao, and M. Qiu, Photonics and Nanostructures: Fundamental and Applications, 3, 134 (2005).

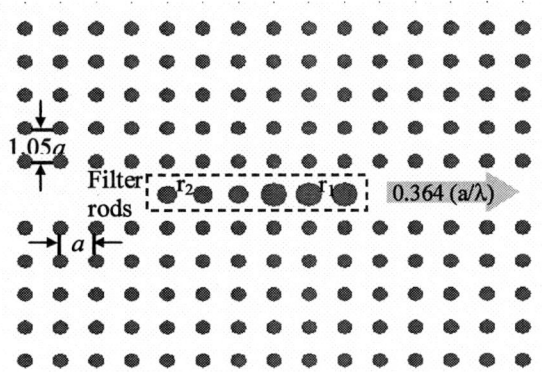

Figure 1: Schematic layout of the rectangular PC with filter rods. $r_1 = 0.354$ a and $r_2 = 0.26a$

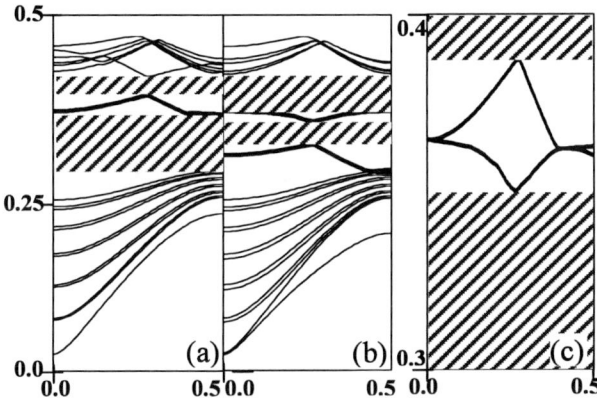

Figure 2: Band structure of the waveguide at different values of r_1 and r_2. (a) $r_1 = 0.354a$ (b) $r2 = 0.26a$ (c) The combined band structure

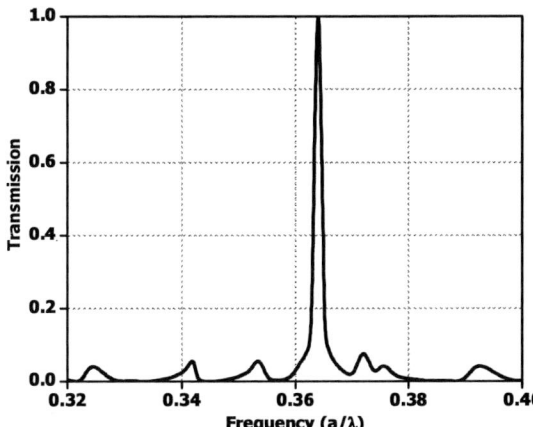

Figure 3: Transmission spectrum of the waveguide filter by band engineering. It is observed that the resonance is at 0.364 (a/λ).

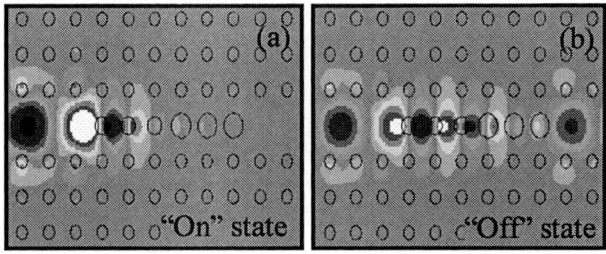

Figure 4: Field pattern distribution of the filtering at "on" and "off" states. (a) Light waves is reflected at 0.37 (a/λ). (b) Light waves transmitted through at 0.364 (a/λ).

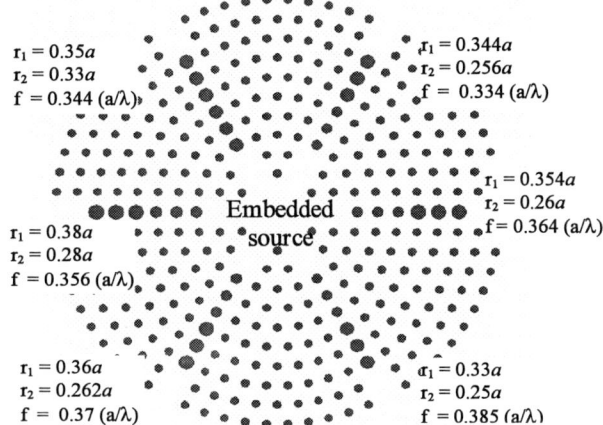

Figure 5: Schematic layout of the six port CPC filter waveguide. Each waveguide consist of a unique set of filter rods. Those with frequency in the filter rods can be transmitted through.

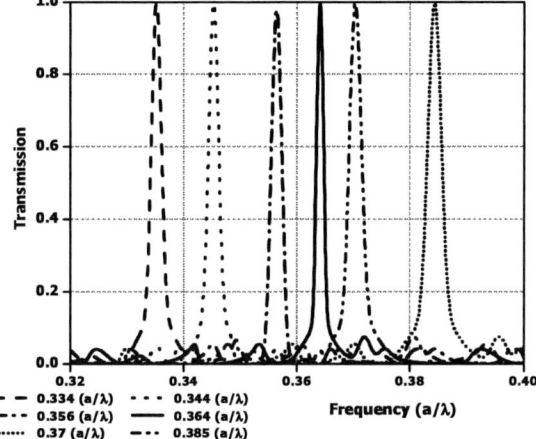

Figure 5: Transmission spectrum of the six port waveguide filter.

Towards integration of glass microlens with silicon comb-drive X-Y microstage

Karolina Laszczyk[1], Sylwester Bargiel[1], Christophe Gorecki[1], Jerzy Kręzel[2]

1) University of Franche-Comté, FEMTO-ST/CNRS, Besançon, France   2) Warsaw University of Technology, Warsaw, Poland

Abstract: We present the design and the fabrication of a silicon X-Y microstage and a new concept of its integration with a glass microlens to obtain an optical 2D scanner for the miniaturized microscope on-chip.

1. Introduction

The integration of a microlens with a silicon microstage is a general problem in the technology of the MOEMS scanners. The fabrication of a microlens with high optical quality onto a fragile movable part of the microstage remains a very difficult task, since compatibility between microlens technology and silicon micromachining is required. To meet this requirement, the microlens is typically formed from liquid polymer materials (e.g. photoresist, optical adhesive) on the prefabricated silicon platform, in the last technological step [1,2]. However, in comparison with polymers, glass offers compatibility with silicon micromachining, much better chemical resistance as well as thermal and mechanical stability. Recently, we have reported the successful fabrication of the borosilicate glass microlens from silicon mold [3]. Here, we propose to integrate the borosilicate glass microlens with the silicon X-Y microstage using silicon-glass thermal bonding. We present preliminary results of the X-Y microstage design and fabrication.

2. Design and technology of X-Y microstage

The architecture of the electrostatically actuated microstage is based on two frames (outer support frame and inner movable frame) and four pairs of straight suspensions (Fig.1a). The outer frame consists of the first pair of comb-drives, which move the inner frame and the platform (2.0 x 2.1mm²) along Y-axis (Fig.1b). The second pair of comb-drives, attached to the inner frame, allows the suspended platform to be moved along X-axis (Fig.1c). Thus, the platform is driven independently in both directions. The electrodes of comb-drives are electrically separated either by insulation gaps (outer frame) or by the insulation bridges (inner frame). The suspensions are 11μm-wide, 30μm-thick and 2mm-long. The inner comb consists of 160 movable fingers, every of which is 11 μm-wide and 70 μm-long.

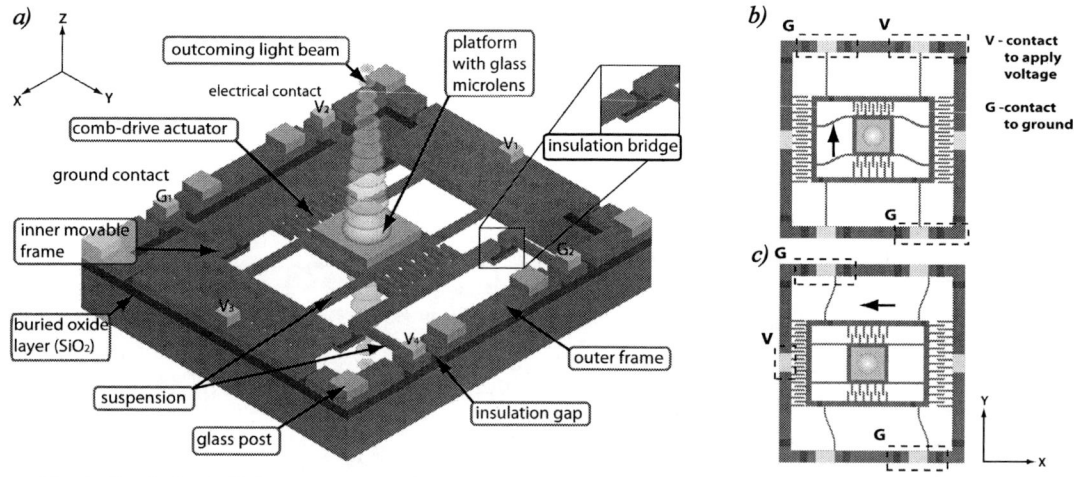

Fig.1. Comb-drive X-Y microstage with glass microlens: a) schematic view, b-c) operational principle.

The movement of the platform depends on generated electrostatic force and on suspension geometry. The calculation of lateral displacement of the microstage was performed in ANSYS using the Finite Element Method. The results of the electromechanical simulations show the maximum displacement of ~27 μm in both X and Y directions for the applied voltage of 150 V (Fig.2a).

The X-Y microstage is fabricated on (100)-oriented SOI wafer with 30 μm-thick device layer. First, the titanium/gold (200Å /2000Å) layers are sputtered on the top side and patterned photolithographically to form the electrical contacts. The microstage structure is then etched into the device layer by DRIE (A601E, Alcatel) using a photoresist mask (SPR-220, Rhom and Haas Electronic Materials) (Fig.2b). In the next step, the cavity with insulation bridges is etched from the back side with two successive DRIE processes. After reaching the buried oxide layer (BOX), the X-Y microstage is released in vapour HF etching.

978-1-4244-1917-3/08/$25.00 ©2008 IEEE   168

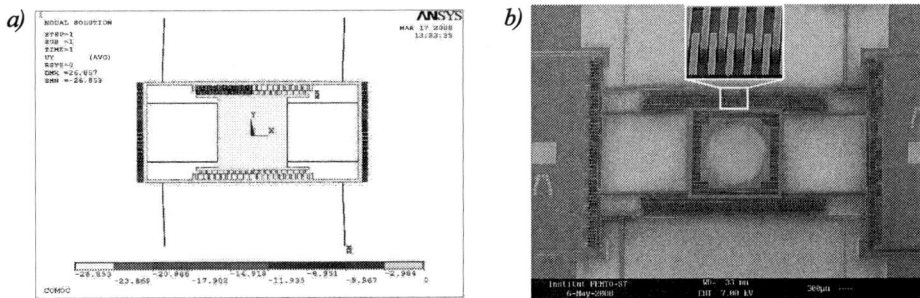

Fig.2 a) FEM simulation of X-axis displacement of microstage (max 27μm at 150V), b) SEM picture of DRIE etched microstage before releasing.

3. Integration concept

The integration concept is based on wafer-scale thermal bonding of the glass microlens, already formed in the silicon mold (process is described in [3]), with prefabricated X-Y microstage (Fig.3a). Before bonding, the top side of the microstage is fully structurized whereas the bottom side is prepared to final releasing DRIE. The key process (silicon-glass thermal bonding) is carried out in vacuum at the temperature in the range of 500-600°C using external force. In the tests, the glass microlens was successfully bonded to the silicon frame as it is shown in Figure 3b. After bonding, the silicon mold is removed in TMAH solution. Finally, the microstage is released with two DRIE processes from the bottom side, followed by etching of BOX in HF vapour.

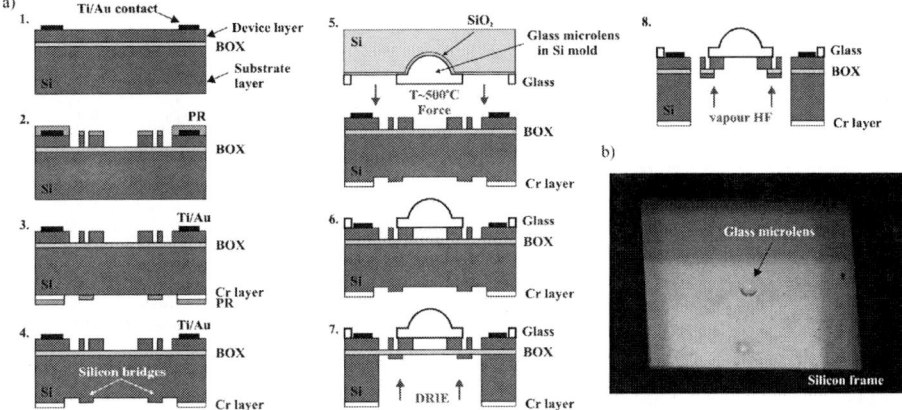

Fig.3. The concept of integration of the prefabricated X-Y microstage with the glass microlens: a) flow chart of the fabrication process, b) picture of the test sample with the glass microlens (φ=300μm) bonded to the silicon frame.

5. Conclusions

We have proposed the concept of integration of the borosilicate glass microlens with the silicon X-Y microstage based on silicon-glass thermal bonding. The concept has been verified by successful fabrication of the simple test structures. We also presented the preliminary results of the X-Y microstage design and fabrication. The next step is the application of the presented method for the fabrication of X-Y microlens scanner for the micromachined scanning confocal microscope on-chip.

Acknowledgments

This work has been partially financed by European Commission in frame of the Marie-Curie Fellowship (FP6-042123). The authors acknowledge the support of the Network of Excellence on Micro-Optics.

References

[1] H.N.Kwon, J-H. Lee, K. Takahashi, H. Toshiyoshi, *Micro XY stages with spider-leg actuators for two-dimensional optical scanning*, Sensors and Actuators A 130-131 (2006), 468-477.

[2] A. Jain, H. Xie, *A Tunable microlens scanner with Large-Vertical-Displacement Actuation*, Photonics Technology Letters, IEEE,Volume 17, Issue 9, (2005), 1971-1973.

[3] S. Bargiel, R. Walczak, J. S. Albero, J. Dziuban, K. Ałkowska, Ł. Nieradko, M. Józwik, Ch. Gorecki, *Towards the miniaturisation of optical microscopes*, XX Eurosensors, Göteborg, Sweden, 17-20 September 2006.

Optical Add/drop Filter Based on Dual Curved Photonic Crystal Resonator

P. Andalib and N. Granpayeh

Faculty of Electrical Engineering, K. N. Toosi University of Technology, Tehran, Iran.

Emails: p.andalib@ee.kntu.ac.ir; granpayeh@kntu.ac.ir

Abstract- **In this paper we propose a novel ultra compact photonic crystal resonator and investigate add/drop filter based on it. Simulation and analysis have been done by finite difference time domain and plane wave expansion method.**

I. INTRODUCTION

Photonic crystals are dielectric or metallic structures with periodic refractive indices. Studying behavior of these structures by analyzing their dispersion diagrams shows that they prevent photon cross-propagation for specific range of frequency called band gap.

Different band gap devices have been proposed and fabricated based on confinement and dispersion in these structures. Self guiding structures, beam bending in self guiding structures [1] beam splitters and unidirectional emitters are examples of dispersion based structures. Beam splitters, beam steering devices [2], micro cavities, micro rings, channel drop filters and deflectors as an optical via are examples of confinement based structures.

Add/drop filters are mostly used in WDM systems. The resonant type of these filters by their narrow band resonance allows efficient wavelength selection in multiplexing and demultiplexing systems and reduces channels wavelength spaces in these systems.

Add/ drop filters consist of two high group velocity photonic crystal waveguides, bus and drop and one or more resonators between them. By applying lightwave to bus, the lightwave with resonance wavelength of resonator tunnels to the resonator or resonators, couples to the drop waveguide and exits from one port of it [3-4].

II. NUMERICAL METHODS

We have used numerical two dimensional (2D) finite difference time domain (FDTD) method to analyze the light guiding, resonance and transmission dropping in our photonic crystal device. Perfectly matched layers (PMLs) are used in all around of the simulation area for modeling open space regions and Yee-cell technique used for discretizing fields in the space and time.

For analyzing wave behavior, group velocity, band gaps range and defect modes we obtain dispersion diagram by plane wave expansion method (PWEM).

III. PHOTONIC CRYSTAL

We have designed this novel resonator with triangular lattice photonic crystal. Background is BSC glass and the rods are GaAs, the refractive indices of which are 1.507 and 3.59, respectively.

Dispersion diagram of this lattice shown in Fig. 1, demonstrates that this lattice has only TM gap. Therefore, this device has been designed for TM mode.

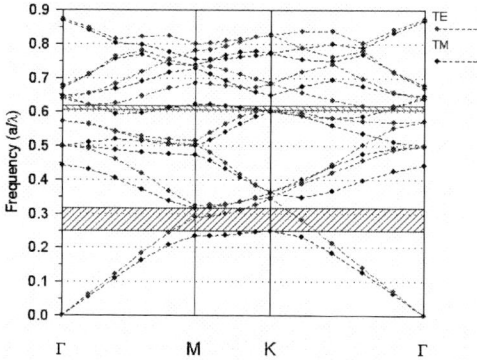

Fig. 1. Dispersion diagram of a triangular lattice Photonic crystal with n_h=3.59, n_l=1.5.

The lattice constant of the photonic crystal is selected to be a=487nm, designing photonic crystal device in the third communication window because of minimum optical loss of glass fibers in this region. The radius of the dielectric rods are r=0.2a for increasing transmissibility of the barrier in order to increase coupling.

IV. RESULTS AND DISCUSSION

Our add/drop filter consists of a photonic crystal resonator that is placed between two open ended waveguides which is demonstrated in inset of Fig. 3. By putting the pump laser at port A, laser power enters photonic crystal waveguide, attracts by the resonator and exits from port D.

Our resonator is two coupled curved Fabry Perot resonator, each is a Fabry Perot cavity with curved structure while Conventional cavities are straight closed waveguide. The length and curvature radius of resonator could be selected due to condition for having more compact photonic crystal optical integrated circuit, specific operational wavelength, enlarging mode spacing or higher quality factor and drop efficiency. Fig. 2 shows electric field pattern of a single curved Fabry-Perot resonator in one of its resonance mode.

978-1-4244-1917-3/08/$25.00 ©2008 IEEE

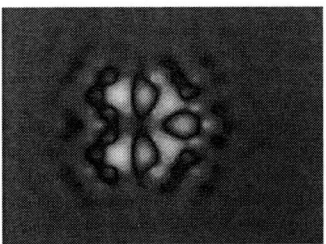

Fig. 2. Electric field pattern of single curved Fabry-Perot resonator in one of its resonant modes.

Dual-curve resonator consists of two curved Fabry-Perot resonator which energy can be exchanged between them by means of evanescent wave coupling. In our proposed resonator, facing two curved cavities to each other causes three contact points with high coupling intensity, instead of one, in coupled ordinary Fabry-Perot or ring resonators [5]. Quality factor, reflected power to input port, drop efficiency and transmission spectra are affected by the barrier thickness between two coupled resonators. Therefore these variations investigated with altering the number of spacer rods. Results show that the best dropping occurs for two rods spacer between two coupled resonators. This result has been expected since Fabry-Perot cavities have larger overlap with the other defects in the angle of 60° with respect to their axes [6]. The two rod separated dual curved, as can obviously be seen in Fig. 3 resonates at λ =1551.3 nm, but forward dropping occurs and lightwave exits from port D, while in single one lightwave drops to backward and exits from port C. Quality factor of this resonator is 153.6 and its drop efficiency is 68%. The electric field distributions of lightwave in both states of in and out of resonance are depicted in Fig. 4. Quality factor of this add/drop filter increases by enhancing period of coupling section between the resonator and two open ended waveguides; however drop efficiency decreases [7].

This resonator shows many advantages and appropriate properties for nonlinear all-optical applications, more investigations are under study.

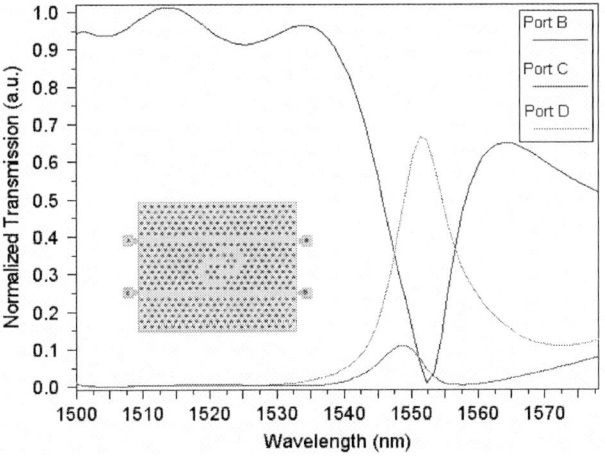

Fig. 3. Dual curve add drop filter transmission spectra.

 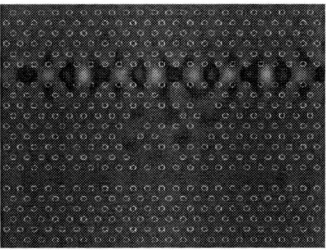

A   B

Fig. 4. Field distribution in, respectively drop and through states.

V. CONCLUSION

In this paper, we proposed a novel photonic crystal resonator constructed by coupling two curved Fabry Perot resonators together. The structure of this device is a dual curve resonator, between two high group velocity waveguides, made by triangular photonic crystal lattice with dielectric rods in glass background. Add/ drop filter based on this resonator is studied by finite difference time domain method and transmission spectra of several add/drop filters with different number of spacer rods is investigated for obtaining optimum coupling situation. Results of this studies showed that the best dropping condition occurs by using dual curved resonator with two spacer rods between them. This resonator shows appropriate properties for nonlinear all optical applications.

REFERENCES

[1] C. Chen, A. Sharkawy, D.M. Pustai, S. Shi, and D.W. Prather, "Optimizing Bending Efficiency of Self-Collimated Beams in Non-channel Planar Photonic Crystal Wave guides", *Optics Express*, vol. 11, pp. 3153, 2003

[2] C. Chen, S. Shi, D.W. Prather, and A. Sharkawy, "Beam Steering with Photonic Crystal Horn Radiators", *Optical Engineering*, vol. 43, pp. 174-180 , Jan 2004.

[3] E. Drouard, H.T. Hattori, C. Grillet, A. Kazmierczak, X. Letartre, P. Rojo-Romeo, and P. Viktorovitch, "Directional channel-drop filter based on a slow Bloch mode photonic crystal waveguide section," *Opt. Express*, vol. 13, No. 8, pp. 3037-3048, April 2005

[4] S. Fan, P.R. Villeneuve, J.D. Joannopoulos, H.A. Haus, "Channel drop filters in photonic crystals," *Opt. Express*, vol. 3, No. 1, pp. 4-11, July 1998

[5] B. Min, J. Kim, H. Yong Park, "Channel drop filters using resonant tunneling processes in two-dimensional triangular lattice photonic crystal slabs," *Opt. Commun.*, vol. 237, pp. 59–63, 2004.

[6] A. Faraon, D. Englund, I. Fushman, J. Vuˇckoviˊc, E. Waks, " Efficient photonic crystal cavity- waveguide couplers," *Phys.*, Oct. 2006.

[7] Z. Qiang, W. Zhou, R.A. Soref, "Optical add-drop filters based on photonic crystal ring resonators," *Opt. Express*, vol. 15, No. 4, pp. 1823-1831, Feb. 2007.

CMOS-SOI-MEMS Transistor (TMOS) for Infrared Imaging

Leonid Gitelman*, Zivit Gutman, Sharon Bar-Lev, Sara Stolyarova
and Yael Nemirovsky, Fellow, IEEE

Department of Electrical Engineering
Technion – Israel Institute of Technology, 32000, Haifa, Israel
Email: leonid.gitelman@gmail.com

1. State of the Art and the TMOS Motivation: State of the art uncooled thermal imaging is based on micromachined microbolometeric sensor arrays [P. W. Kruse, "Uncooled Thermal Imaging", SPIE Press, 2001]. Current bolometers are based on non-standard materials and processes. The result is the increased complexity of the design and cost. Moreover, the non-CMOS materials introduce non-uniformity as well as poor repeatability and reduced stability. Furthermore, the bolometers require relatively high bias currents resulting in significant self heating and allowing only short integration times with respect to the frame time. The large readout bandwidth results in higher noise.

2. TMOS Principle of Operation and Performance: The TMOS is a thermally isolated micromachined CMOS transistor operating at subthreshold [E. Socher, O. Degani and Y. Nemirovsky, "TMOS- Infrared uncooled sensor and focal plane array", Pending Patent, PCT IL 2004/000142]. Fig.1 is a schematic micromachined (released) CMOS transistor implemented on an SOI substrate, using DRIE (Deep Reactive Ion Etch) micromachining backside etching. The transistor operates in the subthreshold regime, so that the *measured* temperature coefficient of the current (TCC) can achieve values of more than 4%/K and the results are corroborated by simulations ["Temperature Sensitivity of SOI-CMOS Transistors for Use in Uncooled Thermal Sensing", IEEE Trans. On Electron Devices, IEEE Trans. On Electron Devices, 52(12), 2784-2790 (2005)], far better and more uniform than the typical 2-3%/K in state of the art microbolometers. In addition, the relative 1/f noise contribution of the TMOS operating at subthreshold is much reduced, since the currents are low and the surface passivation technology of the CMOS is superior (compared to conventional bolometers). The TMOS transistor is selected to be P-channel due to its lower 1/f noise contribution. The measured PSD in subthreshold exhibits a typical 1/f behavior at low frequencies, with measured noise parameter $K_{f,sub}$ ~1e-23[F].

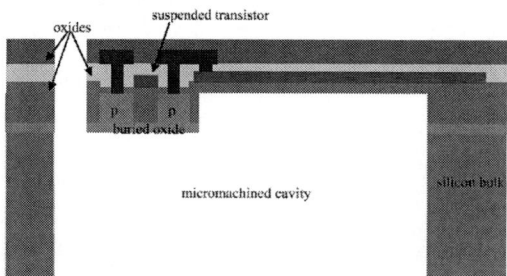

Fig. 1: Schematic cross-section of an SOI-CMOS based TMOS sensor

3. More on TMOS Fabrication: The TMOS is fabricated by MEMS post processing of a standard CMOS-SOI design. Initially, the backside is etched by DRIE using post-processed SiO_2 and Photoresist masks. The buried oxide provides etch stop. The carefully designed metalization layers are used as MASKS for post-processing by RIE of the front surface Inter Level Dielectrics and buried oxide, until the TMOS is completely released (Fig.2). The upper metalization layers, which are used as masks, are then removed. The challenge is to completely release the transistor while retaining its electrical performance. This has been achieved as indicated by the measured I-V characteristics of the released TMOS. At subthreshold, the I-V of the "virgin" MOS transistor and the micromachined released transistor coincide, indicating that the threshold voltage is not affected by the post processing and that the self heating of the thermally isoltated transistor is negligible. At saturation, the current of the released micromachined transistor is higher due to self heating effects, as expected. The image of a released TMOS transistor is shown in Fig.3.

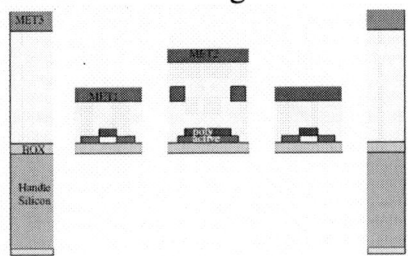

Fig.2: Backside etching of silicon handle using DRIE and BOX as etch stop. Frontside oxide etching using RIE and the standard CMOS metal layers as etch stop. The masking metals (top metals) are finally removed by wet etching

Fig.3: Image of a released TMOS. Pitch size, determined by the black square (black is where there is no material present) is ~ (50microns)^2.

4. TMOS performance: The released element was packaged in 80mTorr vacuum and characterized using a transimpedance amplifier and lock-in amplifier, irradiated by a blackbody of 1000K through a mechanical chopper with controllable frequency. From the measured dependence of the current signal upon chopper frequency, typical frequency response is fitted to a single-pole frequency response and thermal time constant found to be 10msec. Table 1 summarizes the evaluated thermal performance for the device under study.

Pitch Size [μm^2]	50x50
Fill Factor	60%
Interconnect Materials	Poly, Silicon
Resonance Frequency (First Mode) [kHz]	230
Thermal Conductance [W/K]	$1.8 \cdot 10^{-7}$
Thermal Capacitance [J/K]	$4.1 \cdot 10^{-9}$
Thermal Time Constant [msec]	9.2
NEP [nW]	0.098

Table 1.   Evaluated TMOS Performance

In summary, the novel concept of thermally isolated CMOS-SOI-MEMS transistor serving as IR detector (TMOS) is presented and characterized, showing high potential for thermal imaging, allowing long integration time because of negligible self heating.

Mid-Infrared Tunable Resonant Cavity Enhanced Detectors Employing Vertically Moving Comb Drive Actuated MEMS Micromirrors

Niels Quack, Philipp Rust, Stefan Blunier and Jurg Dual
Institute of Mechanical Systems, Department of Mechanical and Process Engineering, ETH Zurich.
Center of Mechanics, Tannenstrasse 3, ETH Zentrum, CLA J35, CH-8092 Zürich, Switzerland.
Tel: +41 44 632 35 63, Fax: +41 44 632 11 45, email: niels.quack@imes.mavt.ethz.ch.

Ferdinand Felder, Martin Arnold, Mohamed Rahim, Matthias Fill and Hans Zogg
Thin Film Physics Group, ETH Zurich, Switzerland, www.tfp.ethz.ch

Abstract

Results on a tunable Resonant Cavity Enhanced Detector (RCED) in the mid-infrared employing a vertically moving, comb-drive actuated micromirror are presented. A wide tuning range of 0.7 μm and a low order configuration have been achieved with a micromirror displacement range of 2.5 μm and a reduction of the optical cavity length respectively.

Keywords: Tunable RCED, infrared detector, MEMS micromirror.

RCED WORKING PRINCIPLE

Resonant Cavity Enhanced Detectors (RCED) are obtained by placing a photodiode inside a Fabry Pérot cavity [1]. The distance between the two cavity mirrors defines the resonances, where the detector is sensitive at. A displacement of one of these mirrors makes it possible to change the cavity length and thus the detection wavelength [2,3].

The hybrid device presented consists of two parts. The complete RCED device working principle is depicted schematically in Figure 1. The lower part shows the detector containing the fixed Distributed Bragg Reflector (DBR) and the p-n photo diode. They are fabricated using lead chalcogenide narrow gap semiconductor materials. The upper part shows the movable MEMS mirror, a displacement of which changes the cavity length and therefore the detection wavelength. Simulations show a shift in detection wavelength from 4.4 μm to 5.6 μm for a mirror movement range of 3 μm, while quantum efficiencies are above 80% [2].

MEMS MIRROR WORKING PRINCIPLE

The movable MEMS mirrors are displaced with a vertically actuated comb-drive mechanism, fabricated in the device layer of a SOI wafer. The mirrors have been realized with a delay mask

process based on a fabrication process presented by Noell *et al.* [4]. The combs attached to the mirror are fabricated much thinner than the fixed combs, resulting in a stiff mirror and limp suspension.

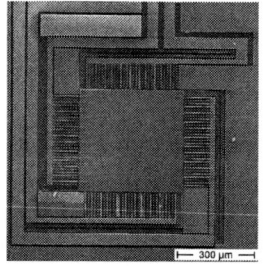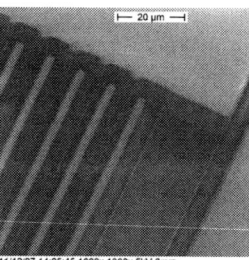

Figure 2: *SEM recordings of fabricated devices. Left: overview of a fabricated micromirror. Right: Close-up combs and suspensions. The lower suspensions and combs on the mirror appear in dark, while the elevated regions of the mirror surface and fixed combs appear bright.*

The asymmetry of the electric field of an applied actuation voltage causes the movable (thin) combs to move vertically towards the center of the fixed (thick) combs. Figure 2 shows SEM images of a fabricated micromirror, where the separation of the thin combs (dark) from the thick combs (bright) is visible.

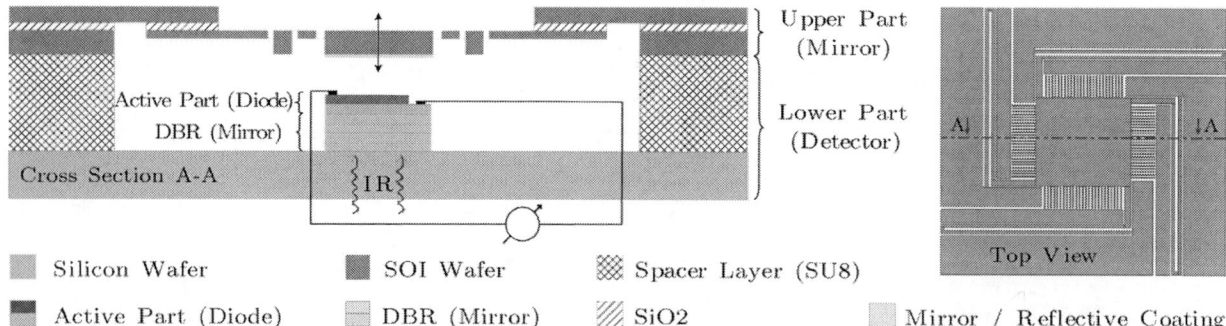

Figure 1: *Tunable RCED working principle. Left: cross section with micromirror (upper part) and the detector (lower part). Right: micromirror top view showing comb drives and mirror suspensions.*

MEMS MIRROR PERFORMANCE

Different suspension geometries have been investigated with FE simulations. The mirror geometry was designed to fit the existing photodiode layout. The mirror size was designed to 400µm x 400µm and the mirror travel range was designed to 3 µm with an actuation voltage of 50V. The comb drive finger gap is 3 µm, finger width 2 µm, length 100 (120) µm, suspension width 10 (5) µm and length 780 (1340) µm for the straight (folded) suspension. Simulation results and measurements for the displacements are shown in Figure 3. Displacements of 2.5 µm have been obtained with actuation voltages of 30V.

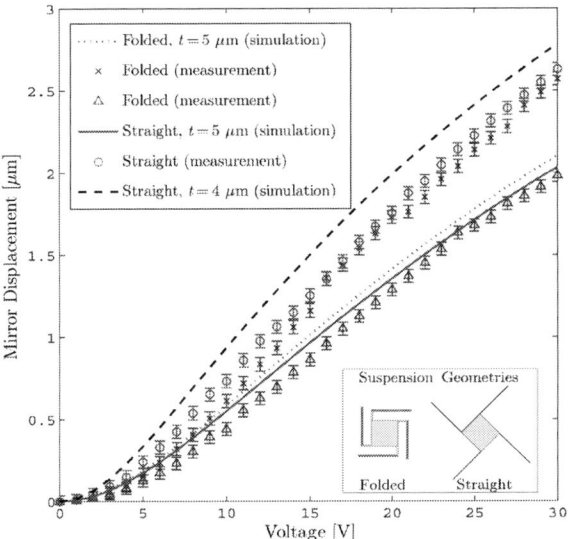

Figure 3: *Simulations and measurements of mirror displacements for actuation voltages applied at all comb drives simultaneously. The discrepancies between the theoretical and measured curve have their origin in thickness (t) variations of the suspensions due to the fabrication process.*

By applying different voltages to the different comb drives, the mirror can be tilted. This is an important feature to adjust the mirror parallel to the photodiode. A total angle variation of 0.23 degrees has been achieved as shown in Figure 4.

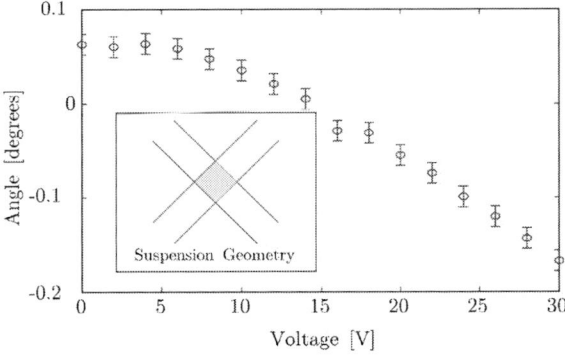

Figure 4: *Mirror tilting by applying a potential difference between two opposite comb drives.*

Contrary to the parallel plate actuated mirror design [3], the comb drive actuated mirror presented is bent towards the diode. This is beneficiary to the detector assembly as stray light loss is minimized. Radius of curvature was measured to be about 1.2m and mirror surface roughness 2nm (*spherical, rms*). Two mechanical resonance modes were measured at frequencies depending on the suspension geometry ~15 kHz and ~40 kHz

TUNABLE DETECTOR PERFORMANCE

First tunable detector results were obtained using a vertically moving comb drive actuated MEMS mirror. At an air cavity length of approximately 10 µm, the detection wavelength was shifted by about 0.7 µm with a total mirror movement of roughly 2.5 µm. The measured detector spectra are shown in Figure 5 (data corrected for common background).

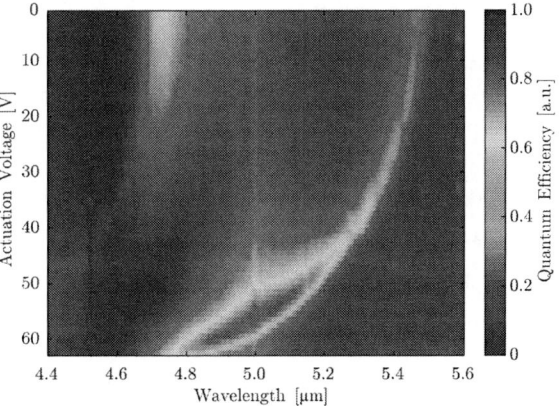

Figure 5: *Detector quantum efficiency for different actuation voltages, corresponding to different MEMS mirror positions. The detection wavelength was shifted from 4.8 µm to 5.5 µm.*

CONCLUSIONS

Integrated tunable Resonant Cavity Enhanced Detectors employing vertically moving comb drive actuated MEMS micromirrors have been demonstrated The tuning range was ~0.7 µm at ~2.5 µm mirror displacement. Next steps are further reduction of the air cavity length and increased mirror travel range in order to achieve higher detector tuning ranges.

REFERENCES

[1] M.S. Unlu, S. Strite, Resonant cavity enhanced photonic devices, J. Appl. Phys. 78, 608, (1995).
[2] F. Felder, M. Arnold, M. Rahim, C. Ebneter, H. Zogg, Tunable lead-chalcogenide on Si resonant cavity enhanced midinfrared detector, Appl. Phys. Lett. 91, 101102 (2007).
[3] Niels Quack, Stefan Blunier and Jurg Dual; Martin Arnold, Ferdinand Felder, Christian Ebneter, Mohamed Rahim and Hans Zogg, Tunable resonant cavity enhanced detectors using vertically actuated MEMS mirrors, J. Opt. A: Pure Appl. Opt. 10 (2008) 044015
[4] W. Noell, A. Hugi, T. Overstolz, R. Stanley, S. Waldis, and N.F. de Rooij, Compact large-stroke piston-tip-tilt actuator and mirror, Proceedings of SPIE, Vol. 6467, 2007.

A dynamic subwavelength pitch grating modulator for continuous Time-Of-Flight ranging with optical mixing

J. Roels [1], W. Van der Tempel [2], D. Van Nieuwenhove [2], R. Grootjans [2], M. Kuijk[2], D. van Thourhout [1], R.Baets [1]

[1] Ghent University –IMEC, Department of Information Technology (INTEC), St-Pietersnieuwstraat 41, 9000 Ghent – Belgium, Contact = joris.roels@intec.ugent.be

[2] Vrije Universiteit Brussel ~ Department of Electronics and Informatics (ETRO) – Lab for micro- and opto-electronics (LAMI), Pleinlaan 2, BE-1050 Brussel

Since continuous Time-Of-Flight ranging systems based on electrical mixing run into fundamental limits we designed an optical MEMS (de)modulator/mixer. With our dynamic (MHz range) subwavelength pitch diffraction grating we demonstrated TOF ranging with optical mixing.

I. Introduction

Continuous Time-Of-Flight (TOF) ranging relies on measuring the time-shift of a modulated carrier wave induced by traveling time. The key component is the electro-optical modulator. Two main research tracks can be identified: modulators in the electrical domain [1] and modulators in the optical domain [2]. The latter allows combining off-the-shelf high performance image sensors with a separate optical modulator, ultimately providing very high image resolution and system performance. In this proceeding we will demonstrate TOF ranging with an optical MEMS-modulator.

II. Mixer/modulator fabrication

In a SOI wafer (top layer monocrystalline Si: thickness t=220nm + buried oxide layer: thickness h=2µm) a diffraction grating with 1.25µm pitch and fill-factor 80% (string width w=1µm) is defined. The diffraction grating is constructed as two intersecting combs each with their own bond pad (Figure 1). Next we deposit a metal stack (150nm Al + 100nm Au) onto the bond pads using a standard lift-off process. To define the freestanding regions (+-25µm string length l) we applied a resist mask after a treatment in a vacuum HMDS oven to ensure good adhesion of the mask. The underetch is performed with wet buffered HF and the samples are dried afterwards using a CO_2 Critical-Point-Drying process to prevent damage due to surface tension.

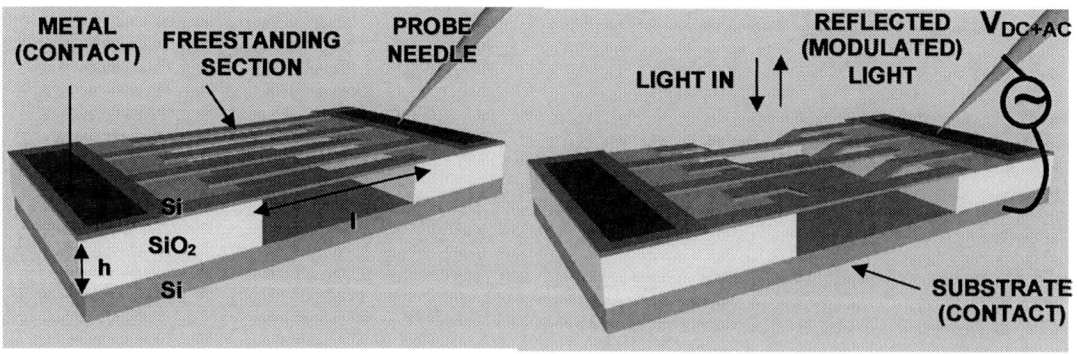

Figure 1: dynamic grating modulator: not-actuated (left) vs. actuated (right)

III. Working principle modulator/mixer

Applying a voltage (V_{AC}=8V, f=2.8MHz and V_{DC}=100V) between grating and substrate creates a closing gap capacitor and a sinusoïdal force with magnitude F_ω=$CV_{AC}V_{DC}/h$ that deflects the grating beams (driving capacitance C=$\varepsilon_0 lw/h$, ε_0 vacuum permittivity, initial gap h =2µm). Hence the reflected light is modulated at the applied frequency f. Using a spring-mass-damper model we find a typical vibration amplitude of 160nm which is sufficient for modulation depths up to 40% (experimental value) for λ =1.54µm. Since the grating pitch is smaller than λ, only one diffracted order is present.

IV. Time-Of-Flight with optical mixing

The modulated laser light ~$\sin(2\pi ft)$ is send through a fiber and reaches the mixer with a phase delay $\varphi(L)$ dependent on the fiber length L. We apply a phase difference ψ between modulated laser and mixer thus the reflected light is proportional with $\sin(2\pi ft-\varphi)*\sin(2\pi ft+\psi)$. The reflected (mixed) light is separated from the incoming light with a circulator and sent to a photodetector. Since the reflected DC optical power P_{DC} has a component ~$\cos(\psi+\varphi)$ we can extract the fiber length by measuring P_{DC} with alternating ψ (0°,90°,180°,270°) and compare with the real fiber length (Figure 2).

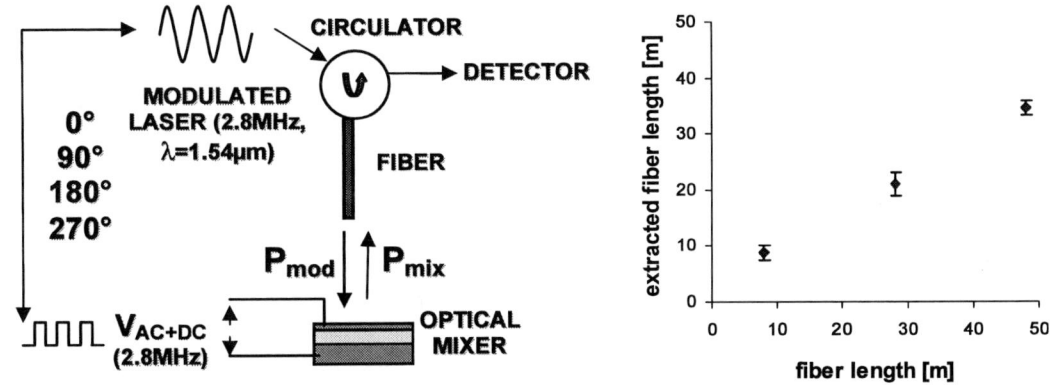

Figure 2: TOF-setup (left) and extracted fiber lengths (right)

V. Discussion + perspectives

Our TOF-technique predicts the fiber length well for the smallest fiber length but underestimates the fiber length for the two longer fibers. We suspect this might be due to higher order components (3f, 5f) that can also mix to the DC-level. In our future work we will eliminate this effect and construct a system that is able to process complete images.

VI. Acknowledgement

This work is supported by the Research Foundation Flanders (FWO-Vlaanderen).

References

[1] R. Schwarte, Dynamcic 3D-vision, International Symposium on Electron Devices for Microwave and Optoelectronics Applications, Vienna, Austria, 2001, pp. 241-248

[2] W. Van der Tempel et al., State-of-the-art modulator approaches for continuous time-of-flight range finding, annual symposium of the IEEE/LEOS Benelux Chapter, Brussels, Belgium, 2007, pp. 63-66

Radiation Heat Transfer Dominated Microbolometers

Anand S. Gawarikar, Ryan P. Shea, Alexandre Mehdaoui and Joseph J. Talghader
Department of Electrical Engineering
University of Minnesota
200 Union St. SE, Minneapolis, MN - 55455
Email: gawarikar@umn.edu, joey@umn.edu

Abstract—Radiation heat transfer limited thermal conductance represents the ultimate lower limit of the thermal isolation achievable in a microbolometer. A microbolometer structure with radiation limited thermal conductance has been fabricated and its operation demonstrated.

I. INTRODUCTION

Extreme thermal isolation is essential for optimal operation of microbolometers. As the thermal conductance, G, of a microbolometer is reduced, its fundamental thermal noise decreases while its responsivity increases as $1/G$. The net thermal conductance is the sum of the thermal conductance of the support legs and the effective thermal conductance of the radiation heat transfer from the bolometer [1]. The noise due to radiation is the fundamental limit on the performance of any thermal detector[2]. For most of the history of microbolometer technology, the radiation heat transfer has been so low as to be negligible compared to the heat transfer through the legs [1]. However, current state-of-the-art devices are approaching sensitivity levels where radiation conductance must be considered. In this paper, we examine microbolometer detector plates with radiation-limited thermal conductance, show that this conductance approximately scales with device area and quantitatively agrees with heat transfer theory.

The thermal conductance of the legs depends on the thermal conductivity of the support leg materials and the dimensions of these legs. The thermal conductance due to radiation, G_{rad} is given by

$$G_{rad} = \frac{dQ_{rad}}{dT} = 8A\sigma\epsilon T^3 \qquad (1)$$

where A is the area of the bolometer plate, σ is the Stefan-Boltzmann constant, ϵ is the emissivity of the bolometer and T is its temperature.

II. MICROBOLOMETER FABRICATION AND TESTING

The microbolometers were fabricated on a silicon substrate using LPCVD SiO_2 as the structural material. The bolometer plate consists of reactively sputtered Vanadium Oxide and electrical contact to this is provided by sputtered Ni-Fe alloy metal lines. The entire structure protected by a PECVD SiO_2 cap layer. Evaporated Au-Cr bond pads provide electrical contact to the buried Ni-Fe metal lines. Finally, the bolometer is isolated from the substrate by bulk micromachining the silicon substrate by deep reactive ion etching. The bolometer

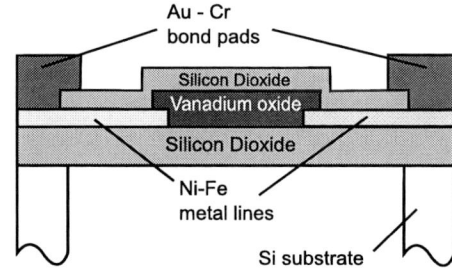

Fig. 1: Schematic diagram of microbolometer cross section. The thicknesses are as follows: Bottom SiO_2 - 150 nm, top SiO_2 - 120 nm, Ni–Fe - 40 nm, Vanadium Oxide - 70 nm, Gold–Chrome - 300 nm

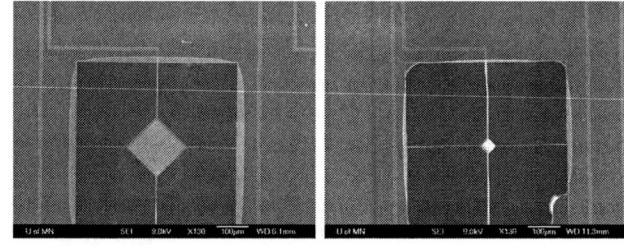

Fig. 2: SEM images of two bolometer devices with different emissive areas. The bolometer in the left pane is 100 μm wide and the one in the right pane is 20 μm wide

legs were designed to have a thermal conductance of about 1×10^{-8} W/K. A schematic diagram of the structure is shown in Fig. 1 and SEM images of the two fabricated devices are shown in Fig. 2.

The thermal conductance of the finished devices was measured in air and in vacuum by electrically heating the bolometer and measuring the resistance change due to increase in temperature [3]. The knowledge of temperature coefficient of resistance is required to extract the thermal conductance by this method and was measured prior to conducting this experiment.

The effect of bolometer plate area and the bolometer emissivity on the thermal conductance was investigated. To analyze the area dependence, the thermal conductance was measured for two devices with different dimensions. To analyze the

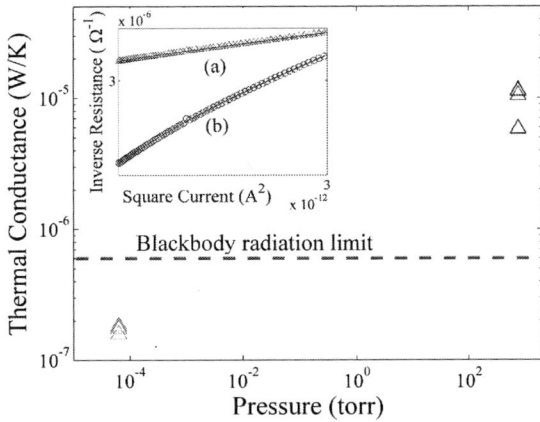

Fig. 3: Thermal conductance of 225 μm wide bolometer devices in air and vacuum. The blackbody radiation limit is also indicated in the figure. The inset shows the square of current vs. inverse of resistance data as measured for two devices of different area. Device (a) is 225 μm wide and device (b) is 80μm wide. The slope of (b) is higher than (a) indicating a lower G

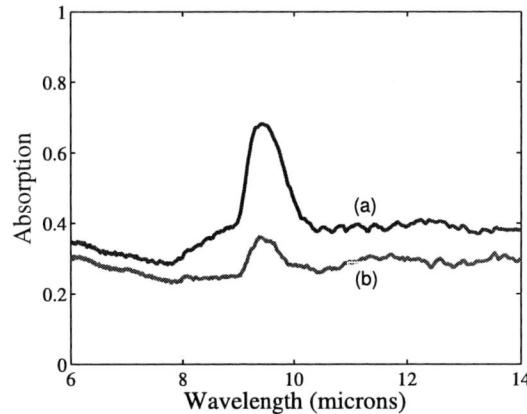

Fig. 4: FTIR absorption spectra for two devices - (a) without oxide etching and (b) after partial oxide etching

emissivity dependence, the SiO_2 on some devices was partially etched using RIE to reduce their infrared (IR) absorption. The IR absorption of these devices was measured using a Thermo-Nicollet FTIR microscope to estimate their emissivity. The thermal conductance was then measured after the oxide etching.

III. RESULTS AND DISCUSSION

Fig. 3 shows the thermal conductance measurement of a few 225 μm wide bolometer devices at two different pressures. The blackbody radiation conductance limit for devices of this area is also shown in the figure. It can be seen that in vacuum, these devices are operating well below the radiation conductance limit of a blackbody. This is primarily due to the low emissivity of the bolometer structures.

To analyze the area dependence, the thermal conductance is also measured for devices of smaller area. According to (1), the thermal conductance of a radiation limited bolometer should change linearly with the emissive area. The inset of Fig. 3 shows the plot of square of measured current vs. inverse of resistance for two different area devices. The G the 80 μm wide device is found to be 4.4×10^{-8} W/K in vacuum. Comparing this conductance to that of the 225 μm wide device, we can see that there is a reduction in the thermal conductance from a value of 1.8×10^{-7} W/K as measured for the larger device. There is a 75% reduction in the thermal conductance corresponding to an 87% reduction in the emissive area. The significant reduction in thermal conductance with area indicates a radiation limited behavior, though this behavior is not completely linear. This discrepancy could be caused by (1) the curvature in the larger bolometer plates which leads to a smaller effective emissive area and (2) the non-negligible effect of the leg thermal conductance for

the smaller device.

The effect of emissivity on the thermal conductance was investigated by partially etching the SiO_2 on a bolometer and measuring the absorption with an FTIR microscope. Fig. 4 shows the FTIR absorption data before and after the etch. The sharp absorption peak around 9.5 microns is due to the presence of SiO_2 [4]. Upon etching this SiO_2 absorption peak decreases significantly (curve b in the plot), indicating a reduction in its thickness. The bulk of the radiation heat transfer at room temperature is concentrated within the 8-12 micron atmospheric window. In this wavelength range the baseline emissivity of the devices is estimated to be 0.3 for devices with thinned oxide. The thermal conductance measured for this etched device is 1.34×10^{-7} W/K. Using this value for G_{rad} in (1), with a side length of 200 μm for area, gives a value of 0.28 for the emissivity. This value is very close to that found from the FTIR spectroscopy. This close correlation between the two measurements indicates that the thermal conductance is being limited by radiation.

ACKNOWLEDGMENTS

This project is supported by Microsystems Technology Office of DARPA and the Army Research Office under contract DAAD19-03-1-0343.

REFERENCES

[1] R. A. Wood, "Monolithic silicon microbolometer arrays," *Semiconductors and semimetals*, vol. 47, pp. 43–121, 1997.

[2] B. E. Cole, R. E. Higashi, and R. A. Wood, "Monolithic two-dimensional arrays of micromachined microstructures for infrared applications," *Proceedings of the IEEE*, vol. 86, pp. 1679–1686, 1998.

[3] P. Eriksson, J. Y. Andersson, and G. Stemme, "Thermal characterization of surface-micromachined silicon nitride membranes for thermal infrared detectors," *Microelectromechanical Systems, Journal of*, vol. 6, no. 1, pp. 55–61, 1997.

[4] Coblentz Society Inc., "Evaluated infrared reference spectra," *NIST Chemistry WebBook, NIST Standard Reference Database Number 69*, 2005. [Online]. Available: http://webbook.nist.gov/chemistry/

NEAR-FIELD SCANNING NANOPHOTONIC MICROSCOPY

John X.J. Zhang, Kazunori Hoshino and Ashwini Gopal

Department of Biomedical Engineering, Microelectronics Research Center

The University of Texas at Austin, Austin, TX 78712-0238

Tel: (512) 475-6872   Fax: (512) 232-4275   Email: John.Zhang@engr.utexas.edu;

Research Website: http://www.bme.utexas.edu/research/zhang/

ABSTRACT

We fabricated nanoscale light emitting diodes (NANO-LED) at the tip of silicon probes. Simultaneous optical and topographical images were acquired using the probes with the Nano-LED in a standard near-field scanning optical microscope.

1. INTRODUCTION

Near-filed scanning optical microscopy (NSOM) overcomes the resolution limitation by utilizing a miniaturized light source which tracks the sample surface. The conventional approach is to use an optical fiber with a nanometer-sized aperture created through the metal coating [1]. Several other methods have been introduced for NSOM, but they usually require an external laser light source. The large energy dissipation at the probe tip and the low signal intensity to the background excitation light have been the main problems. In this paper, we introduce a novel near-field scanning "nanophotonic" microscope through monolithic integration of a nano-scale light emitting diode (Nano-LED) on a silicon cantilever.

The unique feature of this near-field optical nanoscope is that the optical resolution is directly defined by the light source size, not the aperture dimensions as in the conventional NSOM. We tested several types of the Nano-LED probes in a standard NSOM setup and demonstrated simultaneous optical and topographic imaging.

2. NANO-LED DESIGN AND PROBE INTEGRATION

We have demonstrated a series of design, nanofabrication and MEMS technologies to form the light emitting probe tip on silicon, as shown in Figure 1: (1) formation of thin silicon dioxide layer buried between a phosphorus-doped N+ silicon layer and a gallium-doped P+ silicon region, created locally at the tip by a focused ion beam (FIB); (2) electrostatic trapping and excitation of semiconductor nanoparticles, such as CdSe/ZnS core-shell quantum dots at the probe tip; (3) deposition of monolayer nanocrystals on the tip using Langmuir – Blodgett (LB) method. The main body of the scanning probe is made by wet-etching of single crystal silicon wafers. Since the probe body retains the flat top surface of the original wafer, the methods described above for creating light emitting diode can be easily employed.

(1) Silicon oxide based Nano-LED

A silicon-on-insulator (SOI) wafer with a 2μm-thick device layer was used. The device layer was pre-patterned before the etching of the probe body and utilized as the LED electrodes. The device layer was doped by phosphorous oxychloride for 50 min at 900°C. The probe tip and the electrodes were trimmed with a focused ion

beam (FIB) so that a 150 nm-wide gap on the electrode tip converges to the very tip of the probe body. This milling process induces re-deposition of gallium doped p-type silicon, which eventually creates a P-N junction [2] with a thin silicon dioxide layer as the light emitter [3]. The LED was typically driven at 4-8V and emitted light centered at 560nm [4].

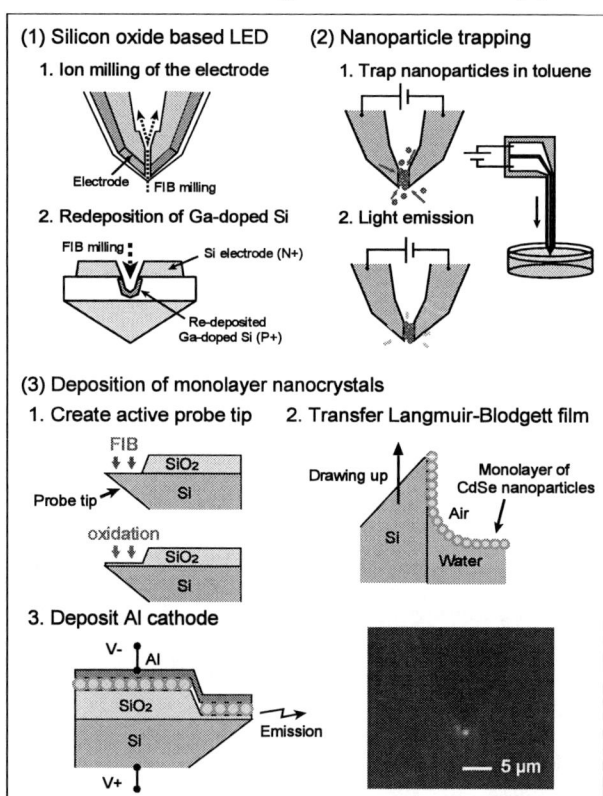

Figure 1: Nano-LED design and fabrication. (1)Silicon oxide based LED; (2)Nanoparticle trapping; (3) Deposition of monolayer nanocrystals.

(2) Electrostatic trapping of semiconductor nanoparticles

Additional semiconductor nanoparticles at probe tip were used as the light source [5]. The same probe as in (1) was used for the type-(2) LED. The electrodes were then immersed in the toluene solution to trap the nanoparticles. When the voltage (typically 5-10V) was applied, the nanoparticles were polarized and attracted to the gap along the electric field gradient.

(3) Deposition of monolayer nanocrystals on the tip

The type-(3) LED utilizes the probe body as the anode. A

standard double-side-polished wafer was used. The top surface of the probe is passivated by silicon oxide. The oxide at the tip is milled by the FIB to reveal active silicon area used as the anode. The silicon anode was then oxidized typically at 800°C for 10 minutes to create a thin energy barrier layer to balance Fermi energy levels. CdSe semiconductor nanoparticles are directly deposited at the probe tip. The Langmuir-Blodget (LB) film was prepared on a trough [6]. A solution of CdSe nanocrystals in chloroform or toluene was added in drops on the water surface. Small patches of particles at the air-water interface were then compressed to form a continuous layer. A silicon probe was dipped into the water and drawn to transfer the LB film.

One unique feature of using nanopatcle-based LED [7] is that the emission wavelength can be tailored by changing the diameter of the deposited nanoparticles. Emission spectra were measured for the three types of the LEDs, as shown in Figure 2. For the cases with CdSe nanoparticles, i.e. type –(2) and type-(3) Nano-LEDs, we found several small emission peaks which is not found in type-(1). The peaks sometimes didn't corresponds to the designed emission peaks, probably because some aggregates of particles formed larger particles at the emitting cite.

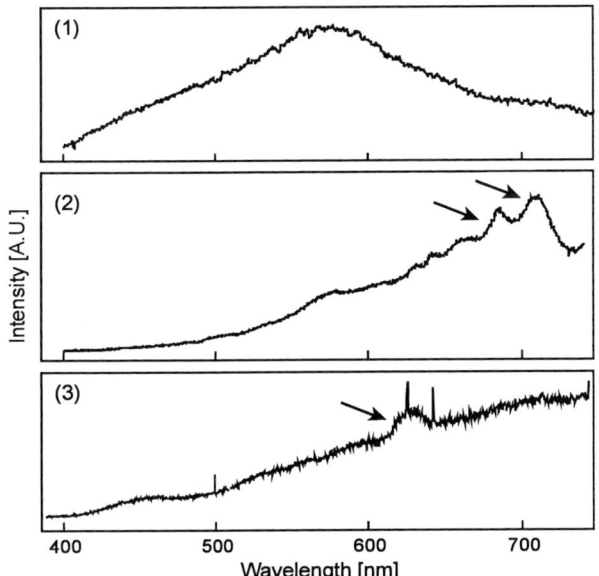

Figure 2: Emission spectra of the three types of Nano-LEDs.

3. IMAGING EXPERIMENTS

The Nano-LED probes were tested in a standard NSOM setup. The silicon probe was attached to a quartz tuning fork and controlled in shear-force feedback to perform simultaneous topographic and optical imaging.

Successful optical and topographical images have been acquired with all three types of Nano-LED probes. With type-(1) and type-(2) probes, the light emission may not happen exactly at the physical tip of the scanning probe as indicated in Figure 1. We observed significant image

shift between obtained topographic and optical images [4]. Because the type-(3) LED utilizes the probe tip as one of the electrodes, the location of the light source corresponds to the tip of the silicon probe. Figure 3 (b) and (c) are the optical and topographic images taken with the type-(3) probe. They match each other well, as shown in Figure 3d and do not show any significant image shift as previously reported[4].

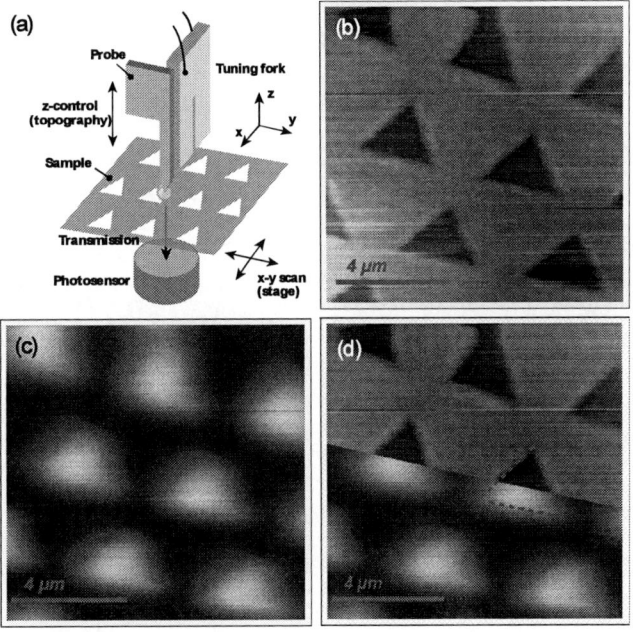

Figure 3:Near-field scanning nanophotonic microscopy. (a) Experimental setup. (b) Topographic and (c) Optical images simultaneously recorded; (d) well matched images of a chromium pattern taken using type-(3) probes.

4. CONCLUSION

We have made a Nanometer-scale LED (Nano-LED) at the tip of silicon scanning probe. We demonstrated fabrication of three types of nano-LEDs and tested them in a standard NSOM setup to demonstrate successful imaging. For the type-(2) ND –(3) probe, emission peak from deposited CdSe nanoparticles were observed. The type-(3) probe recorded simultaneous topographic and optical images with an excellent correlation.

REFERENCES

[1] B. Hecht, B.Sick, U.P. Wild, V. Deckert, R. Zenobi, O.J.F. Martin and D.W. Pohl, Journal of chemical physics 112, 7761-7774 (2000).

[2] A. J. Steckl, H. C. Mogul, and S. Mogren, Journal of Vacuum Science & Technology B, 9, 2718 (1991).

[3] L. Heikkil, T. Kuusela, and H. P. Hedman, Journal of Applied. Physics, 89, 2179 (2001).

[4] K. Hoshino, L.J. Rozanski, D.A. VandenBout and X.J. Zhang, Applied Physics Letters 92, 131106 (2008).

[5] K. Hoshino, L.J. Rozanski, D. A. Vanden Bout and X.J. Zhang, IEEE/ASME Journal of Microelectromechanical Systems 17, 4-10 (2008).

[6] B.O. Dabbousi C.B. Murray, M.F. Rubner, and M.G. Bawendi, Chem. Mater. 6, 216-219 (1994).

[7] S. Coe, W.K. Woo, M. Bawendi, and V. Bulovic, Nature 420, 801-803 (2002).

The Dependence of Poly-crystalline SiC Mid-Infrared Optical Properties on Deposition Conditions

[1]J Provine, [2]Christopher Roper, [3]Jon A. Schuller, [3]Mark L. Brongersma, [2]Roya Maboudian, and [1]Roger T. Howe

[1]Department of Electrical Engineering, Stanford University

[2]Department of Chemical Engineering, University of California at Berkeley

[3]Gaballe Laboratory of Advanced Materials, Stanford University

127X Allen CIS Building, Stanford, CA 94305

Tel +1.650.644.9403, Fax +1.650.644.0464, E-mail jprovine@stanford.edu

Abstract

We report on experimental measurements of the optical properties of thin films of poly-crystalline silicon carbide (poly-SiC) deposited by means of low pressure chemical vapor deposition (LPCVD). Measurements in the mid-IR region of the EM spectrum show strong dependence of both far field transmission and surface modes upon the deposition conditions including doping levels, ratio of dichlorosilane (DCS) to 1,3 disilabutane (DSB), and anneals performed on the film. We observe the link between the relative strength of near field phonon polariton resonances, the appearance of a polariton gap, and the achievement of extraordinary optical transmission (EOT).

Keywords: Nanophotonics, Plasmonics, Silicon Carbide, mid-Infrared

1 INTRODUCTION

The interest in SiC as an optical material has grown rapidly in recent years because its many interesting properties in the mid-IR regime of the EM spectrum allows potential for Negative Index Metamaterials, super-lensing, and extraordinary optical transmission. SiC deposited by LPCVD is attractive because of the controllable film stress, conformal deposition, and variable electrical properties [1]. However, the dependence of the optical properties on deposition condition have not previously been reported.

2 FABRICATION

A variety of poly-SiC thin films were prepared by means of LPCVD at the UC Berkeley Microfabrication Laboratory. A collection of undoped films with varying ratios of DCS to DSB as a precursor for the SiC formation were deposited at 800°C. Additionally, for a fixed ratio of DCS to DSB precursor a variable amount of nitrogen dopant was added to create another series of films. For the doped films differing annealing conditions were used for further differentiation. Table 1 summarizes the films prepared.

A sampling of the films were further processed to produce a square lattice of circular apertures of diameter 5.8μm and pitch of 10.0μm, which we have previously shown to be pattern capable of producing extraordinary optical transmission [2]. The fabrication of the hole array is achieved by contact optical lithography to pattern a hard mask of low temperature oxide which in turn allows the pattern transfer into the SiC via reactive ion etching (RIE). The SiC hole array membrane is released by a gas phase XeF₂ etch. The hole array fabrication is shown schematically in Figure 1.

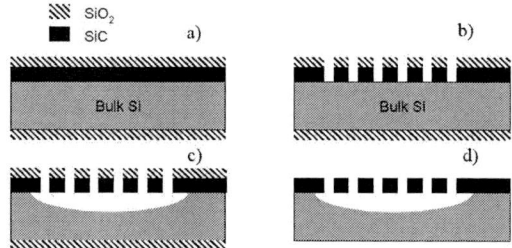

Figure 1. Schematic of hole array fabrication process flow.

Table 1. Thin film LPCVD poly-SiC samples.

Chip ID	DCS fraction	N doping (% atomic)	Anneal temp (8hrs)	Thickness (nm)
R11	0.00	0	No	288
AG10	0.10	0	No	258
AG20	0.14	0	No	255
AH3	0.16	0	No	259
R16	0.18	0	No	274
R11	0.31	0	No	326
U10	0.47	0	No	375
U20	0.31	0	No	279
U25	0.31	0	925°C	192
U22	0.31	0	1050°C	235
W2	0.31	.25	No	272
W6	0.31	.25	925°C	180
W4	0.31	.25	1050°C	229
W9	0.31	.39	No	254
W13	0.31	.39	925°C	254
W11	0.31	.39	1050°C	215
W16	0.31	.51	No	233
W20	0.31	.51	925°C	158
W18	0.31	.51	1050°C	197

3 OPTICAL TESTING

The optical properties of the SiC films are characterized with a Thermo Fisher Scientific Nicolet 6700 FTIR system with microscope attachment. An attenuated total reflection (ATR) crystal was used when only excitation of the surface mode was desired. Figure 2 shows the mid-IR reflectance surface mode spectra of the various thin film for undoped (a) and doped (b) films. The figure of merit for this quality of the films is their similarity to the spectra of single crystal SiC (shown in Fig 2a) which has a high reflectance in the mid-IR except for a sharp dip at ~900cm^{-1} caused by the phonon-polariton resonance.

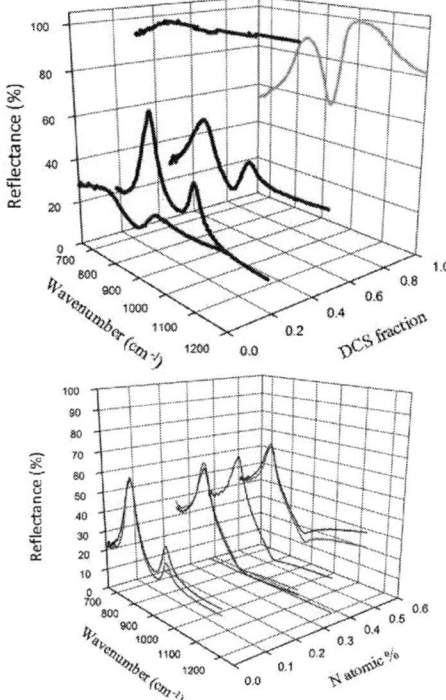

Figure 2: (a)Transmission spectra of (blue) undoped poly-SiC films as a function of DCS fraction and (green) single crystal SiC (arbitrarily placed at 1 on the DCS fraction axis). (b) Transmission spectra of doped poly-SiC films as a function of N atomic percent in the film for (blue) as-deposited films and (red) films annealed at 1050°C in Ar ambient for 8 hours.

Figure 3 shows the far field transmission of two films (R11 and U25) for both unpatterned films and for the hole arrays (in both cases the spectra is referenced to a background of the transmission through the Si substrate in a region with no SiC film). Note that film U25 shows a strong far field transmission minimum characteristic of the polariton gap of crystalline SiC around 12.4μm, while R11 also dips in this region it is less transmissive over the full spectral range.

U25 simultaneously demonstrates a phonon-polariton resonance producing EOT in the polariton gap, while R11 does not.

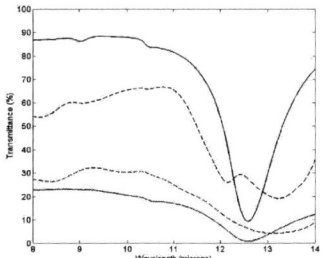

Figure 3: Transmittance in the far field for two thin film SiC samples R11 (blue) and U25 (black). The transmittance of an unpatterned film is shown by the solid lines and the transmittance of a hole array with pitch of 10.0μm and diameter of 5.8μm is shown by the dashed line.

Referring to Table 1, the only difference in the processing of these two films is inclusion of an anneal step for U25 and a difference in thickness. Thicker films are known to generally support stronger polariton resonances, so it can be inferred that the anneal step improved the mid-IR properties of the poly SiC thin film. In general, films that demonstrate strong surface mode responses (ie, overall high reflectivity with a precipitous drop at the phonon-polariton resonance) also have a well defined polariton gap and the ability to produce EOT in the polariton gap for a correctly designed hole array.

4 CONCLUSIONS

We have evaluated the mid-IR optical performance of various types of LPCVD poly-SiC films in both the near and far field. In general, films that perform close to single crystalline SiC in terms of surface mode resonances, also demonstrate a strong polariton gap and the possibility for extraordinary optical transmission in the polariton gap. The ability to tune these surface modes via the deposition conditions provides an opportunity to explore the contributions to optical performance from phonon-polariton modes compared to traditional optical diffraction.

REFERENCES

[1] C. S. Roper, R. T. Howe, and R. Maboudian, "Stress Control of Polycrystalline 3C-SiC Films in a Large-scale LPCVD Reactor using Dichlorosilane and 1,3-Disilabutane as Precursors", *Journal of Micromechanics and Microengineering*, 16 (2006) 2736-2739.

[2] J Provine, P. CAtrysse, C. Roper, R. Maboudian, S. Fan, R. Howe, "Extraordinary Transmission Through a Poly-SiC Membrane with Subwavelength Hole Arrays," IEEE/LEOS Optical MEMS & Nanophotonics, 2007.

[3] E.D. Palik, *Handbook of Optical Constants and Solids*, (Academic, Orlando, Fla., 1985)

Impact of an air barrier on the electron states of etch-released quantum heterostructures

Jan D. Makowski*, Mika J. Saarinen[†], Christopher J. Palmstrøm[‡], Joseph J. Talghader*

*Department of Electrical and Computer Engineering
University of Minnesota, Minneapolis, MN 55455, USA
Email: makowski@umn.edu, joey@umn.edu
[†]Optoelectronics Research Centre
Tampere University of Technology, 33101 Tampere, Finland
[‡]Electrical and Computer Engineering and Materials
University of California Santa Barbara, CA 93106, USA

Combining III-V materials with cantilevers facilitates the integration of photonic and mechanical elements to produce devices with novel properties. Optical cavities are a fundamental component of many photonic systems. Given the similarity of electromagnetic waves to the wavefunction of electrons, a quantum well is a one dimensional resonant cavity for electrons. The properties of the well interfaces affect the energy of the bound electron states. Replacing the semiconductor-semiconductor interface (buried quantum well) on one side of the well with a semiconductor-air interface (exposed quantum well) therefore perturbs the electronic configuration of the well.

This work investigates the effect of the quantum well interface on the photoluminescence (PL) spectrum. Our samples consist of an MBE grown InP/InGaAs heterostructure with two sets of a multi quantum well structure (MQW) and a single quantum well (SQW) symmetrically placed around a sacrificial layer (cf. figure 1). All cantilevers are attached to a common anchor where the heterostructure is still in its pristine state; only the bottom heterostructure is present in the exposed well region next to the cantilevers. The transfer matrix method yields a transition wavelength of 1385 nm for the SQW in the unprocessed buried case and 1362 nm for the exposed case [1]. The electron states in the MQW structure form a mini band at the top of the quantum well and are not affected by the nature of the interface. The transition from the bottom of the mini band occurs at 1100 nm.

Removing the semiconductor cap layer from a quantum well causes the quality of the interface to decrease precipitously. The etch process itself and the ensuing oxidation from air exposure drastically increase the number of surface states where carriers combine non-radiatively. Much research has been dedicated to overcome this problem and hydroxide or sulfide treatments have been initially suggested [2]. In recent years ALD deposited Al_2O_3, however, has proven to be very successful as a passivation layer. During deposition the aluminum precursor reacts with the surface oxide leaving a clean interface [3].

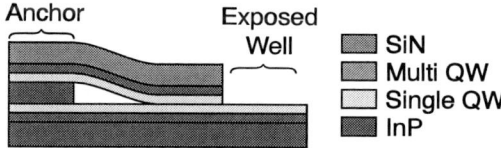

Fig. 1. Schematic of the released heterostructure cantilever.

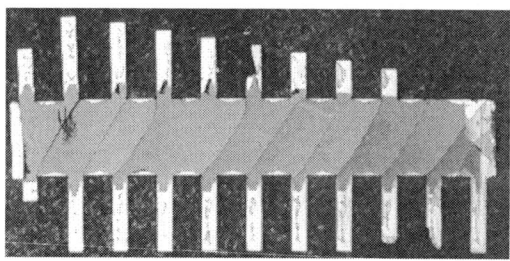

Fig. 2. Micrograph of tape with the removed top structure. The heterostructure is visible as the highly reflective lining on the cantilever.

After patterning the nitride, a chlorine based RIE etch cut through the top MQW. A HCl:Isopropanol 1:1 etch then removed the sacrificial layer and released the cantilevers. In order to test the efficacy of the ALD passivation, one set of four samples was soaked for 10 minutes in 28% NH_4OH. Then they were encapsulated with 100 cycles of ALD Al_2O_3 at 300°C. Another set of four samples did not undergo any further processing after the release etch and had a bare exposed well. In order to measure the PL from the top heterostructure, the cantilevers were removed with double-sided clear tape. Figure 2 shows a set of encapsulated cantilevers — the only set we could retrieve from all samples without severe damage.

We saw a strong photoluminescence from the pristine anchor regions even at room temperatures. At 4.2 K, band filling caused a significant blue shift of the emission line as figure 3 illustrates. Furthermore, at a high excitation intensity of 800 Wcm^{-2} the peak width increased and extended significantly to higher energies as compared to a

978-1-4244-1917-3/08/$25.00 ©2008 IEEE

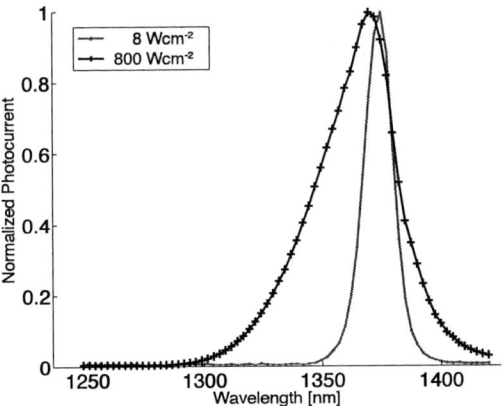

Fig. 3. The photoluminescence (PL) peak from the single quantum well (SQW) experiences significant band filling at high intensities. Lowering the excitation intensity shrinks the peak width and changes the position.

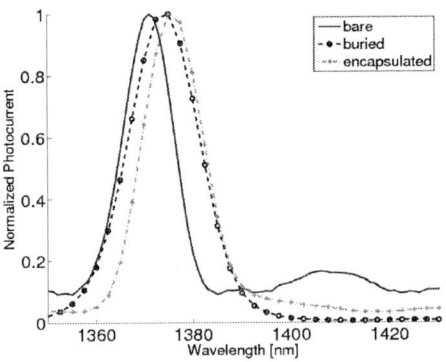

Fig. 4. The peak of the bare exposed quantum well blue shifts 5 nm compared to the anchor peak. The encapsulation appears to prevent any change.

low excitation intensity of 8 Wcm^{-2}. The exposed wells required illumination with the high intensity laser light in order to overcome the non-radiative recombination at surface states and to raise the signal above the background noise. Hence, there is no band filling in this case despite the high excitation intensity. As the peak intensities varied by one order of magnitude we normalized all spectra to emphasize shifts.

The transition peak of the SQW in the encapsulated case had the same wavelength as in the anchor for two samples. For the other two samples we measured small shifts of 1 nm and 2 nm to lower energies. In the bare case the PL peak was blue shifted by 5 nm as seen in figure 4. We repeatedly observed this shift in all measurements from the four bare samples. The observed shift is significantly lower than the theoretical 23 nm. Several processes could be the source of the deviation: partial Fermi-level pinning, interaction of the electron-hole pair with surface states [4], an increase of the exciton binding energy for surface wells [5] or the interference of etch residue and contaminants. The encapsulated wells do

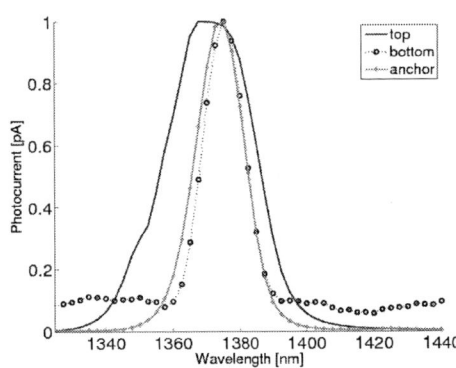

Fig. 5. The PL peak from the top encapsulated SQW is significantly widened at the top and base. The spectrum from the bottom encapsulated well does not differ from the anchor.

not blue shift at all but possibly slightly to the red. Either the effects that limited the shift in the unencapsulated wells are even more pronounced in the encapsulated wells or the aluminum oxide shields the surface well from the vacuum potential without affecting the energy state. The PL peak from the MQW experiences significant band filling in both the buried and the exposed case. Processing affects the peak intensity less than for the SQW peak — just as we expected from a structure that is farther away from the interface.

Removing the nitride and the heterostructure on its underside with tape, allows us to investigate the PL of each heterostructure individually (figure 5). The PL peak of the top structure exhibits a 10 nm wide plateau and its slope significantly extends further to the high energy side than the anchor or the bottom peak. These features could be the result of local variations of the shift between 0 nm and 10 nm which are much smaller than the 10 μm illumination spot of the fiber. In order to assess the efficacy of the surface passivation we compared the ratios of the intensity of the SQW peak to the MQW peak. While the anchor reference had a ratio of 23, all other measurements were randomly scattered between 0.05 and 1.75. At present our ALD encapsulation therefore does not yet display the self-cleaning properties mentioned in the literature.

The authors like to thank the NSF for funding under grant ECCS 0702515. Jan Makowski thanks the University of Minnesota for funding under the Doctoral Dissertation Fellowship program.

REFERENCES

[1] Tsu, R. and Esaki, L., *Appl. Phys. Lett.*, pp. 562–564.
[2] Yablonovitch, E., Cox, H. M., and Gmitter, T. J., *Appl. Phys. Lett.*, pp. 1002–1004.
[3] M. L. Huang, Y. C. Chang, C. H. Chang, Y. J. Lee, P. Chang, J. Kwo, T. B. Wu, and M. Hong, *Appl. Phys. Lett.*, p. 252104.
[4] J. M. Moison, K. Elcess, F. Houzay, J. Y. Marzin, J. M. Gérard, F. Barthe, and M. Bensoussan, *Phys. Rev. B*, pp. 12 945–12 948.
[5] L. V. Kulik, V. D. Kulakovskii, M. Bayer, A. Forchel, N. A. Gippius, and S. G. Tikhodeev, *Phys. Rev. B*, pp. R2335–2338.

Bandgap Tuning of Photonic Crystals by Polymer Swelling

Wolfgang Mönch, Philipp Waibel, and Hans Zappe

Laboratory for Micro-optics, Department of Microsystems Engineering – IMTEK, University of Freiburg
Georges-Köhler-Allee 102, 79110 Freiburg, Germany
moench@imtek.uni-freiburg.de

Abstract—Polymer swelling is studied as a novel method for bandgap tuning of photonic crystals (PCs). Crosslinked polymers swell in the presence of solvents, thereby increasing their volume and changing their refractive index. Consequently, the lattice constant and the refractive index of a polymer PC is modified by swelling and leads to a shift of the optical bandgap. We present experimental studies of swelling of one-dimensional PCs and theoretical calculations of swelling of three-dimensional PCs. It is shown that the frequencies of the optical bandgap of a polymer PC is shifted by about 10 % when exposed to a solvent-nitrogen gas mixture at 80 % of the solvent saturation pressure.

Index Terms—Photonic crystals, polymer swelling, photonic bandgap, tuning mechanism.

I. INTRODUCTION

POLYMER swelling is usually considered as a detrimental effect in microsystems engineering due to its influence on the stability of glue bonds or polymer-based structures. Polymer swelling may be understood as a mutual interdiffusion of small molecules (e. g. organic solvents, water, oligomeres) and a crosslinked polymer network. The polymer network increases its volume upon swelling. At the same time the refractive index is changed due to mixing of the polymer and small molecules.

II. THEORY

We consider here swelling of crosslinked polymer structures in a gaseous mixture of nitrogen and an organic solvent at a partial pressure p and with a saturation pressure of p_0. The volume expansion of a polymer structure may be described by a linear swelling degree α defined by $\alpha = \Lambda/\Lambda_0$, in which Λ is a characteristic length (e. g. lattice constant) of the swollen polymer structure and Λ_0 that in the initial "dry" state. Flory-Huggins type swelling [1] of the polymer structure in this gas mixture is described by

$$\frac{p}{p_0} = \left(1 - \frac{1}{\alpha^d}\right) \cdot \exp\left[\frac{1}{\alpha^d} + \frac{\chi}{\alpha^{2d}}\right] \qquad (1)$$

where d is the number of dimensions of the swelling process and χ is a material-dependent interaction parameter. For calculation of the change in refractive index, the Lorentz-Lorenz theory [2] may be used. Upon comparison of both effects using typical numbers, it can be seen that a polymer network is linearly expanded by 100 %-1000 % upon swelling, while the refractive index is changed only by 10 % at most.

III. EXPERIMENTAL

Bragg mirrors and filters (i. e., one-dimensional photonic crystals (PCs) with and without a defect mode) were fabricated by repeated spin coating / UV-crosslinking of photo-crosslinkable poly(styrene) (PS) and poly(methyl-methacrylate) (PMMA) on glass substrates. The polymer layers were attached to the substrate by covalent bonds. The samples were brought into a solvent vapor-nitrogen gas mixture. Due to the surface attachment, the polymer layer stack is free to swell in only one dimension, i. e. perpendicularly to the substrate surface. Upon swelling, the reflectance / transmittance maximum of the Bragg mirrors / filters is then shifted to higher wavelengths [3]. The partial pressure of the solvent (ethanol, acetone, toulene) as well as optical transmission spectra of the samples in the visible to near IR range were measured at the same time. By parallel measurement of these quantities (see Figure 1), the wavelength of the transmittance / reflectance peak can be correlated then to the partial pressure of the solvent, and the swelling model (here, Flory-Huggins) can be identified [4].

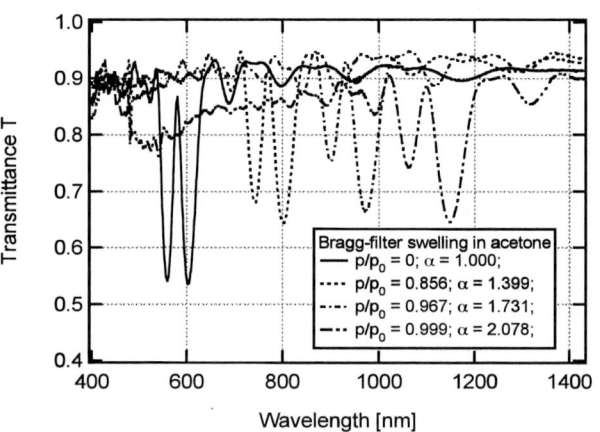

Fig. 1: Measured and simulated transmittance spectra of a PS-PMMA-multilayer Bragg filter at different partial pressure values of acetone in nitrogen.

IV. SIMULATION STUDIES

Based on these results, polymer swelling was investigated as a tuning mechanism for three-dimensional PCs by numerical band structure calculations. The material system and PC structure in these simulations was chosen in such way that the results are accessible to future experimental comparative studies. Crosslinked PS (for the PC) and an acetone-nitrogen gas mixture were chosen as the material system. The assumed PC structure had a rhomboedric unit cell (see Figure 2) with a lattice constant of $\Lambda_0 = 1.12\,\mu m$ and approximately prolate-ellipsoidal voids in the polymer at the lattice sites.

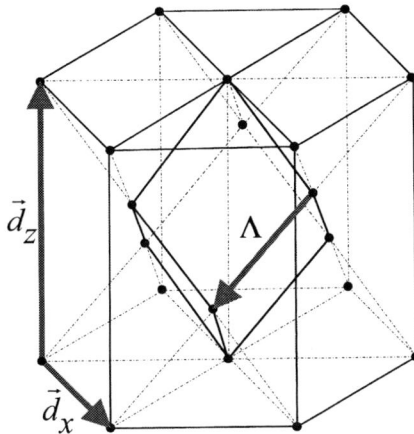

Fig. 2: Rhomboedric crystal structure and its characteristic lattice vectors.

Special care was taken in order to consider structures accessible to fabrication by laser interference lithography (LIL) using the abovementioned photo-crosslinkable PS as a negative photoresist. In first experiments, the described structure could be fabricated in SU-8 as a negative photoresist by four-beam LIL. Here, the output of a HeCd laser was split into four beams that were brought to interference on a silicon substrate coated with SU-8. The incidence angles of the beams are $0°$ for the central beam and $35°$ for the three side beams each, with the side beams arranged symmetrically around the central beam.

Band structure diagrams were calculated for the Γ-Z-direction of the reciprocal lattice by using the software tool "MIT photonic bands" by S. G. Johnson and J. D. Joannopoulos [5]. Surface effects and the finite size of real PCs were not considered in the calculations. From the band diagrams, the midband wavelength λ_C and the bandgap width b (normalized to the midband wavelength) was derived and both quantities were plotted as a function of the normalized partial pressure p/p_0. The result is shown in Figure 3.

As can be seen from this plot, the bandgap tuning characteristics is approximately linear with a small slope at "low" partial pressure values (up to $p/p_0 \approx 0.6$). Consequently, sensitive tuning of the photonic bandgap of polymer PCs is achieved by polymer swelling. The small width of the photonic bandgap can be attributed to the low refractive index ($n \approx 1.58$) of crosslinked PS compared to semiconductors ($n \approx 3.0$-3.5).

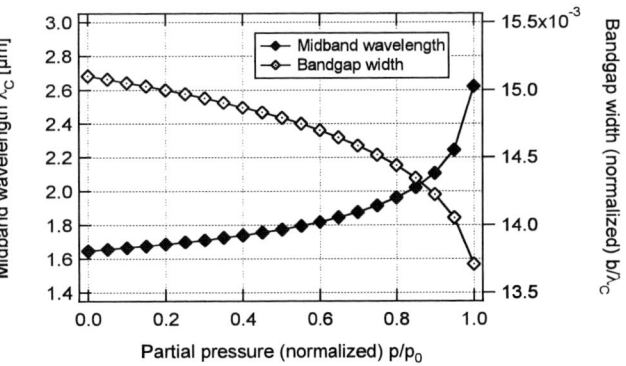

Fig. 3: Midband wavelength and bandgap width plotted as a function of solvent partial pressure (simulation results).

V. SUMMARY

In summary, it was shown by experimental (1d PCs) and simulation (3d PCs) studies that polymer swelling may be used as a mechanism for photonic bandgap tuning.

It should be stressed that polymer swelling is a widespread mechanism that may be accomplished not only by swelling in organic solvent vapor, but also by numerous other mechanisms, e. g. electrochemical changes of the oxidation state of a polymer backbone [6].

VI. ACKNOWLEDGEMENTS

The authors thank their colleagues Dr. Oswald Prucker and Prof. Dr. Jürgen Rühe for stimulating discussions and for synthesis of the polymers.

REFERENCES

[1] P. Flory, *Principles of Polymer Chemistry*. Ithaca, New York: Cornell University Press, 1953.
[2] M. Born and E. Wolf, *Principles of Optics*, 7th ed. Cambridge: Cambridge University Press, 1999.
[3] W. Moench, J. Dehnert, and H. Zappe, "Tunable polymer Fabry-Perot filters," in *Proceedings of the IEEE/LEOS International Conference on Optical MEMS and their Applications*, 2004, pp. 8–9.
[4] W. Mönch, J. Dehnert, E. Jaufmann, and H. Zappe, "Flory-Huggins-swelling of polymer Bragg mirrors," *Applied Physics Letters*, vol. 89, p. 164104, 2006.
[5] http://ab-initio.mit.edu/wiki/index.php/MIT_Photonic_Bands, last checked: 2007-12-13.
[6] A. C. Arsenault, D. P. Puzzo, I. Manners, and G. A. Ozin, "Photonic-crystal full-colour displays," *nature Photonics*, vol. 1, pp. 468–472, 2007.

Formation of a nitrified hafnium oxide buffer layer on silicon substrate and GaN quantum well crystal growth for GaN-Si hybrid optical MEMS

H. Sameshima, M. Wakui, R. Ito, F.R. Hu, K. Hane

Department of Nanomechanics, Tohoku University, Sendai 980-8579, Japan

Tel: +81-22-795-6965, Fax: +81-22-795-6963, E-mail:sameshima@hane.mech.tohoku.ac.jp

Abstract

We study the growth of GaN crystal on Si substrate by molecular beam epitaxy (MBE), in order to integrate GaN light source and MEMS monolithically. Since the lattice constant of HfN is close to that of GaN (only 0.35% mismatch), the crystal growth of GaN on HfN film is superior. On the other hand, HfO_2 film is a good candidate for waveguide, dielectric and sacrificial layer. In this study, HfO_2 film is surface-nitrified by a rf nitrogen plasma source of MBE to generate HfN layer. The morphology of the grown GaN crystal was better on the nitrified HfO_2 layer. The photoluminescence (PL) efficiency of GaN quantum well grown on the nitrified HfO_2 layer was better than that on Si substrate. As a simple hybrid lighting device structure, GaN grating on Si substrate was fabricated and the PL intensity from GaN diffraction grating was measured.

Key words: HfO_2, GaN, quantum well, lightning device, hybrid MEMS

1. INTRODUCTION

Optical MEMS technology is attractive for miniaturizing several optical systems. Using photolithographic process in Si micro machining, most of the optical components except light source can be integrated. In order to integrate a light source in optical MEMS, some novel assembling technologies have been proposed [1, 2].

On the other hand, GaN is a superior material for physicochemical devices such as photonic devices or power devices. Most of the GaN devices are fabricated on sapphire substrate. GaN semiconductor on Si substrate is also attractive not only for electronics on Si substrate, but also for optical MEMS with a light source. The crystal growth of GaN film on Si substrate is now a promising technology [3]. However, it is still necessary to reduce the crystal defect by using a good buffer layer.

In MEMS fabrication, sacrificial layer etching is a key technology. To fabricate GaN suspended structures on Si substrate as shown in Figure 1 a high selectivity is necessary for the sacrificial layer. Simultaneously, for epitaxial growth, the layer must be used as a buffer layer having lattice constant and thermal expansion coefficient close to those of GaN. Hafnium compound materials are studied well for next generation semiconductor devices. Lattice mismatch between GaN of wurtzite structure (0001) and HfN of NaCl structure (111) is only 0.35%. And GaN crystal is grown well on HfN (111) crystal [4]. On the other hand, HfO_2 is a good candidate for gate oxide film of next generation field effect transistor. It is reported that the surface of HfO_2 film can be nitrified by heating HfO_2 film in N plasma [5]. From these reports, the nitrified HfO_2 (N-HfO_2) layer may play both roles of buffer layer and sacrifice layer for GaN-Si hybrid structure.

In this paper, we study the new N-HfO_2 layer for optical MEMS. A GaN film with quantum wells (QWs) was grown on Si substrate with the new buffer layer by MBE. A GaN diffraction grating of 400nm in the period was fabricated on Si substrate and optical characteristics were examined.

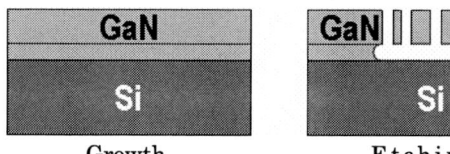

Growth   Etching

Fig.1 Simultaneous roles for buffer and sacrificial layers

2. GaN QUANTUM WELL GROWTH ON Si SUBSTRATE WITH N-HfO_2 LAYER

In our experiments of GaN growth, a MBE machine (Riber 32) was used with a rf plasma source for N atoms and the solid Ga source. A 2cm × 2cm Si (111) wafer was used as the substrate. A 200nm thick HfO_2 film was deposited on Si substrate by electron beam evaporation. After cleaning the HfO_2 / Si substrate at 850 ℃, the surface of the HfO_2 film was nitrified at 800℃ with the N plasma source power of 440W for 3 hours. A 30nm low-temperature GaN film was deposited at 650℃ on the N-HfO_2 layer before the growth of GaN film. The 800nm thick N-type GaN film with Si dopant was grown at 800℃ for 6 hours.

For the GaN light sources, the QW is a key structure to obtain the high efficiency. The QWs consist of six periods of InGaN/GaN (1nm/ 5nm) thin layers. After the QW growth, the 70nm thick P-type GaN layer with Mg dopant was grown at 650℃ for 30 minute.

978-1-4244-1917-3/08/$25.00 ©2008 IEEE   188

3. EVALUATION OF GaN CRYSTALS

The surface of N-HfO₂ layer was evaluated by XPS after the nitrifying. Figure 2 shows the XPS spectra as a function of energy. Since the electron energy peak for Hf-N bonding appears at 400eV, a part of HfO₂ film is actually nitrified. In addition, the cross sectional sample was evaluated by TEM and EDX. The nitrogen atoms were not observed in the deep region of HfO₂ film. Therefore, only the surface of HfO₂ film was nitrified. Figure 3 shows a SEM image of the GaN film with N-HfO₂ layer. Comparing with the GaN crystal without N-HfO₂ layer, the grown column crystal is wider and connected. The PL from the GaN film on the Si substrates with and without N-HfO₂ layer were measured by the irradiation of He-Cd 325nm laser. The PL efficiency of the GaN film with N-HfO₂ layer was larger than that without N-HfO₂ layer. Figure 4 shows the PL spectra of the InGaN/GaN QWs with the temperature as a parameter. The PL intensity is stronger than the case without the N-HfO₂ film.

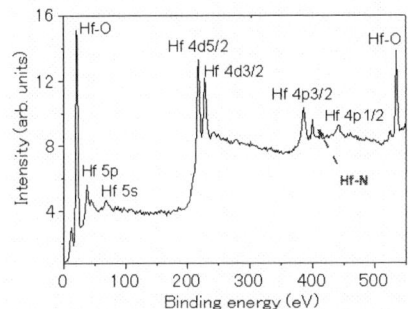

Fig.2 XPS spectra of N-HfO₂ layer

Fig.3 SEM image of the GaN film with N-HfO₂ layer

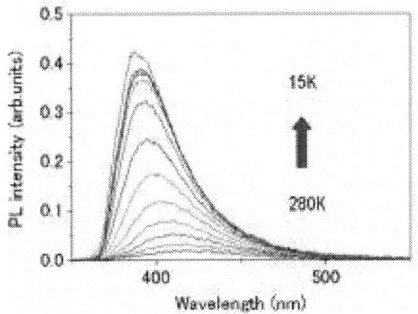

Fig.4 PL spectra of InGaN/GaN QWs

4. GaN DIFFRACTION GRATING FOR EMISSION ENHANCEMENT

Diffraction grating on the surface of a light emitting diode is attractive to increase the light extraction efficiency [6]. Moreover, the self-suspended subwavelength grating is a candidate for tunable filter and laser [7, 8]. The pattering of the 400nm period diffraction grating on the GaN film with N-HfO₂ layer was carried out by using electron beam lithography (JEOL-EB 5000ls). The GaN film with N-HfO₂ layer was etched with Cl atom beam. Figure 5 shows the top and cross sectional SEM images of the grating. The PL intensity of the GaN film with the diffraction grating is more than twice stronger than that of the GaN film without diffraction grating as shown in Figure 6.

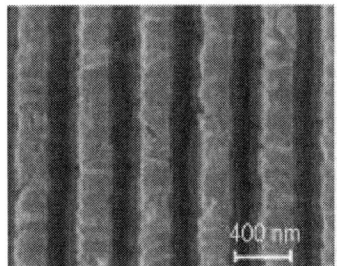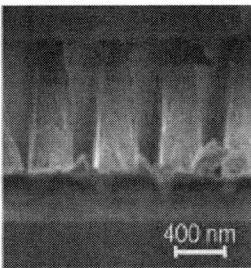

Fig.5 SEM images of the grating top view (left) and cross sectional view

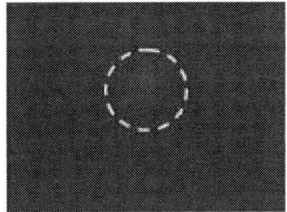

Fig.5 The light emission in 12K
without (left) and with (right) diffraction grating (left)

REFERENCES

[1] M.C.Wu, L.Y.lin, S.S.Lee and C.R.King, Int. J. High Seep Electronics and Systems, 8, 283-97 (1997)

[2] R. Sawada, Transducers '95, 281-5 (1995)

[3] T. Egawa, T.Moku, H. Ishikawa, K.Ohtsuka, and T. Jimbo, Jpn. J. Appl. Phys., 41, L663-5 (2002)

[4] R. Armitage, Qing Yang, H. Feick, J. Gebauer, and E. R. Weber, Appl. Phys. Lett., 81, 1450-2 (2002)

[5] S.J. Wang and J. W. Chai, Appl. Phys. Lett., 88, 192103 (2006)

[6] Y.Kanamori, M.Ishimori, K.Hane, IEEE Photon, Technol. Lett., 14, 1964-6, 2002

[7] Y.Kanamori, N.Matsuyama, K.Hane, IEEE Photon. Technol. Lett., 20008 impress

[8] N.Matsuyama, Y.Kanamori, J.S.Ye, K.Hane, J. Opt. A : Pure Appl. Opt., 9, 940-4, 2007

This page intentionally left blank.

Author Index

A

Adams, David P. 68
Ahmed, I. 130
Alaca, Erdem 46
Aljasem, Khaled 44
Andalib, P. 154, 170
Arakawa, Yasuhiko 1
Arslan, Aslihan 140
Asano, Takashi 3
Ataman, Çaglar 140

B

Baets, R. 176
Barbastathis, George 9
Bargiel, Sylwester 168
Bar-Lev, Sharon 172
Bergeron, Sacha 35
Blunier, Stefan 174
Bonacinay, Luigi 39
Brongersma, Mark L. 182
Byun, Young Tae 29

C

Caica, Michael 114
Chae, Sun-ki 37
Chang, Yao-tien 108
Chao, Cha-Hsin 124
Chau, Fook Siong 98, 142
Checkovskiy, Aleksandr 70
Chen, Chia-Yun 84
Chen, L. S. 150
Chen, Rongshun 106, 164
Chen, Wen-Chien 152, 164
Chen, X. 56
Cheo, Kelvin K.L. 98
Chew, XiongYeu 142
Chino, Daisuke 62
Chiou, Eric Pei-Yu. 19
Chiou, Jin-Chern 102, 110
Chiou, Sheng-jie 108
Chiu, Yi 102, 110
Chou, Chen-Yu 5, 100
Christensen, Marc P. 25
Christy, Andre 82
Chung, Youngchul 29
Contag, C. H. 42
Cowan, William D. 68
Cunningham, Brian T. 50

D

Dagli, Nadir 29
Davis, Wyatt O. 31
de Rooij, Nico 39, 114
Dell, J. 66
Demir, Hilmi Volkan 11

D

Dharmarasu, Nethaji 134
Dickensheets, David L. 54, 138
Douahi, A. 162
Du, Yu 98
Dual, Jurg 174
Dunbar, Erwin 138
Dziuban, P. 162

E

Endo, Takashi 70
Enoksson, Peter 140
Evans, Gary A. 25
Extermanny, Jerome 39
Ezoe, Y. 104, 122

F

Fang, Weileun 102, 158, 164
Fang, Weiluen 152
Faraone, L. 66
Feng, Hanhua 98
Fischer, Holger 82
Fujishima, Masayuki 120
Fujita, H. 48
Fujita, Hiroyuki 70, 116
Fujita, Masayuki 3
Furch, B. 126

G

Gabriel, N. T. 64
Gao, Lanyu 19
Garai, E. 42
Gawarikar, Anand S. 178
Gerken, Martina 112
Geyer, Ulf 112
Giordano, V. 162
Gitelman, Leonid 172
Gopal, Ashwini 180
Gorecki, C. 162
Gorecki, Christophe 168
Granpayeh, N. 154, 170
Grootjans, R. 176
Grossetete, Grant D. 68
Guldimann, B. 126
Gutman, Zivit 172

H

Haffner, Jan 88
Hah, Dooyoung 108
Haist, Tobias 88
Hane, K. 23, 188
Hane, Kazuhiro 92, 96, 120
Hasegawa, M. 162
Hast, Jukka 116
Hayashi, T. 104
Hedsten, Karin 140

Author Index

Hermerschmidt, Andreas 88
Herzig, Hans-Peter.....................54, 58
Higo, A. 48
Higo, Akio 27
Higurashi, Eiji.....................13, 62
Hiitola-Keinänen, Johanna.....................116
Hill, Daniel 52
Hillmer, Hartmut.....................134
Hinkov, Vladimir 78
Ho, S. T.....................130
Höfler, Heinrich 78
Hokari, Ryohei.....................92
Holmstrom, Sven 140
Hoogerwerf, A. 33
Horsley, David A. 90
Hoshino, Kazunori.....................180
Hosseinkhannazer, H. 128
Howe, Roger T.....................182
Hsiao, Sheng-Yi.....................102, 158
Hsieh, Tien-liang 108
Hsu, Chia Chen 160
Hsu, Chia-Pao.....................152
Hsu, Hsan-Yin 7, 74
Hu, F.R. 188
Huang, Jing-Shun 5, 100
Huang, Wei-Zhi 110
Huang, Y. Y. 130
Huanga, Jing-Shun.....................118
Huib, Sheng-Hao 118
Hung, Shih-Che 124
Huntoon, Nathan R. 25
Huttunen, Olli-Heikki 116

I

Ito, R. 188

J

Jamshidi, Arash.....................74
Jeong, Ki-Hun.....................37
Jun, Min-Ho.....................146
Jung, Hyukjin.....................37
Jung, Il Woong.....................76, 86

K

Kanamori, Y. 23
Kanno, Isaku.....................132
Kavakli, Halil.....................46
Keating, A.....................66
Khoo, E. H.....................166
Khoo', E. H.....................130
Kim, Doo Gun 29
Kim, Man Geun 146
Kim, Suhyun.....................29
Kimura, Yoshinori.....................13
Kino, G. S.....................42
Kiselevy, Denis.....................39

Klose, Thomas.....................78
Kojima, Takanori.....................3
Kopola, Harri.....................116
Koshiishi, M. 104, 122
Kostiuk, L. W.....................128
Kotera, Hidetoshi.....................132
Krause, Holger.....................15
Krezel, Jerzy.....................168
Kubota, Masanori 27
Kuijk, M. 176
Kusserow, Thomas 134

L

Lai, Ngoc Diep 160
Lammle, David 94
Lanzoni, Patrick.....................114
Laszczyk, Karolina 168
Lee, Chih-Chun 102, 158
Lee, Chun-Yu 5, 100
Lee, Jong-Hyun 146
Lee, Sung-Kil 146
Leea, Chun-Yu 118
Lemmer, Uli 112
Leone, Matthew.....................138
Leveque, Gaetan 82
Li, C......................56
Li, E. P......................130
Li, E. P......................166
Li, Ling-Han.....................27
Liao, Chun-da 144
Lin, Ching-Fuh 5, 100, 118, 124
Lin, Jian Hung 160
Lin, Keng-Shuen 106
Linden, Stefan 80
Liu, A. Q......................166
Liu, J. T. C......................42
Lo, Cheng-Yao 116
Lo, Wing Kin.....................72
Lockhart, R......................33
Lukes, Sarah 138

M

Maaninen, Arto.....................116
Maboudian, Roya 182
MacFarlane, Duncan L......................25
Mader, Daniel.....................60, 94
Maeda, R. 104, 122
Maeda, Y. 104
Mah, M. L. 64
Makowski, Jan D......................184
Mallick, Shrestha Basu.....................76, 86
Mandella, M. J. 42
Manfred, M. E. 64
Manzardo, Omar.....................58
Martin, Olivier J.F. 82
Masson, J. 17
McMullin, J. N. 128

Author Index

Mehdaoui, Alexandre 178
Milanovic, Veljko 72
Miller, Josh 31
Milne, J. 66
Mita, M. 104, 122
Mitsuda, K. 104, 122
Mitsuishi, I. 104, 122
Miura, Hideo 120
Monch, Wolfgang 15, 186
Morii, Hiroki 156

N

Nakada, M. 48
Nakagawa, Wataru.................................54
Nakano, Yoshiaki 27
Nazirizadeh, Yousef 112
Neale, Steven L......................................7
Neale, Steven 74
Nemirovsky, Yael 172
Nichol, Anthony J....................................9
Niedermann, P. 33
Nieradko, L. 162
Noda, Susumu...3
Noell, Wilfried.................................39, 114

O

Ocakli, H. Ilker 46
Ochi, Toshihiko 156
Offrein, B.J. 21
Ohira, Yasutaka 70
Ohta, Aaron T. 7, 74
Onoe, Atsushi 13
Oohira, Fumikazu 156
Osten, Wolfgang 88
Overstolz, T. 33
Özber, Natali...46
Öztürk, Alibey.......................................46

P

Palmstrømz, Christopher J.......................184
Park, Hyunkyu 90
Peng, C. L. 150
Perdigues, J..126
Peshave, Manasi 25
Petäjä, Jarno 116
Peter, Y.-A. 17
Peter, Yves-Alain....................................35
Phinney, Leslie M. 68
Piyawattanametha, W. 42
Provine, J...182

Q

Quack, Niels 174

R

Ra, H..42
Roels, J. 176
Roper, Christopher 182
Rust, Philipp 174

S

Saarineny, Mika J. 184
Sameshima, H. 188
Sandner, Thilo 78
Sasaki, Minoru...............................120, 156
Sasaki, Takashi.......................................96
Sawada, Renshi 13, 62
Saý di, Samir 35
Scharf, Toralf..58
Schenk, Harald 78
Schriemer, Henry.....................................136
Schuler, L. 66
Schuller, Jon A. 182
Schwarzer, Stefan....................................78
Seifert, Andreas 44, 60, 94
Seren, H. Rahmi 140
Shaik, Riyaz P. 15
Shea, Ryan P. 178
Shieh, Han-Ping D.............................102, 110
Shih, Hsi-Fu..102
Shirata, T. 104
Shiu, Shu-Chia 124
Sodnik, Z. 126
Solgaard, O. 42
Solgaard, Olav 76, 86
Song, Jun 148
Sonksen, Julian 134
Spahn, Olga Blum 68
Sprague, Randy 31
Stanley, R.P. 33
Stauferc, Urs 54
St-Gelais, R. 17
Stolyarova, Sara 172
Stumpf, Wolfgang C.....................................3
Suc, Wei-Fang 118
Suga, Tadatomo 62
Sugiyama, Masakazu 27
Sultana, Nahid 25
Suterb, Kaspar 54

T

Takahashi, K......................................23, 48
Takano, T. 104, 122
Talghader, J. J. 64
Talghader, Joseph J. 178, 184
Tanaka, Yoshinori......................................3
Tang, Tsung-Lin 152, 164
Tormen, M. 33
Toshiyoshi, H. 48
Toshiyoshi, Hiroshi 70, 116

Author Index

Tsai, C. Gary...150
Tsai, Jui-che...108, 144
Tsang, H. K...56
Tsuda, Shogo...132

U

Urey, Hakan..46, 140

V

Vahala, Kerry...41
Valley, Justin K. 7, 74
Van der Tempel, W..176
Van Nieuwenhove, D.......................................176
van Thourhout, D..176
Venancio, L. 126

W

Waibel, Philipp 94, 186
Wakui, M. 188
Waldis, Severin.......................................39, 114
Wallace, K. 126
Webery, Stefan M..39
Wegener, Martin...80
Wei, Kenneth..19
Wheeldon, Jeffrey F..136
Wildenhain, Michael...78
Wölfelschneider, Harald.....................................78
Wolfy, Jean-Pierre..39
Wong, C. Channy 68
Wu, Jhong-Wei..110
Wu, Ming C. 7, 74, 108
Wu, Ting-Hsiang 19

Y

Yamanoi, Toshio...70
Yamasaki, N. Y...104
Yeh, J. Andrew 150
Yen, Ta-Jen..84
Yip, Ming-Chuen..152
Yu, Hongbin 142
Yukihara, E. G. 64
Yun, Sungsik 146
Yuzuriha, Asumi..156

Z

Zamkotsian, Frederic 114
Zamora, Ricardo 134
Zappe, Hans.........................15, 44, 60, 94, 186
Zhang, John X.J..180
Zhang, Qingxin...98
Zhou, Guangya...98, 142
Zhou, Wei...25
Zhu, Ning...148